INTEGRATED LAND-USE AND TRANSPORTATION MODELS: BEHAVIOURAL FOUNDATIONS

Related Books

HENSHER & BUTTON (Eds.)
Handbooks in Transport 6-Volume Set

AXHAUSEN (Ed.)
Travel Behaviour Research

GÄRLING et al (Eds.)
Theoretical Foundations of Travel Choice Modeling

HENSHER (Ed.)
Travel Behaviour Research: The Leading Edge

MAHMASSANI (Ed.)
Travel Behavior Research Opportunities and Applications Challenges

TIMMERMANS (Ed.)
Progress in Activity-Based Analysis

Related Journals

Land Use Policy
Editor: A. Mather

Transportation Research Part A: Policy and Practice
Editor: Phil Goodwin

Transportation Research Part B: Methodological
Editor: Fred Mannering

Transport Policy
Editors: Moshe Ben-Akiva, Yoshitsugu Hayashi & John Preston

For a full list of Elsevier transportation research publications visit:
www.elsevier.com/transportation

INTEGRATED LAND-USE AND TRANSPORTATION MODELS: BEHAVIOURAL FOUNDATIONS

Edited by

MARTIN E.H. LEE-GOSSELIN
Université Laval, Québec, Canada

SEAN T. DOHERTY
Wilfrid Laurier University, Waterloo, Canada

United Kingdom – North America – Japan –
India – Malaysia – China

Emerald Group Publishing Limited
Howard House, Wagon Lane, Bingley BD16 1WA, UK

First edition 2005.

British Library Cataloguing in Publication Data
A catalogue record for this book is available from the British Library

ISBN: 978-0-080-44669-1

Printed and bound in Great Britain by
CPI Antony Rowe, Chippenham and Eastbourne

ISOQAR QUALITY ASSURED ISO 9001 UKAS QUALITY MANAGEMENT 026
Certificate number 1985

Awarded in recognition of Emerald's production department's adherence to quality systems and processes when preparing scholarly journals for print

INVESTOR IN PEOPLE

CONTENTS

LIST OF FIGURES

LIST OF TABLES

ACKNOWLEDGEMENTS

This book was a product of the PROCESSUS network research programme, funded by the Social Sciences and Humanities Research Council of Canada (SSHRC), the Network of Centres of Excellence in geomatics (GEOIDE), and the Quebec Ministry of Transport (for more details, please see the Introduction).

The editors undertook the preparation of this volume with the much appreciated assistance of an Editorial Board consisting of Pavlos Kanaroglou, Eric Miller and Paul Villeneuve. Between us, we persuaded a large number of colleagues to perform independent reviews: we wish to thank these anonymous people for their many suggestions, which in some cases guided substantial revisions of the draft papers.

We also much benefited from the thoughtful guidance of Chris Pringle and Zoë Youd of Elsevier Limited, Oxford, throughout this undertaking.

The editors would like to thank each of the contributing authors for their careful collaboration during the multiple stages of review, correction and/or modification, as applicable.

The preparation of camera-ready copy requires the fastidious attention of someone with infinite patience, and the ability to talk politely to the more esoteric functions of word-processing software, something the editors did not always succeed in doing. That indispensable colleague was Manouane Théberge of the Groupe de recherche interdisciplinaire mobilité, environnement et sécurité (GRIMES), Université Laval. We offer our sincerest thanks for her impeccable work and good humour throughout the preparation period.

The costs of technical services for this book were shared between the SSHRC grant and the GRIMES.

Balancing family and work is never easy, and this book could only increase this tension, especially as the editors were often on different continents. Our heartfelt thanks to our spouses and children, and to our employers, l'École Supérieure d'Aménagement du territoire et de Développement regional of Université Laval, and the Department of Geography and Environmental Studies, Wilfrid Laurier University, for their generous understanding.

INTRODUCTION

It may seem odd to see a book with the words "behavioural foundations" come from a group of authors that includes scarcely anyone with a traditional behavioural background. If anything, we would like this to serve as an invitation to set us straight – we believe something isn't quite right with the behavioural assumptions going into the gamut of land-use and transport models, and we'd like to get the ball rolling on some innovative and creative new solutions.

In this vein, we brought an international group of modellers, data-designers, geomatics experts policy analysts, and some behavioural scientists, to Quebec City in 2002. We asked them to unplug their computers and share ideas under the broad topic of "The Behavioural Foundations of Integrated Land-Use And Transportation Models: Assumptions and New Conceptual Frameworks". This Colloquium was a landmark event in a major research programme on those behavioural foundations that had been underway since 1999 in a network of six (later eight) Canadian universities, with the collaboration of eight (later ten) institutions in six other countries[1]. Known as the PROCESSUS[2] Network, it overlapped and built upon the work of the ILUTE[3] Consortium, founded by Eric Miller of the University of Toronto in 1994 to pursue collaborative research on major simulation models for urban areas.

The behavioural research programme, consisting of 20 linked projects, was made possible by unprecedented funding for this class of urban problems through a Major Collaborative Research Initiative grant from the Social Sciences and Humanities Research Council of Canada, by an important synergy with the spatial sciences through project funding from the GEOIDE network (Geomatics for Informed Decisions), supported by the Network of Centers of Excellence Programme of the Canadian federal research councils, and by additional financial support from the Ministère des Transports du Québec for PROCESSUS' empirical and database work in the Quebec City region.

It was at the Québec Colloquium that most of the chapters of this book were presented in a preliminary form. They were selected for the book with a view to establishing an open forum to share our experiences and to plot our course. Our overriding objective for both the research programme and the book was to question the behavioural foundations of ILUTE-type models, and to determine how such decision-support tools could help Canadian cities plan a future in

---

[1] The Canadian members were Université Laval (network host), the University of Toronto, the University of Calgary, Université de Québec/INRS-Urbanisation, Culture et Société, Wilfrid Laurier University, McMaster University, McGill University and St.Mary's University. The international collaborating institutions were Imperial College, London (UK), the University of Westminster (UK), Göteborg University (Sweden), L'École Navale de Brest (France), IVT-ETH, Zurich (Switzerland), Massachusetts Institute of Technology (USA), Boston University (USA), Georgia Institute of Technology (USA), the University of Sydney (Australia) and The Urban Transport Institute (Australia).

[2] PROCESSUS Network: PROCesses of behaviour underlying Equity and Sustainability in Systems of Urban access and their Simulation. Réseau PROCESSUS : PROcessus Comportementaux Essentiels aux SystèmeS d'accès Urbain durables et équitables et à leur Simulation

[3] ILUTE: Integrated Land-Use, Transportation and Environment, a framework for modelling.

which the access that they offer to activities and services is environmentally sustainable, healthy and equitable. By behavioural foundations, we mean any framework or set of assumptions concerning the mechanisms and processes of human behaviour and decision-making over time, space and across multiple actors. As shown throughout the papers in this book, these may be adopted from a variety of disciplines, such as from: economics (e.g. random utility maximization), geography (e.g. space-time prisms), psychology (e.g. decision-making processes/rules), sociology (e.g. social networks, family dynamics), and others.

The behavioural foundations are explored in the context of land-use and transportation decision-making in built environments – by individuals, households, and businesses. These range in scale from daily decisions of households concerning what to do, where and when to do it, and by what travel mode, to longer term decisions concerning residential choice, vehicle holdings or land consumption by business entities. For it is the summation of all these individual micro-level decisions that culminates in the aggregate flows of traffic and urban growth, and has important implications for the environment, equity and sustainability.

Why are the behavioural foundations so important? Put simply, this class of model is only as good as the behavioural assumptions upon which it is based. One of the major uses of integrated urban models is to forecast human behavioural changes in response to emerging policy, technologies, and growth scenarios. Many of these, such as travel demand management policies or growth in telecommunications, inherently invoke changes at a micro decision-making level: if the behavioural mechanisms for these changes are not incorporated in the model, how much faith can you put in their output?

This book contributes in three key thematic areas, covered with varying degrees of overlap by all the authors: (I) conceptual issues and approaches; (II) key behavioural components and related data needs; and (III) operational modelling issues and advances.

Much of the material comprising the conceptual theme resulted from a challenge to those who normally focus on operationalising models through mathematical formulation and computer simulation: they were asked to clearly identify and communicate the underlying bases and assumptions of their models and approaches. We felt it was critical to establish the open forum with researchers who can offer new concepts, theories, frameworks, and data collection methods that could potentially enhance or question these bases. Surely, if we are to solve the emerging problems in our field, we need to make it accessible and understandable to a wide range of disciplines, and encourage intense collaboration.

Many of the papers in this volume explore new data collection tools that offer the potential to address past assumptions about the key behavioural components more empirically, and provide a basis for fundamentally new model types. For example, computerized surveys have emerged that can interactively prompt and track aspects of behaviour that are difficult to capture on paper, such as the sequence of data entries. At the same time, in-depth household interview methods have been extended, for example to engage respondents in interpreting the room for manoeuvre that they perceived to have been present when organising themselves and others in time and space during a recently observed week of activities. Stated preference/adaptation surveys and retrospective recall have evolved to more thoroughly explore future and past response scenarios. Other tools, such as Global Positioning Systems (GPS) and 3-D visualization, are being used to reduce respondent burden and to potentially provide an opportunity for more in-depth exploration of other issues. In all, a shift in focus from observed outcomes alone (e.g. trip patterns, aggregate goods flows) to the inclusion of underlying decision processes is accruing.

The operational model development issues stemming from these behavioural foundations touch on a longstanding criticism of behaviourally rich disaggregated modelling approaches: the challenge of accounting for all the behavioural mechanisms and causal effects that are present. Rather than recoil at the apparent complexity, this book demonstrates how we can embrace the challenge, conceptualize the issues, and offer new solutions. Some of these capitalize on recent technologies such as Geographic Information Systems (GIS), agent-based simulation, and distributed computing, while others extend and integrate the established and the new, such as "hybrid" models and the proposed interdependence of sub-models within an ILUTE "suite".

In all, this book was intended as an opportunity for the authors to openly expand their ideas, without a requirement for empirical support at this stage (although this is often provided where available), thus allowing many questions to emerge and creative approaches to be suggested. This serves both to stimulate and challenge other researchers, and provide an important basis for students and younger researchers to launch new projects – something that we feel is critical to forming a basis for substantive growth in the field.

The *conceptual issues and approaches* theme opens with contributions from two leaders in the field (Axhausen and Miller) who provide their latest perspectives on emerging approaches to understanding travel behaviour. In particular, Axhausen provides both a general introduction to travel behaviour issues and their historical evolution, highlights the most prominent issues and conceptual approaches (activity generation, activity spaces, scheduling, projects, etc.), and introduces several new focal areas for future research (e.g. social networks). Eric Miller, a recognized leader in the development of urban simulation models, takes a much more applied approach, providing a detailed sketch of how such concepts could be translated into analytical modelling advances and the challenges involved. The presentation is particularly comprehensive as travel decision making is treated as derived from the wider notions of activities, projects, schedules, and resources/budgets, all within a household multi-day context.

Roorda *et al.* take up the challenge posed by Axhausen and Miller with respect to activity scheduling behaviour, and take a specific look at the many assumptions that are often made when it comes time to model, and what new types of data could be used to address, expand and/or avoid such assumptions. For example, when activities and trips are inserted into an existing schedule, assumptions are made concerning how various activities may be modified as a result, and suggestions for exploring the validity of these assumptions using empirical data made.

Ramadier *et al.* complete the conceptual theme with a review of social psychological theories of the dynamic relationship between an individual in her social setting and the built environment. They present a conceptual framework intended to guide survey methods that can observe the rules and heuristics with which agent-based models could more realistically represent the decisions that underlie spatio-temporal behaviour. A key implication for data collection is that rules can best be observed functioning in a two-stage "negotiation" cognitive process that is upstream from activity and travel patterns, a notion that fits well with the interplay of household, person and project "scheduler" roles proposed in the Miller chapter. They also argue for new efforts to simultaneously measure spatial and temporal flexibility, both as it is perceived, and as it may be inferred from executed activities.

The four chapters in the section on *behavioural components and data needs* expand on many of the key issues related to the representation of activity/travel decision-making that are put forth in the first chapters. McCray *et al.* focus on the adequacy of data collection methods to capture the articulation between what people do in time and space and what else they might do – and especially in the case of disadvantaged groups, those desired activities for which access is too costly or personally disquieting. In their wide-ranging chapter, they examine the implications of the evolution of activity patterns (including their extension into cyberspace), and the challenges of making surveys inclusive of all groups, including those that are hard to reach, such as the urban poor. To keep up with the phenomenological changes, and to overcome the selection biases that may accompany some types of survey methods, they conclude that flexible multi-instrument packages are often indispensable, especially at the specification stage of new model development. They give three case examples of such methods, two of which were guided by the conceptual approach in the Ramadier *et al.* chapter.

The Jotisankasa and Polak chapter gives a thorough review of recent efforts to incorporate in travel behaviour models a dimension that is often absent: the mechanisms of traveller learning. They develop in-depth the example of the cognition of travel time (including the effects of departure time) and route choice. They compare modelling approaches with and without updating mechanisms for perceptions of network attributes, and examine other aspects of learning. The chapter identifies six modelling issue areas in which research is needed to more fully incorporate learning mechanisms into integrated land-use and transport models: the learning component, choice modelling, habitual choice, choice set formation, data collection methods, and the evolution and convergence of the systems.

On of the key components that is changing the dynamics of activity scheduling is the rapid diffusion of mobile telephony. Richardson and Seethaler present a portrait of this phenomenon from several countries, including some characteristics of users and uses. They describe their efforts to find data on the interaction between mobile phone use and travel, which is too recent a development to have been covered, for example, by national travel surveys. As an interim measure, they used an Australian national time-use survey to compare a number of characteristics of the use of land-line phones and of tripmaking, including durations and temporal profiles of both for different population groups. This provides some leads for future analysis and some principles for new types of data collection.

To conclude the data discussion theme, Stopher describes a number of experiments to simulate household travel data. These methods start from suitable demographic data sources, and employ parameter distributions from travel surveys that are carried out in other geographic areas, but which support comparable demographic definitions. Such simulations are attractive cost-savers. He presents statistical evidence that, with some cautions, useful datasets can be developed, and through the microsimulation of households, the datasets can be projected into the future, assuming that underlying distributions of travel characteristics remain stable. Research into a number of enhancements are suggested, including the introduction of random components to make the variance of characteristics of expanded households more realistic, the simulation of trip tours rather than trips, and the Bayesian updating of the simulations.

The chapter by Buliung *et al.* opens the set of four papers on *modelling issues and advances*. It moves the focus more directly on to integrated land-use and transportation models, including a thorough historical review, tying in how emerging travel behaviour approaches (dealt with in previous chapters) and geocomputing technologies are being incorporated into such models and their expansion. They pay particular attention to the need for interfaces that can be readily understood by model users who are not GIS experts. Ray and Claramunt subsequently demonstrate how a distributed computing environment offers a new avenue for the simulation of dynamic systems, and the exploration of the underlying properties and behaviour of large real-world systems. Despite remarkable recent increases in computing power, they illustrate from an example of passenger flows in an airport that multi-agent simulations can easily push the limits of memory and processing of a single computer. Distributed computing platforms seem well suited to future extensions of such simulations, especially those incorporating the complexity of collaboration between agents.

Hunt and Abraham follow with a chapter devoted more exclusively to the structure of land-use models, emphasizing how the many system elements (businesses, households, etc.) behave and interact. Their approach to simulating markets for goods, services, labour and space involves allocating flows between zones using nested logit models, according to exchange prices and transport (dis)utilities. It has been applied not only to urban regions, which are the main preoccupation of ILUTE frameworks, but also to larger geographies, namely the American states of Oregon and Ohio.

The book ends with a chapter by Bolduc *et al.* that explores recent extensions of econometrics to represent idiosyncratic aspects of choice processes, and to allow for apparently irrational decisions. They describe the development of Hybrid Choice Models (HCM) that build on basic Random Utility Theory, but include key elements of the cognitive processes believed to underlie choice, including aspects of history and context, the formation of perceptions, heterogeneity and unobserved constraints. Until recently, HCM methods have been demonstrated only on small-scale models with two or three alternatives: their chapter explores the application of HCM to large scale models with numerous dimensions and flexible formulations of the choice process, such as would be needed for urban simulations.

In an ideal world, all disciplines would truly collaborate together, surveys would have 100% response rates, all data would be shared, all models would be open source, and behavioural assumptions would not be needed (since the models would capture all aspects of them!). But questions remain. How will we know if we are on the right track? How far off are models that

that would reasonably satisfy the behavioural standards/frameworks outlined in this book? How will we know we got there? We encourage everyone to keep asking these questions, and look forward to future stages of the work, including the second International Colloquium to be held in Toronto in 2005.

Sean T. Doherty Martin E.H. Lee-Gosselin

Integrated Land-Use and Transportation Models: Behavioural Foundations
M. Lee-Gosselin and S. Doherty (Editors)

1

A Dynamic Understanding of Travel Demand: A Sketch

K.W. Axhausen, IVT, ETH Hönggerberg (HIL), Zürich, CH

Abstract

This paper highlights two under-researched areas in transport planning: a) the structure of travellers' social networks in connection with their basic mobility choices; and b) the day-to-day dynamics of activity generation. This discussion is set against the background of overall economic changes during the last 50 years, which have funded the behavioural changes observed.

The discussion is conceptual, as there is not enough empirical evidence available to conclusively demonstrate the trends identified. The paper argues that the spatial dispersion of the social networks in combination with the residential location choices were, and will be, the main factor in the observed increases in miles travelled. The conceptual framework for the day-to-day dynamics is based on the rhythmic patterns of need accumulation and their satisfaction through activities and long-term projects.

Averages and Dynamics

Transport planning claims to understand the dynamics of the systems it analyses: transport infrastructures and travel behaviour. A closer look reveals that certain dynamics are well understood, others less so, some not at all. The reasons for these partial understandings are the previous and current professional tasks of the discipline and the preferred modelling tools.

Understanding the systems as being in equilibrium is not a motivation to search for change (See also Manski, Forthcoming). No one doubts the usefulness of the mathematical tools provided by the equilibrium concept, but the differences between two different equilibria do not necessarily tell the analyst anything about the path of the implied system and behavioural changes.

In this paper, we cannot set out a theory for everything; we cannot even provide a description of every substantial dynamic process in the transport system. We want to highlight two specific areas against the background of the overall economic change of the last 100 years. The paper is speculative, despite of using graphs to illustrate its argument at various stages. It has to be speculative, as there is little empirical base on which to draw for the two main issues it wants to address: a) changes in the spatial distributions of social networks and b) intrapersonal dynamics of daily travel behaviour.

Transport planning describes behaviour during particular times and in specific locations. It uses the parameters estimated from it to forecast changes, which may occur with new structures of generalised costs, new population levels and distributions, and activity opportunities. Notions of constant time budgets and money budgets, as well as constant tastes, help to justify this approach (Zahavi, 1979; Becker, 1976). While constant interpersonal time budgets have been reported in the literature, money budget shares have not been constant over the last 50 or 100 years (see below), which raises doubts about the approach. In addition, even 25 years after Zahavi's work, no credible mechanism has been suggested to explain the observed stabilities in average travel times over the years. These stabilities are surprising given the overall change in social and economic conditions.

In the first part of the paper, we will inform readers about the scale of those changes mentioned above by highlighting a small number of key developments. In the second part, we will discuss likely impacts of the anticipated further changes on the physical distribution of travellers' social networks. Finally, in the third part, we will outline a framework for the daily dynamics of travel behaviour which translate these bigger changes into daily practise.

A Brief Look Back

The key development of the last 100 years is the translation of the accumulated differences between productivity and population growth (See Figure 1.1) into mass consumption. Key sectors in this process were housing and travel demand. In housing, the trend to smaller households combined with the use of cheaper suburban land made, for example, about $50m^2$ available for each Swiss by 2000[1]; an increase of 66% since 1950 (Rumley, 1984).

[1] The numbers for Germany are substantially lower: 41 m^2/person in 1998 (Statistisches Bundesamt, 2000) and substantially higher for US metropolitan areas: 59 m^2/person for blacks and 73 m^2/person for whites (Kahn, 2001).

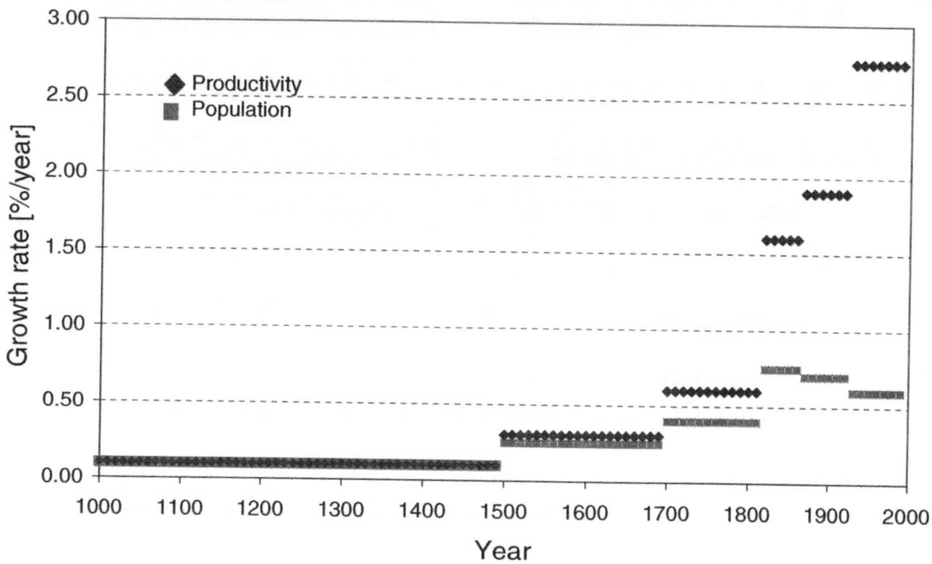

Source: adopted from Galor and Weil (2000)

Figure 1.1 Western European productivity and population growth rates since 1000

In terms of travel demand, the Swiss, like most other populations of the OECD area, have achieved almost complete motorisation, defined here as (almost) universal licence ownership and (almost) universal car availability of licence holders. Figure 1.2 shows that women raised during and directly after World War II will be the last group to be excluded from this. A share of non-licence holders will remain, but very few out of a reasoned voluntary stance. The licence has become too much of a badge of adulthood to be ignored. Car availability among licence holders is impressive for all age groups. It is clear that certain younger age segments (to some extent) do not have the resources, which delays the universality of access[2]. Virtually everyone has access to a parallel technology, telephony. Its process of adoption started earlier than that of the car, ran parallel with it for most of the post-war period and is now in a new phase, during which households are slowly starting to replace the joint household land-line access with mobile phones for each member.

The access has been translated into use, and the associated speed gains translated – in the main - into longer distances (Figure 1.3 and Figure 1.4). The discussion about the translation rate is still inconclusive, but the majority opinion suggests a value of 1.0, i.e. a constant travel time budget (e.g. Schäfer, 2000). The pattern is consistent with the on-going process of residential dispersion visible in all OECD countries.

[2] The population of licence holders without access would be an interesting subject of in-depth study.

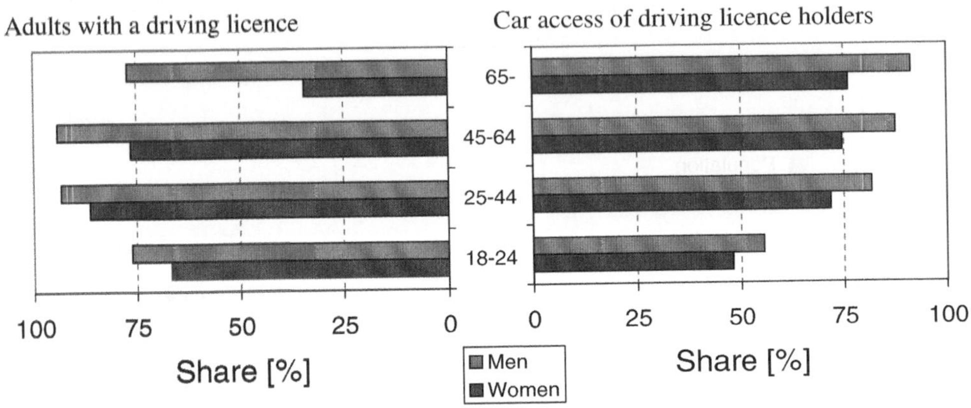

Data source: Mikrozensus Verkehr 2000 (ARE and BfS; 2001)

Figure 1.2 Access to motorised travel in Switzerland by age

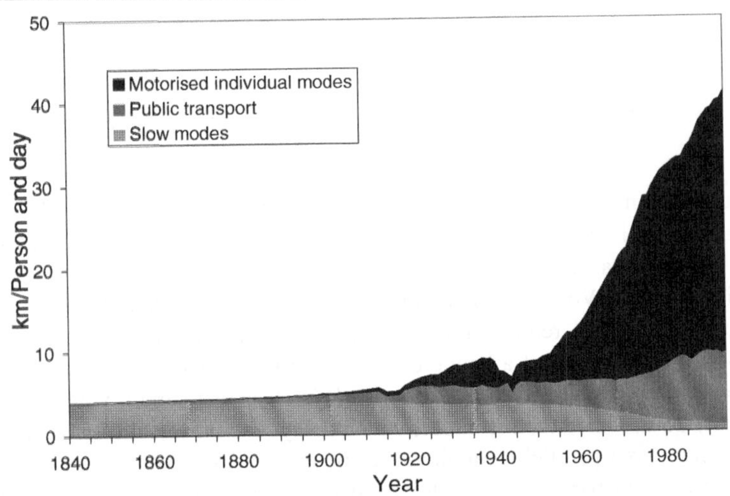

Source: adopted from Gruber (1998)

Figure 1.3 France: Estimated daily personal travel [km/day]

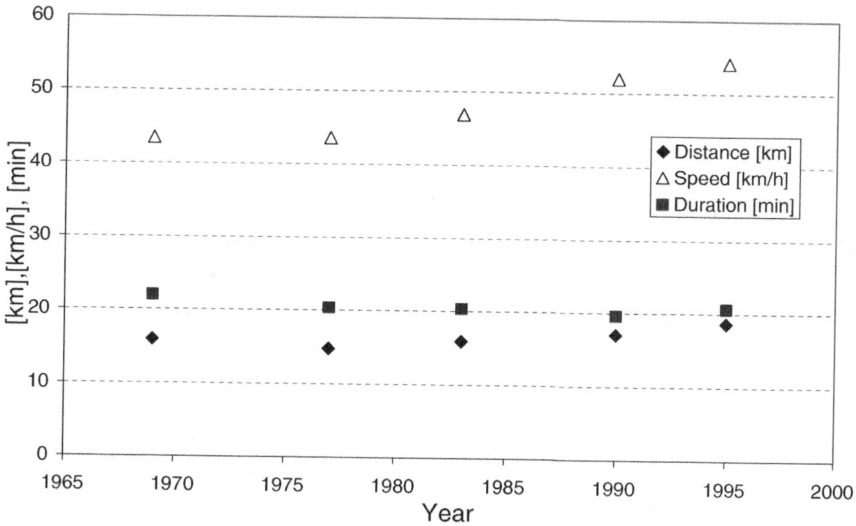

Source: http://www.publicpurpose.com/ut-6995commute.htm using US NTSurvey data

Figure 1.4 USA: Development of daily commute

These increases in housing consumption, distances travelled and telephone contact have led to a rearrangement of increasing household budgets (Figure 1.5). Expenditure for food dropped massively as a result of industrialised retailing and agriculture, while the shares for leisure, education, travel, and communication (respectively) have approximately doubled since 1913, for example in Switzerland. The larger shares for leisure and education reflect the longer life expectancy in Western Europe, the longer time spent in school, the shorter work week and longer retirement. The adoption of the car is the main reason for the increase in the share for travel and communication.

The motorised, connected, and spatially dispersed population of today with its access to cheap travel must have different patterns of work and sociability than the population of 100 or even 50 years ago. The next section will discuss one element of this change: likely changes in spatial structures of the overall populations' social networks.

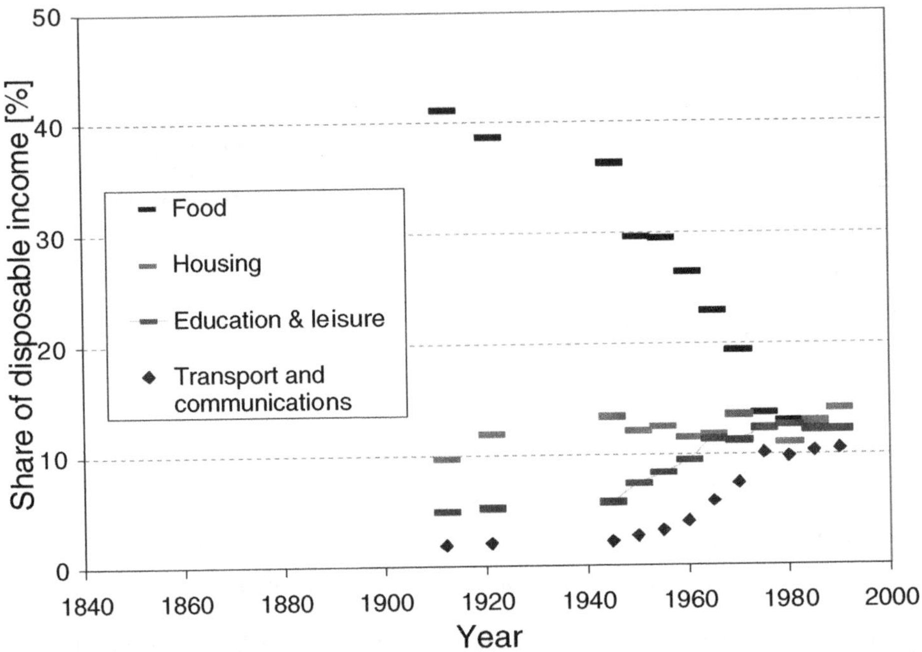

Source: adopted from Widmer (2001), 19

Figure 1.5 Switzerland: Shift in household consumption patterns

Activity Space and Social Space

Daily life revolves around friends, family, work, school, and shopping (See for example Wellmann, 1996 and his website[3], or Törnqvist, 1970 and Burt, 2000 for similar work on the contact systems of firms). The distribution in space of these commitments/opportunities shapes the activity space of a person in his or her everyday environment. The size of this activity space determines a person's consumption of transport services, obviously including those which are produced by the person walking or driving. If we accept this idea, then increasing distances between those poles of a person's activity system has long reaching and permanent effects: a dynamic which cannot be reversed quickly, given the investments people make into their work, their family, and friendships.

The personal contacts of a certain quality from the social networks of a person. A person belongs to the network, if the contact has a particular intensity, quality and permanence. It can include persons at a remove, if the contacts with and between intermediaries is strong enough. Clearly, this will vary by the social context of the contact: work, studying, kindergarten, bible study group, voluntary fire brigade, etc. It is therefore useful to think of a person's network to

[3] See www.chass.utoronto.ca/~wellman.

consist of different, functionally defined (sub)-networks. Clearly, some persons will be a member of more than one of those defining the overlap between them. The definition above implies also, that many, maybe most of our daily interactions occur with persons, who are not members of our social network, even if know them, as for example the cashier at the supermarket, the local policeman, or students in a teacher's class etc.

Wellman (2000, 2001) and Hampton and Wellman (2001) argue that society has been transformed from something based on *little boxes* to a network society; this could be defined as a change from a situation where a person's social networks were strongly overlapping and spatially coherent to a situation where the membership overlap will be small and the members spatially dispersed. The reasons behind this are the breakdown of the walking environments dominant before the car and the associated weakening of the closed social milieus of yesterday: We do not live next door to our work colleagues anymore, our aunt does not serve us in the corner store, nor is the foreman/union steward/local councillor godfather to our children[4].

Spatial structure of social networks

Wellman (1996) provides some information about the distances between network members' residences based on an earlier Toronto survey. He highlights the large number of contacts living relatively close by or working with the respondents, but mentions that these are not necessarily the emotionally/socially important contacts, except – obviously - those in the respondents' households. Unfortunately, his work has no time dimension so that no trends can be identified. Still, based on the general spatial trends one can assume the following:

- The spatial distribution of social networks' members will be wider than in the past, i.e. the average distances between the home locations will have grown
- Average social networks will be less coherent, i.e. fewer people will share multiple affiliations today than in the past
- The memberships will also overlap less in spatial terms, i.e. vis-à-vis their residential locations and activity spaces[5]

In transport terms these trends and assumptions imply longer distances when people want to meet face-to-face. The current unwillingness to replace face-to-face contact might fade over time, but currently most analysts agree that such contact is required to build up the trust required for on-going, productive relationships, in both private and work domains (Sobel, 2002).

Distance is the first perspective from a transport point of view, but frequency the second. Again, the literature is silent on this question, but one can argue for the following assumptions:

---

[4] It should be noted that this spatial density of the social networks never applied to aristocratic or later intellectual families.

[5] Current telecommunication technologies allow in particular the maintenance of contacts with friends and acquaintances from non-local earlier residential locations, who might have no overlap at all with the local social networks of a person.

- People have a larger set of active contacts today than in the past, e.g. 50 years ago.
- The contacts are spread across more social networks than in the past.
- The maintenance of each contact requires more effort at the same level of quality than in the past.

Maintaining contact requires time and effort. One needs time to inform the other, to learn about the other, and to readjust to the other and his/her changed circumstances and contexts. Modern telecommunication and travel have become substantially cheaper, which reduces the generalised cost of personal contact (See Figure 1.6 for a transport example). In addition, cheap ways exist today to duplicate messages, e.g. xeroxed Christmas letters or email mailing lists. While these do not have the same quality as a handwritten letter as a token of interest in the other, they are a sign of life, and help the other side to stay informed. Still, the effort of maintaining the contact lies nearly exclusively with the person him (her)self, as less dense social networks might not convey information as reliably as in the past: the grapevine on which a wide range of information and assessments travelled has withered to an unknown extent[6,7]. Also, the opportunity for chance encounters is smaller when the overlap of the activity spaces of the network members becomes smaller. The on-average reduced amount of overlap in contexts and social networks should increase the amount of information needed to be conveyed for each party to understand the other's position and situation on each renewal of contact.

The developments described above could offset each other; thus no clear trend can be postulated for the distribution of the contact frequencies among the members of a social network. The reduction of the generalised costs of travel and telecommunications[8] speak for more contacts overall, and relatively more contacts with persons further away, as long distance travel and communication had a steeper drop in their generalised costs than local travel and telecommunication. The increase in the required total duration of the direct contacts (face-to-face, personal email, telephone conversation) implies a concentration on fewer people than in the past combined with more superficial contacts with a larger number of people overall.

Given the acknowledged trend away from strongly localised social networks, the two arguments indicate that the activity space of an average person should be larger now than in the past. It is necessary to point out that this activity space consists of two elements: the local activity space[9] and the non-local contact space. The characteristics of each can be chosen by individuals depending on their preferences/habits. Examples might include: the owner of a platinum frequent-flyer card who walks to everything locally, the suburban car-owning couple,

---

[6] Commercially distributed gossip about the rich and famous, on the other hand, has increased by orders of magnitude.

[7] The role of the internet as a supplier of personal information and particular of assessment of reputations is still unclear, although the experiences with cumulative ratings of sellers on web-based auction sites are an interesting indication of future possibilities.

[8] The weighted sums of the relevant monetary costs, the time taken and the comfort of the trip or contact. In the case of telecommunication this includes: direct and fixed charges; search time for the other person; quality of connection and its bandwidth; for the complementarity between travel and telecommunication see Claisse and Rowe, 1993.

[9] The local activity space will be duplicated for persons and groups which follow a regime of season migration, a practise which has always been important (See Grieco, 1988 and 1996 or Weber, 1976 for European examples).

who travel regularly to their weekend home in the region, or the cyclist who visits her partner 400 km away every weekend by train.

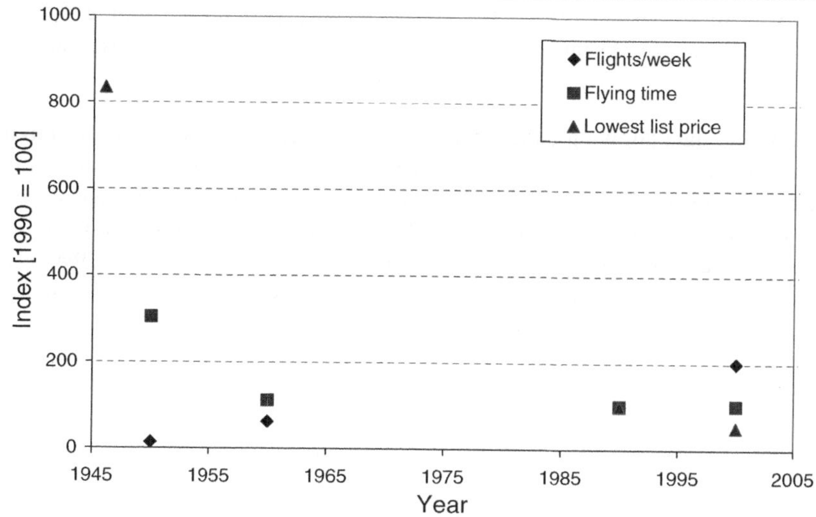

Source: adapted from Widmer (2002)

Figure 1.6 Example for the change in the generalised cost of travel: Zürich – New York

The further dynamics of these trends are difficult to forecast. The passing of the cohorts born before World War II means that the full motorisation will soon be the rule in Europe, as it already is in North America: licence ownership rates of 85-90% combined with car access rates of licence owners of around 80-90% (See Figure 1.2 above for an example). This will increase car-based travel per head further, but the relative contribution to social totals will be small, as the big growth spurt due to initial motorisation is past in the already industrialised countries. The size of the impact will depend on further rate of suburbanisation, i.e. the development of the average speeds of motorised travel. While this normally means car based-travel, regional rail-systems can have substantial impacts as well. In addition, high-speed rail systems can act as super-regional *stopping trains*, if the price system favours commuting use.

Activity spaces

There is an increasing amount of literature, which highlights the importance of lifestyle patterns for the structure of a people's travel behaviour (e.g. Götz, Jahn und Schultz, 1998). Central to the spatial definition of a lifestyle is the combination (Simma and Axhausen, 2001; Axhausen, Scoot, König and Jürgens, 2001) of:

- Work/school location
- Housing location and the style of neighbourhood
- Mobility tool ownership (bicycle ownership, car ownership/access, public transport season ticket).

Figure 1.8 shows the impact of the last three factors on the size of the activity space. The measure of the activity space is very crude (two-dimensional 95% confidence intervals of the geo referenced locations visited during the six-week diary period, see Figure 1.7; for the Mobidrive data source see Axhausen, Zimmermann, Schönfelder, Rindsfüser and Haupt, 2002).

There is currently no data allowing us to trace the development in the sizes of the activity and contact spaces of the public. As discussed above, motorisation, suburbanisation, and the network society should have increased it substantially over the last 50 years, but we have no idea whether it will continue to grow and, if so, how quickly. Has airline deregulation since 1980 and the vigorous competition in the telecommunication markets since 1990 given its growth a new push? We do not know, but we should, as these trends have substantial social externalities.

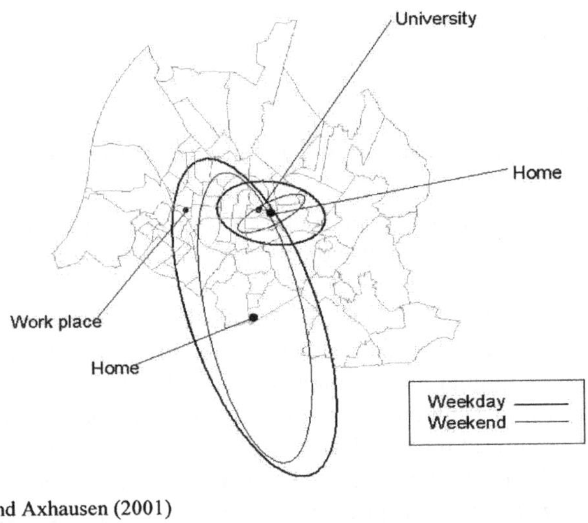

Source: Schönfelder and Axhausen (2001)

Figure 1.7 Local activity spaces of two persons by type of day (Measured as 95% confidence intervals of all locations visited during a six-week diary)

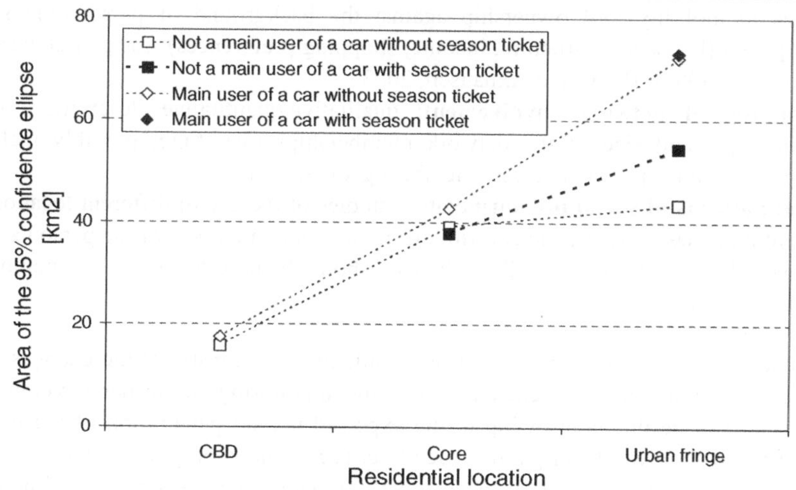

Source: Axhausen, Scott, König and Jürgens (2001)

Figure 1.8 Variation of the size of the activity space by home location, car and public transport season ticket ownership

We cannot review this topic here, but it is clear that the current patterns of sociability leave many people worried (see Putnam, 2000[10]; Sobel, 2002; Franzen, 2000 or for a selective review Axhausen, 2000). Does the stronger selectivity[11] of social networks today improve the quality of life? Does reduced local networking lead to more reliance on commercial/government services in times of crises, such as acute illness, or permanent care? Is less generalised trust (see Seligman, 2000 for a discussion) a corollary of less dense local networks? (e.g. the pervasive feelings of a lack of safety reported by many urban and suburban residents.)

Research issues

Home, work, mobility tool ownership, and social network membership interact and depend on each other. The further trajectory of these interactions will, to a large extent, determine the shape of personal travel in the future. Unfortunately, the current focus of travel behaviour research is on daily travel behaviour with little attention given to longer-term commitments of people and households. Some effort should therefore be redirected to the following types of studies:

[10] Putnam's focus on the decrease in organised sociability misses the range of informal, but still regular contact patterns, which might have taken up the time freed up.

[11] The ability to select one's social contacts given the spatial range the car makes available to most households.

- **Mobility biographies**[12] documenting changes in home and school/work location plus changes in mobility tool ownership against the background of personal/household development (household structure, income, shopping preferences); for partial examples see Roorda, 1998 or Beige and Axhausen, 2003.
- **Biographies of network involvement**: qualitative/quantitative description of the number, type, and size of the network memberships over time; possibly including attempts to localise past and current members geographically.
- **Spatial patterns of social life**: quantitative studies of the use of different locations for different purposes and for contacts with different people/members of the person's social networks. If possible, a snowball mechanism should be used to capture all members of a social network.

The first two are needed to reconstruct, (at least to attempt to reconstruct) the changes of the last 50 years, while the third is needed for on-going monitoring of further developments. These are expected to be dramatic owing to the expected market penetration of the internet, anticipated further withdrawal from public space (e.g. use of malls or gated residential areas), further decreases in the cost of long-distance travel, projected increases in second and third residences, extrapolated increases in residential mobility or, alternatively, long-distance commuting (weekend-commuting)[13].

DAY-TO-DAY DYNAMICS

Just as it has generally ignored the biographical dimension of travel behaviour, transport planning methodology has also, as a rule, ignored the day-to-day dynamics of traveller behaviour at the individual level. There were good reasons for an engineering subject, such as transport planning, to limit its field of enquiry to what was perceived to be necessary to address the questions at hand (Weiner, 1988). However, this does not really excuse the lack of curiosity obvious in the larger-than-required gaps in data availability and theoretical understanding of this issue. It was only the interest in transport telematics, which has put the questions of day-to-day variability on the table during the last decade, that made it clear that learning and short-term behavioural change are crucial to the assessment of the business and economic case for this bundle of technologies.

Conceptual framework

The conceptual framework suggested below can therefore only be a sketch, as substantial issues have not yet been covered by empirical work. They will be flagged below. The building blocks are the following concepts (See also Axhausen, 1998):

---

[12] It would be helpful, if travel diaries would routinely capture all these elements in cross-section, for example season ticket ownership is frequently missing, as is information about exact locations of work, school, home, and regular shopping.

[13] To my best knowledge, the growing acceptance of weekly long-distance commuting in place of a residential move has not been documented in the transport literature. Long-distance commuting also raises the issue of divided activity spaces and activity repertoires.

- **action stream, activity**, and **activity calendar** to describe the structure of the actions in the short term.
- **activity space** and **knowledge space** to structure the **mental mapping** of the environment.
- **activity repertoire, needs, projects, commitments** and **self image** to characterise the medium and long-term.

Respondents, usually narrators, will generally report daily life at a level of abstraction transport planners and time use researchers call **activities** (Szalai, 1972; Ås, 1978; Grønmo and Harvey, 1982; Jones, Dix, Clarke and Heggie, 1983): they refer to a group of actions, which can be understood as a coherent whole: supervising the children, going shopping, working, etc. The detailed **action stream** of individual acts, with all its normally confusing chaos, can be recalled by a respondent, but only with great effort; the chaos is because the performance of multiple activities is often intertwined, especially away from a structured work environment. The individual acts of the action stream can be named and described, but they make full sense only as part of the larger activity. For example, a phone call is part of organising an event, chopping the carrots contributes to the soup being prepared, and removing a weed readies a garden for the summer.

Activities have both purpose and meaning: the purpose identifying the instrumental goal of the activity, and the meaning indicating the expressive goal (See Axhausen, 1998 or Hargreaves – Heap, 1989). Also, the immediate utility of an activity should be distinguished from a contribution towards self-realisation given a person's **self-image**. The **activity repertoire** is the set of activity types (specific actions streams at particular - types of - locations and times), which a person knows of, or has performed in the past. The size of the repertoire changes over time as it is enlarged through experimentation and learning or reduced through forgetting and diminished physical or mental capabilities.

At any one time, a person has a subset of these activity types flagged as desirable or necessary to do in his or her **activity calendar**. The calendar can include multiple instances of the same type scheduled for different days, e.g. work, picking up the children from school, going grocery shopping. This notion has been put forward by a number of authors (Nuttin, 1984; Doherty, Axhausen, Gärling and Miller; 2002; Gärling, Kalén, Romanus, Selart and Vilhelmsson, 1998). The empirical work until now was based on prompted recall, which raises the question whether the respondents think about their daily life in this way. Doherty and Miller (2000), for example, asked their respondents to fill out an agenda for the next week with the activities they had already planned or committed themselves to. While the respondents had no problems with the task, the question remains, if this thinking in "agendas" is natural for all respondents (see, for example, the discussion in Kiel, Post and Dollase, 2000 on "time-planning" and "non-time-planning" persons).

The concept of the calendar, even without its psychological relevance completely proven, is appealing as the locus of the decision and optimisation task of daily activity scheduling. It is constantly refilled with activities deriving from the **long-term commitments** of a person: school, work, child care, and care for others, and household work. The time requirements for the basic **personal needs** of sleep, eating, and personal care have to be integrated. The time requirements of these two types have to be balanced against the freely chosen ones deriving from **projects,** individually planned activities or spur-of-the-moment choices. A project is any group of activities, which have a joint goal. This can be as simple, as preparing dinner for friends or as complex as building a house. Depending on the complexity of a project, it will be divided into parts or sub-projects. The time horizon of a project varies, some lasting a day or two, others stretching over years. For very long-duration projects the concept starts to blend with the idea of *life goals* (Nuttin, 1984).

The idea of the project, as well as of the long term commitment involves again the social networks discussed above, as both normally require co-ordination with others. This additional complexity will be ignored for the moment.

The calendar, as seen here, is the instrument which allows the person to schedule time and other resources to perform his or her plans. Activities will be added, modified and sometimes removed before they are finally executed in a particular form. The variables describing the final form of activity are clear:

- Type of activity
- Location
- Timing
- Duration
- Size and composition of the company (i.e., others who are involved)
- Money expenditure, physical resources used, social resources used[14]

The variables describing the logic of the scheduling and rescheduling are less clear, in spite of the recent work in the area (see Ettema, Borgers and Timmermans, 1995). Nor is there a clear understanding how the calendar is filled up, how much buffer time is provided, and how the "when and why" of any activity are decided.

Recent work using long-duration travel diaries has shown that there are strong temporal rhythms for certain types of activities. The trade-offs made between activities and their characteristics in detail are still unclear (Schönfelder and Axhausen, 2001; Winston, 1992). It is known that the basic physical needs of a person are driven by diurnal rhythms. These in turn will produce repeating patterns for those activities associated with satisfying those needs (shopping, returning to a place for sleep, etc.). The rhythmical patterns identified for activities satisfying higher level needs, such as personal leisure, time with friends, contributions to civic activities, imply that the behavioural dynamics of the medium term (week to year) are driven by processes of need accumulation and satisfaction. Longer-term projects can be understood to address even higher level needs or aspirations by organising complex series of activities. There is very little awareness in transport behaviour research about the number and type of

---

[14] From the perspective of this section, the two most important social resources are the time one has saved with others and the favours one can draw on to achieve particular objectives.

projects in which people engage in at any time, or about the way they are formulated or selected.

While it is attractive for the analyst to believe that the calendar leads to something approaching an overall scheduling optimum for any time frame considered, e.g. a day or week, empirical work has not shown that this is the case. People leave substantial parts of the day unplanned to allow for unreliability in the activity and transport system and for unexpected activity opportunities. These times will be filled at short notice, clearly with activities from the activity calendar, but without an overall optimisation across a longer time period. At this time, we have no idea if this scheduling strategy is optimal given the unreliability of the systems and the probabilities of chance occurrences. What is clear is the stress experienced by persons who live *overbooked* lives. This group is at the mercy of the reliability of the transport and communication systems, and they experience high rescheduling costs when they do want to take advantage of unforeseen opportunities.

The long-term diaries mentioned above also allow tracking of the **activity space** of a person; the area which the person uses for his or her activities. The analysis of the Mobidrive data allows only the understanding of the local part of the respondents' activity spaces. One has to add the non-local parts, which derive from visits to friends, excursions, holidays, and previous residential locations. Furthermore, the activity space is only a subset of the **knowledge space**, which covers all locations a person knows of through word of mouth, reading, advertising, or watching television or films.

In decision making, the main difference between the activity space and the remaining knowledge space is the trust level a person can have in his or her **mental mapping** of the generalised costs of those locations and activities. The generalised costs include the effort to get there (generalised costs of travel) and resources used for the activities. For those places and activities already visited, perhaps frequently visited, the traveller can trust his or her own experiences; there is an understanding of the distribution of possible outcomes. Depending on his/her risk averseness, the traveller can compare alternatives reliably. For the rest of the knowledge space, the traveller has to rely on second-hand information, which has to be weighted by assessments of the reliability of the sources.

The updating process of the mental map for the various cost elements is essentially unknown. Most previous work assumed a priori learning mechanisms, but did not derive them from empirical data. For an exception and a detailed literature review see Oladeinde, 2000.

Research questions

The conceptual framework sketched above sees two main processes at work in the activity system: first, the rhythmic patterns of need build up and satisfaction over the short term, and second, the pursuit of longer-term projects, which generally will be one-offs during a person's lifetime. The way these will be formulated and realised depends, in the medium term, on the longer-term commitments discussed in the section above. Given the longer-term projects and those resource allocations (home, work, car, season tickets) the person will select activities from his or her activity calendar which is constantly being refilled. The exact form of the activities will depend on the **personal world** of the traveller. Mental map, activity repertoire, and knowledge space are three ways of seeing this personal world. The mental map focuses on spatial relations and the generalised costs of travel; the activity repertoire describes the kind and form of possible activities and their generalised costs and qualities; and knowledge space emphasises the possible locations of different types of activities.

The main research questions concern then the following issues:

- **Activity calendar**: How is it being filled? How much unscheduled time do people retain until what point in time? How do people reschedule activities? How do they judge the quality of on-going activities?
- **Updating of the personal world**: What does it take to change the mental maps? What mechanisms are used to integrate different sources of information? Do people actively attempt to enlarge their personal world? If so, how? How are general expectations formed?
- **Projects**: Do people use the same concepts as the one suggested here? How do they select projects? What changes happen to projects and why? Under what circumstances are projects abandoned? What number do people pursue at any one time?

The answers to these research questions would allow us to get away from the simplistic notions of optimisation embedded in the current state of the art transport models (Ben-Akiva, Bowman and Gopinath, 1996). They would also allow us to avoid non-causal pattern recognitions approaches, as those suggested in the state-of-the-art Albatross system (Arentze and Timmermans, 2000).

OUTLOOK

The paper has attempted to direct attention to two areas overlooked in previous transport behaviour research: spatial and social structure of travellers' social networks, and an understanding of day-to-day dynamics of travel. It has argued that an understanding of the first issue is crucial for the assessment of the further dynamics of overall transport consumption. For a given number of joint activities the locations of friends, family and colleagues decide the amount of miles travelled, especially for leisure.

The spatial distribution of the social network is tied in with the basic spatial/travel choices of a household (home, work and school locations bound together by the mobility tools owned). These basic choices also frame the day-to-day dynamics of travel given the overall time-space regimes of society (Hägerstrand, 1970). In turn, the experiences of daily life will lead to reassessments of these choices and possible change.

The discussion of day-to-day dynamics has centred on the process of activity generation and its interaction with the personal world of the traveller. Long-term commitments, higher level needs, and projects generate activities, which are coordinated by travellers through their activity calendars.

This conceptual framework requires both empirical testing through observation and surveys and application testing with suitable micro simulation tools (For references see Nagel and Axhausen, 2001). The first test is required to determine whether the framework is valid as a description of the behavioural dynamics. The second test is necessary to prove that the framework offers a better way of forecasting and assessing transport policy.

ACKNOWLEDGEMENTS

The paper is based on discussions and presentations at a number of occasions: the workshop organised by the Mobility Network in Cambridge (February 2002), the Focus Group 3 workshop of the STELLA Project in Bonn (April 2002) and finally the MCRI-GEOIDE International Colloquium in Quebec (June 2002). The author would like to thank the organisers for the opportunity and the feedback.

REFERENCES

ARE and BfS (2001). Mobilität in der Schweiz: Ergebnisse des Mikrozensus 2000 zum Verkehrsverhalten, Bundesamt für Raumentwicklung (ARE) and Bundesamt für Statistik (BfS), Bern[15].

Arentze, T. and H.J.P. Timmermans (Eds.) (2000). *Albatross: A learning-based transportation oriented simulation system*, EIRASS, Eindhoven.

Ås, D. (1978). Studies of time-use: problems and prospects, *Acta Sociologica*, **21**(1), 125-141.

Axhausen K.W., D.M. Scott, A. König and C. Jürgens (2001). Locations, commitments and activity spaces. Paper presented at the Survive Project Workshop, Bonn, December 2001, *Arbeitbericht Verkehrs- und Raumplanung*, **96**, IVT, ETH Zürich, Zürich.

Axhausen, K.W. (1998). Can we ever obtain the data we would like to have? In: *Theoretical Foundations of Travel Choice Modelling* (T. Gärling, T. Laitila and K. Westin, eds.), pp. 305-323 Elsevier, Oxford.

Axhausen, K.W. (2000). Geographies of somewhere: A review of urban literature. *Urban Studies*, **37**(10), 1849-1864.

Axhausen, K.W., A. Zimmermann, S. Schönfelder, G. Rindsfüser and T. Haupt (2002). Observing the rhythms of daily life: A six-week travel diary. *Transportation*, **29**(2), 95-124.

Becker, G.S. (1976). *The Economic Approach to Human Behavior*. University of Chicago Press, Chicago.

Beige, S. and K.W. Axhausen (2003). Besitz von Mobilitätsressourcen und deren Nutzung sowie Änderungen des Wohnortes. Research report UNIVOX 2003: IG Transport, *Arbeitsbericht Verkehrs- und Raumplanung*, **168**, Institut für Verkehrsplanung und Transportsysteme (IVT), ETH Zürich.

Ben-Akiva, M.E., J.L. Bowman and D. Gopinath (1996). Travel demand model system for the information era. *Transportation*, **23**(3), 241-266.

Burt, R.S. (2000). The network structure of social capital. In: *Research in Organization Behaviour* (R.I. Sutton and B.M. Straw, eds.), **22.** JAI Press, Greenwich[16].

---

[15] http://www.admin.ch/gvf/inhalte/grundlagen/mikrozensus_2000.pdf

Claisse G. and F. Rowe (1993). Domestic telephone habits and daily mobility. *Transportation Research*, **27A**(4), 277-290.

Doherty, S.T. and E.J. Miller (2000). A computerised household activity scheduling survey. *Transportation*, **27**(1), 75-97.

Doherty, S.T., K.W. Axhausen, T. Gärling and E.J. Miller (2002). A conceptual model of the weekly household activity-travel scheduling process. In: *Travel Behaviour: Spatial Patterns, Congestion and Modelling* (E. Stern, I. Salamon and P.H.L. Bovy, eds.), 233-264. Edward Elgar, Cheltenham.

Ettema, D., A. Borgers and H. Timmermans (1995). SMASH (Simulation model of activity scheduling heuristics): empirical test and simulation issues. Paper presented at the International Conference "Activity based approaches: Activity scheduling and the analysis of activity patterns", Eindhoven, May 1995.

Franzen, A. (2000). Does the Internet make us lonely? *European Sociological Review*, **16**(4), 427-438.

Galor, O. and D.N. Weil (2000). Population, technology, and growth: From Malthusian stagnation to the demographic transition and beyond. *American Economic Review*, **90**(4), 806-828.

Gärling, T., T. Kalén, J. Romanus, M. Selart and B. Vilhelmsson (1998). Computer simulation of household activity scheduling. *Environment and Planning A*, **30**, 665-679.

Götz, K., T. Jahn and I. Schultz, (1998). Mobilitätsstile: Ein sozial-ökologischer Untersuchungsansatz. *Forschungsbericht Stadtverträgliche Mobilität*, **7**, Forschungsverbund City: mobil, Frankfurt am Main.

Grieco, M. (1988). Keeping It in the Family: Social Networks and Employment Chance. Tavistock Publications, London.

Grieco, M. (1996). Workers' Dilemmas: Urban and Social Networks and Labour Circulations. Routledge, London.

Grønmo, S. and A.S. Harvey (1982). Time-diaries and trip-diaries: a comparison. Paper presented at the 10th World Congress of Sociology, Mexico City.

Gruber, A. (1998). *Technology and Global Change*. Cambridge University Press, Cambridge.

Hägerstrand, T. (1970). What about people in Regional Science? *Papers of the Regional Science Association*, **24**, 7-21.

Hampton, K. and B. Wellman (2001). Long distance community in the network society: Contact and support beyond Netville. *American Behavioral Scientist*, **45**(3), 477-496.

Hargreaves-Heap, S. (1989). *Rationality in Economics*. Blackwell, Oxford.

Jones, P.M. and M.I. Clarke (1988). The significance and measurement of variability in travel behaviour: a discussion paper. *Transportation*, **15**(2), 65-87.

Jones, P.M., M.C. Dix, M.I. Clarke and I.G. Heggie (1983). *Understanding Travel Behaviour*. Gower, Aldershot.

Kahn, M.E. (2001). Does sprawl reduce the black/white housing consumption gap? *Housing Policy Debate*, **12**(1), 77-86.

[16] http://gsbwww.uchicago.edu/fac/ronald.burt/research/

Kiel, P., C. Post and R. Dollase (2000). "Ich plane meine Zeit nicht" – Vergleich der Motive bewusster versus abgelehnter Zeitplanung bei selbstdeklarierten Planerinnen und Nichtplanerinnen. In: *Temporal Muster: Die ideale Reihenfolge der Aktivitäten* (R. Dollase, K. Hammerich and W. Tokarski, eds.), pp. 53-64. Leske und Budrich, Opladen.

Manski, C.F. (Forthcoming). Economic analysis of social interactions. *Journal of Economic Perspectives.*

Nagel, K. and K.W. Axhausen (2001). Workshop report: Microsimulation. In: *The Leading Edge in Travel Behaviour Research* (D.A. Hensher, ed.), pp. 239-246. Pergamon, Oxford

Nuttin, J. (1984). *Motivation, Planning and Action.* Leuven University Press, Leuven.

Oladeinde, F.A. (2000). A model of traveller learning with applications to traffic information systems. Doctoral dissertation, University of London, London.

Putnam, R.D. (2000). *Bowling Alone: The collapse and revival of American community.* Schuster and Schuster, New York.

Roorda, M.J. (1998). Toronto area car ownership study: A longitudinal survey and a preliminary analysis of results. MAS dissertation, Department of Civil Engineering, University of Toronto, Toronto.

Rumley, P.A. (1984). Aménagement du territoire et utilisation du sol. Dissertation, ORL, ETH Zürich, Zürich.

Schäfer. A (2000). Regularities in travel demand: An international perspective. *Journal of Transportation and Statistics,* **3**(3), 1-32.

Schönfelder, S. and K.W. Axhausen (2001). Mobi*drive* - Längsschnitterhebungen zum individuellen Verkehrsverhalten: Perspektiven für raum-zeitliche Analysen. In: *Tagungsband CORP 2001* (M. Schrenk, ed.), **2**, 315-321. Technische Universität, Wien.

Seligman, A.B. (2000). *The Problem of Trust.* Princeton University Press, Princeton.

Simma, A. and K.W. Axhausen (2001). Structures of commitment and mode use: A comparison of Switzerland, Germany and Great Britain. *Transport Policy,* **8**(4), 279-288.

Sobel, J. (2002). Can we trust social capital? *Journal of Economic Literature,* **40**(2), 139-154.

Statistisches Bundesamt (2000). *50 Jahre Wohnen in Deutschland.* Metzler-Poeschl, Stuttgart.

Szalai. A. (ed.) (1972). *The Use of Time.* Mouton, The Hague.

Törnqvist, G. (1970). Contact systems and regional development. *Lund Studies in Geography: Series B Human Geography,* **35**, Royal University of Lund, Lund.

Weber, E.J. (1976). Peasants into Frenchmen: The Modernization of Rural France 1870-1914. Stanford University Press, Stanford.

Weiner, E. (1988). *Urban Transportation Planning in the United States: An Historical Overview.* Report DOT-T-88-26, US Department of Transportation, Washington, D.C.

Wellman, B. (1996). Are personal communities local? A dumptarian reconsideration. *Social Networks,* **18**(4), 347-354.

Wellman, B. (2000). Changing connectivity: A future history of Y2.03K. *Sociological Research Online,* **4**(4), http://www.socresonline.org.uk/4/4/wellman.html.

Wellman, B. (2001). The Persistence and Transformation of Community: From Neighbourhood Groups to Social Networks. Report to the Law Commission of Canada, Toronto.

Widmer, J.P. (2001). Ausgewählte Schweizer Zeitreihen zur Verkehrsentwicklung. *Materialien zur Vorlesung Verkehrsplanung*, 1.02, IVT, ETH Zürich, Zürich.

Widmer, J.P. (2002). Accessibility by air passenger transport related to Switzerland. *Arbeitsberichte Verkehrs- und Raumplanung*, **112**, IVT, ETH Zürich, Zürich.

Winston, G.C. (1992). *The timing of economic activities*. Cambridge University Press, Cambridge.

2

PROPOSITIONS FOR MODELLING HOUSEHOLD DECISION-MAKING

Eric J. Miller, University of Toronto, Canada

INTRODUCTION

This paper is an attempt to think through a unified theory/model of household and household member decision-making. Ideally, this model covers the gamut of decisions of interest within an integrated land use, transportation, environment (ILUTE) type of comprehensive urban simulation model, from "long-run" residential and employment location choice to "short-run" activity/travel decisions within an individual day or week.

The focus of this paper is conceptual; that is, the objective is to sketch out a conceptual theory/model of household decision-making. While it is hoped that this conceptual model is, in fact, practical to implement (and, indeed, a simplified prototype has been developed, see Miller *et al.* (2002, 2005); Miller and Roorda (2003) and Roorda *et al.* (2005)), the intent is not to deal with implementation-level details. Rather, the hope is to describe the overall structure of such a model in sufficient detail to assess its potential as a framework for the development of improved models of household decision-making. In particular, the paper is concerned with developing consistent, explicit and comprehensive **semantics** characterizing the problem of household decision-making, along with detailing the overall **processes** that control and shape these decisions.

In developing this conceptual framework, several assumptions are made at the outset and maintained throughout the paper. These include the following:

1. The model must be **household-based**, both because many decisions of interest within ILUTE-type models are made at the household level (residential location, etc.), and in order to capture the household-level interactions which constrain and otherwise influence individual behaviour within the system.
2. The model is **microsimulation-based**, in that the behaviour of individual households and the individual persons within these households is explicitly simulated over time.
3. As implied by the wording in point 2, the model is dynamic, in which decisions occur over time (both short and long time frames), with the household's attributes and behaviour evolving over time accordingly.
4. The model is **agent-based**, implemented within the **object-oriented** programming paradigm.

What ILUTE-type models fundamentally try to do is model person- and household-level decisions concerning the acquisition and usage of the tangible resources that are available to people in their daily lives. These consist of three major types:

1. time;
2. money;
3. goods and services.

The next three sections of the paper discuss each of these three fundamental types of "resources" in greater detail. The section "Choice of Planning Time Period" discusses the choice of the time period over which planning the consumption/usage of these resources is likely to occur.

We live our lives in terms of a sequence of **activities** in which we engage. A central theme of this paper is that these activities do not consist of a random jumble of events, but rather they arise out of coherent, logically interconnected sets of actions which are labelled **projects**. The following section introduces and discusses the concept of the project in detail, while the section "Activities" then similarly defines the notion of an activity in some depth. Given these definitions, the succeeding section briefly discusses household-level decision-making[1]. This leads to a section that discusses individual person activity/travel scheduling behaviour. Although not central to the conceptual framework being developed in this paper, to help focus the discussion, it is assumed that the period of analysis of activity/travel scheduling is an arbitrary week in the life of the persons being modelled. A final section briefly summarizes the paper.

TIME, ACTIVITIES, EPISODES AND SCHEDULES

Each person must spend each moment of each day doing something (even if this "something" is "nothing"). The things which we do, we call **activities**. A fundamental constraint on all human endeavours therefore is:

[1] This brief discussion is sufficient for present purposes. In particular, it provides a conceptual basis for more detailed treatments of household-level decision-making, such as is discussed in Miller and Haroun (2000), Miller and Salvini (2001) and Mohammadian (2001).

$$\sum_{a \in \mathbf{A}} t_{ia} = T \quad [1]$$

where:

t_{ia}	=	time allocated by person i to activity a during time period T ($t_{ia} \geq 0$)[2]
T	=	analysis or planning time period (typically one day, one week, etc.)
A	=	universal set of all possible activities

Activities are drawn from a universal set of possible activities, **A**. A more detailed definition and discussion of this activity set is provided below. Note that, at this level of discussion, travel is simply one of the many activities in which people engage.

The time spent on a given activity during period T occurs within one or more **episodes**, where each episode is of finite **duration**. Thus:

$$\sum_{e \in \mathbf{E}_{ia}} t_{ie} = t_{ia} \quad [2]$$

where:

$\mathbf{E}_{ia}$	=	set of episodes of activity a for person i in period T
t_{ie}	=	duration of episode e ($t_{ie} > 0$)

In addition to duration (t_{ie}) and type (a), each episode has, at a minimum, the following two attributes:

1. **location** (l_{ie}). For "normal" activities, the episode location is the point in space in which the activity occurs. For travel, "location" involves two points in space: the trip origin and the trip destination.
2. **episode start time** (τ_{ie}). $\tau_{ie} = 0$ corresponds to the beginning of the analysis time period and so τ_{ie} is bounded on the interval [0,T]. The episode end time is also a (derived) attribute of the episode and is simply equal to ($\tau_{ie} + t_{ie}$).

Travel episodes have at least two additional attributes, **travel mode** (m_{ie}) and **route** (r_{ie}). Travel episode duration is equivalent to the trip's **travel time.**

Given these attributes, episodes can only exist within a **schedule**, which consists of a sequence of episodes, in which the end time of the kth episode in the schedule is the start time for episode k+1. A schedule (and hence its constituent episodes) exists first as a **provisional schedule** or

---

[2] Throughout this paper, it is assumed that people engage in exactly one activity at a time. Multi-tasking, however, obviously often occurs (e.g., I typically watch television while doing the dishes). The issue of multi-tasking, however, is too complicated to take on within the current discussion. For present purposes we will elaborate a conceptual framework that we hope/trust will be extensible to include multi-tasking.

plan; i.e. a set of planned episodes in sequence which the person is intending/expecting to undertake[3]. While in the planning stage, all episode attributes may not be known with certainty. For example, one may have planned a shopping trip for Wednesday afternoon, but the exact location of the trip may not yet be known. The provisional schedule, however, is eventually executed in terms of participating in activity episodes of actual duration, location, etc. The **executed schedule** is often referred to by modellers as the **activity pattern** for the person for period T.

More formally, we may define a schedule for person i as the set of activity and travel episodes $\boldsymbol{\Psi}_i = \{e \in \mathbf{E}_{ia}\ \forall\ a \in \mathbf{A}\}$. An executed schedule (activity pattern) obviously satisfies equations [1] and [2] by definition (i.e. we had to have spent every minute of period T doing something). A planned schedule very often will not satisfy equation [1] as an equality constraint, but must satisfy the equivalent inequality constraint:

$$\sum_{a \in \mathbf{A}} t_{ia} \leq T \qquad [1.1]$$

That is, the total time for all currently scheduled activities cannot exceed the planning time period[4].

GOODS AND SERVICES

A **good** is a physical item which can be owned and used by the household. It is either purchased by the household from another agent or manufactured by the household from other goods. A **service** is an activity undertaken by another agent on behalf of the household in order to meet some household need. Services usually are purchased from the agent performing the service, but in some instances (charity, help from a neighbour, etc.) can be received free of charge.

It is convenient to divide goods/services into two major categories: **consumption goods** (food, fuel, etc.) and **durable goods** (houses, cars, etc.)[5]. That is:

$$\mathbf{G} = \mathbf{G}_c \cup \mathbf{G}_d \qquad [3]$$

where:

$\mathbf{G}$	=	set of all goods
$\mathbf{G}_c$	=	set of consumption goods
$\mathbf{G}_d$	=	set of durable goods

[3] In the extreme case of a totally spontaneous activity, it need not be the case that the activity first exists within a plan. This does not invalidate the general concept of a provisional schedule.

[4] Like airlines, it is possible that people may sometimes "over book" themselves on a provisional basis, and then subsequently resolve the conflicts at some later point in time.

[5] Henceforth, for simplicity of discussion, we will simply use the term "goods" to stand for both goods and services.

Reasons for doing this include:

1. Benefit presumably is derived by consuming the consumption good (i.e. a one-time, more or less "instantaneous" benefit), whereas benefit from durable goods flows over an extended period of time (the lifetime of the good or the period during which the good is owned by the household, whichever is less).
2. Consumption goods similarly involve a one-time cost — the cost of purchasing the good. Durable goods have both a one-time capital cost (the purchase cost), plus, potentially, on-going "operating" costs (fuel, maintenance, taxes, etc.), as well as potentially on-going debt-servicing costs (for "big-ticket" items such as houses and cars, for which households often need to finance the capital cost).
3. To the extent that durable goods require maintenance (which is an activity), they can give rise to "projects" (to be defined/discussed further below).

Alternatively, one can argue that all goods fall along a continuum with respect to durability, ranging from very short (fresh fruit) to very long (houses) and that they are eventually consumed. Thus, a more generic, continuous representation of goods in which all goods have a capital/acquisition cost, an operating/usage cost, and a duration over which they are consumed/used may be more accurate. For example, is clothing a consumption good or a durable good? Most clothing lasts for potentially a considerable period of time (although it eventually wears out or otherwise loses its appeal), and involves some on-going maintenance costs (dry cleaning, mending, etc.).

In a practical ILUTE application, however, it is likely that a relatively simplistic representation in which a few "big-ticket", "very durable" goods such as housing and automobiles are treated as durables, while all other goods are treated as consumables, is both sufficient and appropriate. Similarly, it is reasonable to assume that most (although possibly not all) services fall under consumption goods. For simplicity of discussion, this is the approach which is assumed in this paper.

At least three reasons exist for raising the issue of goods and services within this discussion. First, a considerable portion of household members' time is spent in the acquisition of goods and services—shopping for both consumption and durable goods, and obtaining a myriad of services (health services, personal grooming, home maintenance, etc.). Thus, any theory/model of household activity behaviour needs to be concerned with household consumption patterns and decision-making concerning this consumption.

Second, the acquisition of big-ticket durable goods such as houses and cars represents an extremely important set of decisions within an ILUTE-type model for at least the following reasons:

1. The investment in housing and cars typically represents the largest financial transactions in which the household will ever be involved, and consumes a considerable portion of total household financial resources[6].
2. On-going "operating" expenses associated with the use of these durables (maintenance, taxes, energy, etc.) also represent significant, on-going expenditures for the household.
3. These capital and operating expenditures on durable goods must be "traded off" against expenditures on consumption goods and other activities. That is, money spent on houses and cars cannot be spent on food, clothing, entertainment, etc. This clearly constrains decisions concerning the purchase of goods and the participation in activities.
4. At the same time, long-run decisions concerning the acquisition of durables condition short-run activity decisions in a variety of ways. Choice of housing location literally "positions" the household in space, and thereby determines the spatial context within which all activity/travel decisions are made. Similarly, choice of the number and types of vehicles owned/leased by the household establishes the "mobility context" for travel decision-making[7].

Third, it is important to explore the different impact on and treatment within household decision-making of consumption and durable goods. In particular, little concern seems to have been expressed in the literature about how households decide at a discrete point of time (e.g. "now") to acquire durable goods (houses, cars, home entertainment equipment, etc.) for which both the benefits and costs accrue to the household over an extended period of time. That is, a **time-stream** of benefits and costs exists which must somehow be assessed in order to determine whether a current expenditure (e.g. buy a second car) is "worthwhile", compared to the alternative uses to which this money might be put.

In both private and public sector project evaluation, the use of time-value of money concepts is well developed for "collapsing" a time-stream of either costs or (monetarized) benefits to either a "net present value" (NPV, i.e. the amount of money which one would need to have "in hand" now to be equivalent to the time-stream of costs or benefits) or an "annualised value" (AV, i.e. the time-stream of equal annual cash amounts which is equivalent to the actual time-stream). In both cases, the result is a single value (i.e. the NPV or AV) which summarizes the entire cash time-stream and which can be used in choosing one project over another (e.g. by comparing NPV's).

While not for a minute suggesting that households actually perform NPV-type calculations to assess future time-streams of costs and benefits, it is argued that "something like this" must implicitly occur if households are to be able decide how to allocate their resources among

---

6 A recent U.S. study found that American households on average spend 19% of their income on housing and 18% on transportation (McCann, 2000).

7 In addition, purchase of a "home entertainment centre" may have direct impacts on deciding between in-home and out-of-home entertainment options. Purchase of a riding lawn-mower means one probably will not hire the kid from next door to mow the lawn. Ownership of a cottage will have a major influence on weekend activity plans. And so on.

short-run (immediate cost and benefit) and long-run (time-stream of costs and benefits) expenditure/consumption options.

MONEY, EXPENDITURE, INCOME AND BUDGETS

The Expenditure Budget

Each household has a certain amount of money available to spend during period T. If one includes savings/investments, then all available money must be spent during this time period. Money can be spent on activities, goods/services, savings, taxes and debt service. Thus, a second fundamental constraint on household activities is:

$$\sum_{i \in \mathbf{N}_h} \sum_{e \in \mathbf{\Psi}_i} c_{ie}(t_{ie}) + \sum_{g \in \mathbf{G}} p_g x_g + Z_h + S^+_h + D^-_h + R(D_h) = Y_h \qquad [4]$$

where:

$\mathbf{N}_h$ = set of persons in household h
$c_{ie}(t_{ie})$ = cost of episode e for person i ($c_{ie}(t_{ie}) \geq 0$)
x_g = amount of good/service g purchased during period T ($x_g \geq 0$)
p_g = unit price of good/service g ($p_g \geq 0$)
$\mathbf{G}$ = universal set of goods and services available to purchase during period T
Z_h = taxes paid during period T ($Z_h \geq 0$)
S^+_h = net contributions to savings made at the end of period T ($S^+_h \geq 0$)
D^-_h = principal repayment on debt D_h at the end of period T ($D^-_h \geq 0$)
$R(D_h)$ = interest payment on debt D_h ($R_h \geq 0$)
D_h = total household debt (mortgage, car loans, etc.) at the beginning of period T ($D_h \geq 0$)
Y_h = total money available to the household for allocation during period T ($Y_h \geq 0$)

For now, "taxes" refers principally to income taxes. Sales taxes, property taxes, etc. (i.e. taxes associated with consumption of a given good/service) are assumed to be included in the price of the good. In a full accounting system for the urban economy, the tax components of these "prices" may well be explicitly tracked. For our purposes, equation [4] will suffice.

Care needs to be taken in some cases to differentiate properly between the cost of an activity episode and the purchase of goods/services. The obvious example of this is shopping, in which goods are purchased, but other costs associated with the episode (travel costs, parking, etc.) should be assigned to the episode itself. How to handle other activities might be less straightforward. For example, if one goes to a movie, is the price of the movie part of the episode cost ($c_{ie}(t_{ie})$) or is it the cost of "consuming" a movie ($p_{movie}x_{movie}$)? As long as double-counting does not happen and all costs are accounted for in one form or another, the choice between these two options is probably fairly arbitrary. Design decisions in such cases may well depend upon pragmatic considerations driven by other components of the model design.

Also note that expenditures on goods and services can include items that are not associated with any explicit activity episode during time period T. For example, housing operating costs (energy, water, municipal taxes, etc.) are on-going expenditures that may occur during period T but which do not require any (significant) time allocations on the part of household members to occur.

The money available to the household to spend in period T, Y, is given by:

$$Y = \sum_{j \in \mathbf{J}_h} W_j(t_{wj}) + \omega_h + S^-_h + I(S_h) + D^+_h \qquad [5]$$

where:

$\mathbf{J}_h$	=	set of workers in household h ($\mathbf{J}_h \subseteq \mathbf{N}_h$)
$W_j(t_{wj})$	=	income of worker j during period T ($W_j \geq 0$)
t_{wj}	=	time spent by worker j working during period T (i.e. a = w_j for his activity)
ω_h	=	total government transfer payments (unemployment insurance, etc.) ($\omega_h \geq 0$)
S^-_h	=	net deductions from savings at the beginning of period T ($0 \leq S^-_h \leq S_h$)
S_h	=	total savings at the beginning of period t ($S_h \geq 0$)
$I(S_h)$	=	investment income received at during period T ($I \geq 0$)
D^+_h	=	new debt incurred during period T

Equations [4]-[5] represent the household's **budget**. As with a person's time schedule, at the end of a time period these equations must hold exactly. Also as with schedules, households can look ahead and plan their expenditure for a future time period. In this case, the **budget plan** need not satisfy this equality constraint but rather the inequality constraint that total planned expenditures (the left hand side of equation [4]) can not exceed total expected income (the right hand side of equation [5]).

Note that equations [4]-[5] can always hold by incurring new debt (D^+_h) in each time period. One generally assumes, however, that households do not incur debt to cover on-going operating costs (i.e. D^+_h is not a routine, on-going entry in the budget, but rather occurs occasionally, typically associated with the acquisition of major durable goods, as discussed in the next sub-section). Certainly this is not sustainable over the long run because it will eventually lead to bankruptcy. Equations [4] and [5] do not, however, preclude this possibility, which is fine, since bankruptcies do, in fact, occur upon occasion in the real world.

For the sake of simplicity, it is assumed that deductions from savings occur at the end of the time period, that interest is paid at the end of the time period on savings held at the beginning of the time period, and that any new contributions to savings occur at the end of the time period. Similarly, new debt is assumed to occur at the end of period T, so that no interest is paid on this new debt during T. Given these definitions, total savings and debt at the beginning of the next period are simply:

$$S_{h,next} = S_h - S^-_h + S^+_h \qquad [6.1]$$

$$D_{h,next} = D_h - D^-_h + D^+_h \qquad [6.2]$$

and:

$$S^-_h * S^+_h = 0 \qquad [7]$$

that is, S^-_h and S^+_h cannot both be non-zero in the same time period.

In addition to savings and debt, **equity** is also an important concept with respect to major household durable goods, specifically housing and vehicles. Equity is the current market value of the durable good (e.g. house or car), less any outstanding debt attached to this good. Formally, we may define equity as:

$$Q_{hg}(t) = p_g(t) - D_{gh}(t) \qquad [8]$$

where:

$Q_{hg}(t)$	=	equity which household h has in durable good g at time t
$p_g(t)$	=	current market value of good g at time t
$D_{gh}(t)$	=	debt attached to good g for household h at time t (e.g. balance of principal owed in a house mortgage or car loan)

Equity is different from (and not included in) household savings, S_h, since it is a "potential" savings, not an "actual" one. That is, interest (I) does not accrue on equity as it does on savings. Further, equity can only be "actualised" by selling the durable good. Equity is of importance at the time at which durable goods are purchased and sold. In particular, since the purchase of one house (or car) often involves the sale of the currently owned house (or car), equity in the current good is realised at that time and, is usually applied towards the capital cost of the new good being purchased. This issue is dealt with further in the next sub-section.

The market value of the durable good, $p_g(t)$, varies over time in response to market processes as well as a function of the age and quality of the good. In general, vehicles depreciate over time (i.e. $p_g(t)$ decreases with increasing t), whereas housing prices often (but not always) appreciate over time. Thus, purchase of a dwelling unit can represent an investment which provides a positive rate of return (over and above the "psychic" returns or benefit/utility gained through occupancy of the dwelling unit), and this may be a factor in determining housing purchasing behaviour.

Budgeting for Major Durable Goods

When a household is deciding on the purchase of a major durable good, such as a house, equations [4] and [5] need to be simultaneously considered within two planning periods. The first is the point in time at which the purchase actually occurs. At this time, the purchase of the house will show up in the left-hand side of equation [4] as a very large p_{new} term (i.e. the purchase price of the new dwelling unit), balanced by the sum of S^-_h and D^+_h (i.e. deductions from savings – the "down payment" plus new debt in the form of a mortgage on the balance required) on the right-hand side of equation [5]. As noted in the previous sub-section, it is at this time that any equity in a current dwelling unit enters the calculations (assuming that this current dwelling unit is being sold at the same time that the new one is being purchased)[8]. More precisely, the relationship between housing prices, savings, equity and debt which must hold is given by the following expressions:

$$p_{new} = S^-_h(t+\epsilon) + D^+_{new,h}(t+\epsilon) \qquad [9.1]$$

$$S^-_h(t+\epsilon) \leq S_h(t+\epsilon) \qquad [9.2]$$

$$S_h(t+\epsilon) = S_h(t) + Q_{hg}(t+\epsilon) \qquad [9.3]$$

$$Q_{hg}(t+\epsilon) = p_{old}(t+\epsilon) - D^-_{old,h}(t) \qquad [9.4]$$

where:

t = time immediate before sale of the current ("old") dwelling unit and purchase of the new dwelling unit

$t+\epsilon$ = time at which old house is sold and new house is purchased

$D^-_{old,h}(t)$ = principal owed on mortgage for the old house, which is retired at time of sale

$D^+_{new,h}(t+\epsilon)$ = principal owed on new mortgage for the new house, which is taken on at time of purchase

The second time periods of importance are subsequent, on-going budget periods, T, over the lifetime of the ownership of the durable good. During these time periods there will typically be large monthly D^- and R(D) terms (i.e. mortgage or car loan payments which include both

---

[8] This need not always be the case. Sometimes a second unit of the durable good may be purchased without selling the first unit (e.g., a second car is purchased, while retaining the currently owned vehicle). In this case, the equity embedded in the first unit, of course, is not realised. In the case of housing, the order in which the new house is purchased and the old house is sold may vary, leading to various "risk" issues in the decision-making of buyers and sellers within this process.

principal repayment (D^-) and interest ($R(D)$) terms) on the left-hand side of equation [4] which must be accounted for[9].

It is these latter "period 2" calculations which in many ways are critical to the determination of the feasibility/attractiveness of buying the durable good, since it is these calculations which must balance in a sustainable way (i.e. the "period 1" calculation, in principle, can always hold with a big enough D^+ term), and, if they balance, they define constraints on future period benefits/utilities which can be obtained through the purchase of other goods/services.

On the benefits side, the situation is, in most respects, the same. There is little in the way of initial utility at the time of purchase of a house – the utility "flows" over time through usage of the house. The problem is, we are not used to computing "utility time streams". This is suggestive, however, that an "annualised" utility is what is required here in terms of assessing the attractiveness of the durable good, in the same way that "period 2" budget calculations are on an "annualised" basis.

Thus, to summarize, "big-ticket" goods which involve considerable expenditure (and that may involve taking on debt in order to finance the purchase) should be treated as pure durables, with an effective annualised cost and benefit/utility which is applied on an annual basis over the lifetime of the good and/or debt (note that these two may well have different lifetimes). The impact on household savings and debt can be determined using equations [4] and [5] applied in "period 1 mode", i.e. at the time of the purchase. But feasibility/decision-making is then based on equations [4] and [5] applied on an on-going basis in "period 2 mode", in which the post-purchase savings and debt money flows are used. Put another way, purchase of a "big-ticket" item shows up in equations [4] and [5] as ($D^- + R(D)$) terms in post-purchase time periods, rather than as $p_g\ x_g$ terms in the purchase time period (as is the case with consumption goods).

Major durable items in ILUTE obviously include housing and cars. Other possibilities that might conceivably enter an ILUTE-type model include: education (private school, post-secondary, etc.); major vacation travel (as opposed to utilitarian, day-to-day travel); and health care (depending on the society and its funding arrangements for health care).

Choice of Planning Time Period (T)

Equations [1] - [7] in principle "scale up and down" for any time period T. However, they clearly make more sense for certain definitions of T than others. The choice of T depends on the analysis to be undertaken (by the modeller) but also on the decision being made by the person/household.

The key issue as we move up and down the temporal scale in our models is to maintain a consistent treatment of time, money and acquisition/use of goods/services. In addition, multiple T's may well exist simultaneously within the model at one time—i.e. one process

---

9 In addition, various operating costs (maintenance, insurance, taxes, etc.) will generally exist with the $p_g x_g$ terms.

might be planning today's activities, while another is planning over the next year, etc. At a minimum, however, it is likely that there will be at least two time-scales operative within an ILUTE-type model: one of approximately a year in length, during which "long-term" decisions are considered (residential location, auto ownership, employment status, etc.); and one ranging from a single day to a single week, during which shorter-run activity/travel decisions are explicitly simulated. Both of these planning time frames are assumed (as appropriate) throughout the remainder of this paper.

PROJECTS

Axhausen (1998) defines a **project** as a coordinated set of activities tied together by a common goal or outcome. A simple example of a project might be a dinner party, for which individual (but inter-connected) activities might include: planning the party, shopping for food and drink, preparing the meal, cleaning the house, and the actual dinner party. The concept of the project is central to the conceptual model of household decision-making being presented in this paper. This section discusses this concept of the project in three parts: First, the concept is defined in detail; the role of projects within the scheduling process is then discussed; and finally a suggested list of "primary" project types which might provide the starting point for ILUTE-type modelling based on the project concept is introduced.

Definitions and Examples

Formally, a project is defined as a logically connected set of activities that contribute towards a common objective. Projects can spawn sub-projects (and so on and so on) if the sub-project is sufficiently important (in terms of its use of resources) and coherent in terms of having a definable objective in its own right.

Attributes for project $p \in \mathbf{P}_i$ (where $\mathbf{P}_i$ is the set of projects currently undertaken by person i) include:

1. **Type.**
2. **Objective.** For now, this is a verbal statement of what the project is trying to accomplish. It is conceivable that the objective might be tied to utility/priority calculations.
3. **Project task list ($\mathbf{A}_p$).** This is the set of activities that must be executed within this project. This set will generally be ordered by the **sequence** in which the activities must be undertaken in order to logically and efficiently accomplish the project objective, although there may be examples of projects in which activities need not be undertaken in strict sequence.
4. **Project agenda ($\mathbf{E}_p$).** This is the set of specific activity episodes that project p is currently looking to schedule (i.e. actually plan and then execute).

The distinction between the task list and the agenda is perhaps a subtle one, but it is absolutely crucial in the definition of the project. As defined here[10], the agenda is a list of specific activity episodes that are actively being considered for insertion into the current planning period's schedule. They are explicit objects (in the programming sense of the word) which only persist within the model until they have been scheduled and executed. They are primarily defined by their **attributes** (duration, etc.) and possess little behaviour *per se*; i.e. episodes do not "act", they are acted upon. If the simulation model were to be run over a sequence of planning periods, then the agenda's contents (i.e. the set of episodes) would vary over time as current episodes are scheduled and deleted and as new episodes are added by the project to the list.

The task list, on the other hand, is persistent throughout the life of the project. Its elements are activities in which the person **might** engage at some point in the project, but may not necessarily do so within the current planning period. Activities are also objects (in the programming sense); but while they possess attributes, they are primarily defined in terms of their **behaviour**. That is, an activity possesses the information and the procedures/methods needed to generate specific instances of activity episodes. Thus, the "knowledge" about how activity episodes are generated (how often, constraints on duration, etc.) are contained (encapsulated) within the activity. Properties of activities are discussed in greater detail below.

As an example of these concepts, consider the following project:

Project p:

Type: Home Renovation

Objective: Tile the basement rec room floor with Mexican tiles

Task List:

1. Obtain tools and materials required.
2. Lay tiles
3. Seal the tiles.

Agenda:

1. Buy tiles and mortar (requires trip by car to the tile store; this store has been previously identified as the best place available to buy this type of tile).
1. Buy tools required to lay tiles (requires trip to a home improvement store).
2. Rent tile cutter (requires trip by car to the tool rental store, immediately prior to when laying the tiles, since the rental charge precludes keeping the tile cutter longer than necessary; store opens 8:00 a.m.).

[10] As in many other instances within this paper, this term is being used in a specific, perhaps non-standard fashion. "Agenda" is a popular term within the activity scheduling literature. It is often not well defined, and often seems to include elements of both what are here defined as the task list and the agenda. Separating and clarifying these two concepts is, it is argued, a useful step forward.

3. Lay tiles (requires several hours, exact number of hours uncertain; must be done in one episode, both due to the need to mix mortar which must be used at once and due to the need to minimize tile cutter rental fees).
4. Return tile cutter (requires trip by car; store closes 6:00 p.m.).
5. Apply sealer to tiles (requires 3 coats; approximately one hour per application; each coat must dry 24-hours before the next coat is applied).

The numbers indicate the logical sequence of activities or episodes. When the same number is assigned to two or more activities/episodes, it implies that there is no logical need to do one before the other (e.g. I can buy the tools before the tile, if this makes sense from the point of view of my overall schedule, but both of these actions must be taken before I can actually tile the floor).

Note that many of the episodes come with constraints/criteria which are important for scheduling these episodes. These include:

1. sequencing information, as already discussed;
2. expected duration of the activity (if known);
3. travel mode requirements (in this case, a car is required for the trips involving bulky items which must be transported, whereas the tile-laying tools could, if desired be transported via transit); and
4. trip start/end time constraints (can't go the rental store before 8:00 a.m., etc.).

Given this project, it is clear that a day with a large block of free time is required to get the tile cutter, tile the floor and return the tile cutter—preferably the entire block from 8:00 a.m. to 6:00 p.m., given the uncertainty of how long the tile laying will take and the rental store's opening and closing times. Assuming that the person involved works Monday to Friday, this means that these tasks must be scheduled on a Saturday or Sunday (assuming the rental store is open Sundays). When such a block of time can be found depends on the person's competing projects, the "priority" of this project, etc.

Note that the decision to tile the basement floor with this type of tile, along with the identification of the store from which the tile is to be bought, have come out of a "meta-project" dealing with renovating the basement, which has "spawned" this sub-project. Thus, a project's task list in the general case will consist of a collection of (sub-)projects and activities. Indeed, this suggests that a more formal definition of an activity is that it is an elementary project, whose task list consists of exactly one type of action.

Following on with the previous example, the meta-project might look something like:

Project p':

Type: Home Renovation

Objective: Renovate the basement

Task List:

0. Plan the basement renovation
1. Lower basement floor
2. Build stud walls
3. Install wiring
3. Install bathroom plumbing
4. Drywall
5. Finish the basement rec room floor
5. Build shelves in the cold cellar
...

The concept of the project appears to be a potentially exceptionally powerful one. **Indeed, a fundamental proposition of this paper is that <u>all</u> activities in which we participate (both in the long and the short run) stem from projects.**

If this hypothesis is correct, then a fundamental part of the activity scheduling process is the meta-decision of what projects to "take on". "Taking on" a project involves a (typically conscious) **commitment** to pursue the project. This, in turn, means a commitment to allocate time (and, often, money) to the project. It is hypothesized that a project is not taken on unless the person expects to have sufficient resources (time and money) to execute it. If this true, this again emphasizes the importance of projects within the scheduling process in that:

1. it involves assessing the current plan to see if time exists to do the project (similarly, the budget must be assessed to see if funds will be available for the project); and
2. it may involve "booking"/scheduling project activities to ensure that the required time will be available.

Projects can be either "one-time" projects (such as tiling the basement floor), or recurring/on-going (such as work or school). As illustrated above, one-time projects will have a list of activities/episodes that are generally executed once and then deleted from (or, perhaps "checked off") the list (e.g. one only needs to buy the tiles once), although there may be iterative sub-projects within a one-time project (e.g. coating the tiled floor requires iterating through the activity three times with three separate episodes of the activity). Recurring projects will also have a list of activities with some sequencing and iteration rules/constraints, but also with the potential for some "one-time"/new/random activities "mixed in" (e.g. a person's work week, while perhaps generally possessing a fair degree of routine will have "variants" mixed in within any specific week).

Projects and Scheduling

Scheduling can go on "within" a project, provisionally ignoring other projects/parts of a person's schedule. For example, I may schedule my "work week" largely independently of my home-based activities. This within-project scheduling can be **encapsulated** (in the object-oriented sense) within the project, both in terms of the scheduling act (which can "run" as an independent process) and in terms of many of the details of the resulting project plan. What is required, however, is an **interface** (again in the object-oriented sense) between this project's time (and money) requirements with the person's overall plan. It is the task of a person's overall scheduler to mediate among conflicting requirements from all the project plans (e.g. two plans call for an episode to occur Saturday afternoon between 2 and 5), and "feed back" time allocations to projects (e.g. project p_1 gets the Saturday afternoon time slot, project p_2 doesn't, which will trigger a rescheduling of p_2's plan).

As an example of this, note that in the case of the basement tiling project, all that the person's overall scheduler needs to know about the combination of episodes 2, 3 and 4 is that this project requires a full block of time from 8:00 a.m. to 6:00 p.m. and that a car will need to be available at the beginning and end of this time period. Details of what goes on within this time block are not required for the person scheduler to do its job. They are, of course, of importance to the project scheduler, both during its planning phase (so that it can determine its **requirements** which it will present to the person scheduler for decisions about whether and when to allocate these resources to the project), and during execution (e.g. deciding when to have lunch during the day, given how the tiling is proceeding). As another example, my wife does not need to know the details of how my work day proceeds in order to schedule her day—she only needs to know when I am leaving the house in the morning, when I will be returning in the evening and whether I need the car that day or not.

In discussing the requirements of my work project for a given day above, note that these are initially generated by the work project, but they are must be "confirmed/approved" by the household, which mediates between the requirements generated by the person schedulers for each person in the household in terms of resolving conflicts in household resource requirements among the household members' personal schedules (demands on the household's vehicle(s), needs to satisfy various household level activities/projects, etc.).

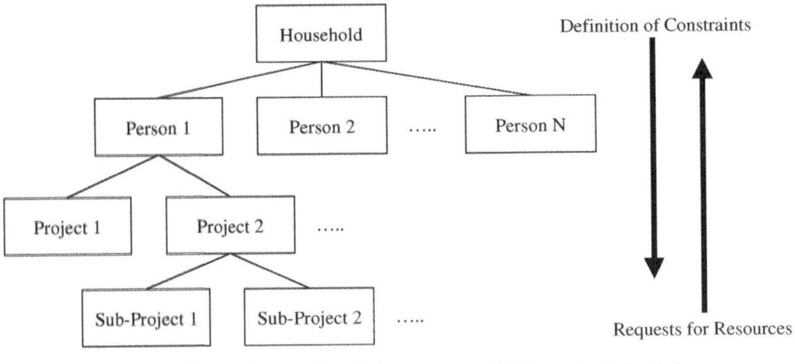

Figure 2.1 Hierarchy of Decision - Making & Scheduling

Putting these observations together, we see that there is a hierarchy of interconnected scheduling activities, as illustrated in Figure 2.1. Although not shown in the figure, households can also have projects. Activities generated by these projects will either involve joint activities among household members (e.g. a family outing) or be allocated to one or more family members for execution (e.g. somebody has to mow the lawn).

As already illustrated and discussed above, provisional decisions at each level independently generate "requirements"/"provisional schedules" which are passed up to a higher-level decision-maker. The agent at the upper level mediates between conflicts among lower-level plans and feeds back constraints to the lower-level agent. This decision-making in turn, may be provisional, requiring approval from yet a higher-level decision-maker to which this agent reports.

Observations/comments about this process include the following.

1. A person's life is the highest level project (i.e. the person scheduler is scheduling the "project" which is the person's life).
2. Higher-level projects have authority to dictate constraints, etc. to lower-level processes (e.g. the basement renovation project decides when the cold cellar shelves will be built).
3. The household similarly mediates between person schedules within the household. It is the highest level decision-maker of all. This does not necessarily mean, however, that persons will value/abide by household level constraints at all times and in all circumstances—presumably divorce is an extreme example of people deciding to opt out of household-level commitments.

Projects involving commitments to other persons/agents (e.g. firms/organisations) tend to be generally more binding/compelling than commitments made only to one-self, for at least two reasons. First, the commitment ("contract") means that the other agent(s) involved are counting on the person to fulfill the commitment (the child expects to be taken to soccer practice, the firm expects the work to get done in a timely fashion, the squash partner expects you to show up at the agreed upon time to play a game, etc.). Second, the successful execution of the project often requires agreement on the allocation/sharing of resources (e.g. who gets the car) and/or cooperation in a joint venture (e.g. agreeing on when to play squash, working together on a proposal at work). This is presumably the basis for placing the household at the top of the scheduling hierarchy, and one of the reasons why projects like work and school (which involve considerable collaboration and committing to contracts for delivery of services) are viewed in most activity models as "mandatory" or "priority" activities.

All of the above sounds extremely mechanistic, rational, deterministically optimal, etc. It need not be. The information used, the rules applied, etc. will determine how optimal, rational, deterministic, etc. the outcomes are. What the above represents is an attempt to put the scheduling process into a coherent, potentially programmable representation that is consistent with observed behaviour—which, it is argued, the foregoing is.

Primary Projects

It is hypothesized that a relatively small number of "primary project types" exists that incorporate all activities in which people engage in attempting to meet their fundamental needs. An obvious (but perhaps partial and certainly provisional) list of such **primary projects** (including representative elements of project task lists) is presented in Table 2.1.

In Table 2.1, "work" includes all types of activity undertaken in order to obtain money. This includes working for salaries/commissions, running your own business, and the time and energy involved in making investments. It involves participating in the exchange of goods and services with others.

"Education" includes participation in any sort of formal education program (i.e. one which is run by another agent such as a school, a "home learning" enterprise, etc.) which involves the explicit commitment of time (and often money) to a course of study. Any "grey area" between this and self-directed learning (which under the categorization shown above would fall somewhere under personal recreation) is, again, not likely to be problematic within an ILUTE modelling context.

Note that "shopping" is not shown as a primary project. Rather, it is an activity/sub-project which is spawned by many of the primary projects (indeed, any project that requires the acquisition of goods for its accomplishment). In particular, note that, while not explicitly delineated in all cases, most projects involve the acquisition and maintenance of durable goods to assist in and/or improve the quality (utility) of activities within the project. Thus, for example, acquisition of television sets, VCR/DVD players, stereo systems, sports equipment, etc. are all implicit in the personal/household recreation project. Similarly, "information acquisition" (window-shopping, inspecting open houses, browsing the internet, etc.) is an activity/sub-project spawned by many of the primary projects.

Similarly, travel is not a primary project, and is again an activity that is spawned by any activity which requires the person to move to a new location in order to participate in the activity. Thus, the generation of a travel episode is fundamentally different from that of an activity episode. Note, however, that "mobility" has been defined as a primary project, where by "mobility" is meant the acquisition and maintenance of household and personal resources which facilitate travel.

Table 2.1 Suggested List of "Primary" Projects

- community service
- education
- family
    - decision to have a(nother) child
    - child care
    - facilitating children's activities (driving them to activities, etc.)
    - visit extended family
    - care for extended family members
- health care (visit doctors, dentists, etc.)
- household evolution
    - leave parents' home to form one's own household
    - form non-family unit (e.g., roommates)
    - marriage
    - divorce
- member of an organised religion
- mobility
    - purchasing/leasing vehicles
    - vehicle maintenance
    - acquisition of driver's licence
    - acquisition of a transit pass
- personal maintenance (bathing, sleeping, clothing oneself, etc.)
- recreation (personal/household)
    - sports (team or individual)
    - hobbies
    - entertainment (movies, theatre, etc.)
    - reading
    - "family outings"
- shelter
    - purchasing/renting a dwelling unit (house, apartment, etc.)
    - home maintenance (cleaning, repairs, yard/garden work, etc.)
    - home improvement (renovations)
    - acquisition of home furnishings/decorations
- socialising (with non-household members)
- sustenance
    - food acquisition (purchasing, growing)
    - food preparation
    - eating
    - maintenance (washing dishes, etc.)
    - acquisition of durable goods for food storage, preparation, cleanup (refrigerator, stove, dishwasher, etc.)
- vacation
- work

The definition of what constitutes a primary project is not very clear from the list shown above, and certainly this list is relatively arbitrary at this point in time. The projects selected above as being primary seem, however, to meet one or more of the following criteria:

1. The project corresponds to a major personal or household decision within ILUTE. In making such a decision, the person/household is explicitly committing to the associated project (e.g. work, school, shelter).
2. The "project" is an integral part of human existence (e.g. shelter, personal maintenance, sustenance). Indeed, it can be argued that the above list is applicable across all cultures, providing one makes necessary adjustments concerning the "modes" within which these projects might be engaged within different cultures/societies (the concept of activity "mode" is discussed in greater detail in the next section).
3. The project represents a major, generic activity type (e.g. personal/household recreation) that does not have an obvious, useful "meta-type".
4. Collectively, the primary projects are capable of incorporating all activities of interest (at least those of concern within ILUTE).
5. They are all recurring/on-going projects which, if not all undertaken at all times by all households, are all potentially available for undertaking at all times by all households.

While this specific list may not be the "last word" on this, it certainly serves for present purposes. Probably the key question is whether there exists activities of interest within ILUTE which cannot fit into this list. If so, then obviously the list needs to be expanded to include them. Conversely, if the list can still be advantageously condensed or rearranged, then it should be.

ACTIVITIES

To this point in the discussion "activity" has been a very loosely defined concept. It is useful, however, to think a little more deeply about activities, as opposed to either the projects they reside in or the episodes which represent their physical manifestation. Two major attributes of activities are introduced in this section: **type** and **mode**.

It is argued in the following sub-sections that activities fundamentally fall into exactly four types, each of which has specific implications for scheduling. In discussing these four types of activities it is assumed that one is interested in constructing an activity scheduling model which deals with an entire week of household activities. Given this assumption, it is convenient to differentiate between activities that are **pre-planned** (i.e. that is at least partially planned prior to the commencement of the week in question) and **within-week** (i.e. are planned during the week in question). In particular, it is assumed that in modelling the week-based scheduling problem that one can speak of a **skeleton schedule** (Doherty *et al.*, 2001) that is (at least nominally) planned prior to the week in question and which provides the starting point for the more detailed, dynamic within-week scheduling that ultimately determines the household members' activity patterns that are actually executed. Given this, one must determine which activities are pre-planned onto the skeleton schedule and which are scheduled within-week.

The concept of "activity mode" is then introduced as an attractive concept for dealing with the myriad of activities potentially available for participation in by household members. Finally, given the preceding discussion, some preliminary thoughts on a "universal activity set" are presented.

Activity Type I: On-going Contract with an External Agent

A number of projects that individuals may take on involve activities that involve commitments ("contracts") with non-household agents. Such activities share common characteristics, including:

- a high level of pre-planning;
- a high level of general regularity/repetitiveness from week to week, particular with respect to temporal attributes[11];
- a high level of scheduling priority is attached to activity episodes, given the commitment to the external agent;
- often a severe penalty exists for consistent non-participation (lose job; dropped from the team; etc.); similarly these episodes will tend to be quite "resistant" to modification/rescheduling, except under quite high priority cases;
- typically tight/"mandatory" episode generation/scheduling constraints exist, usually across all episode dimensions (frequency, start time, duration, location); and
- typically high reward/utility is derived from participation (financial, psychic, physical health, etc.), hence making the high level of commitment "worthwhile".

At least three sorts of activities fall into this category:

1. "regular" work; i.e. "in-office", "usual/routine" work episodes; excludes specific "visit client" type activities which are included in Type II activities, discussed below; includes self-employment, in which case the "external agent" in this case is one's own firm;
2. "regular" school (includes any formal out-of-home "contracted" instruction such as music/dance lessons; excludes "one-of" activities such as "staying late" after school); and
3. participation in formal, regular group activities, which includes a variety of specific types, including: sports teams, cultural groups (theatrical group, band, etc.), religious/community groups, etc.

Again, the key feature of this type of activity is the on-going commitment to an external agent. In order to make this commitment, the activity in general **must** be pre-planned in that this commitment requires the allocation of personal and household resources (time, perhaps money, vehicle usage, etc.), and the expected feasibility of being able to allocate these resources to the activity must be assessed prior to undertaking the commitment (I must be able to get to the job

[11] They usually tend to be routine with respect to spatial attributes as well, but from the pre-planning point of view, the temporal dimension is the more critical one.

in the suburbs; if my son plays hockey I must be willing to allocate the time it will take to get him to the practices and games; etc.).

Similarly, to ensure that this feasibility is maintained in the actual execution of a given week's activity pattern, such activities' episodes **must** receive priority in scheduling.

Finally, pre-planning, priority scheduling is also a feasible thing to do since:

(a) a "typical"/"average" set of episodes and episode attributes are generally known (i.e. they are intrinsic in the activity), at least to a sufficient level of detail/reliability to establish a weekly skeleton schedule that defines the starting point for determining the actual, detailed schedule for the week actually being simulated (the worker generally works 9-5, Monday to Friday; the baseball games are Tuesday, Wednesday and/or Thursday nights; the worker needs a car for business meetings, even if the specific meetings for a given week are not known well in advance; etc); and

(b) the commitment is an on-going, largely repetitive one from week to week (go to work every weekday every week; play soccer 3 times every week; etc.), rather than a one-time engagement.

Another way of putting this is that the activity attributes are generally determined by the external agent being contracted with (the business establishment, the school, the team, etc.). The person decides whether to accept/commit to the activity or not, but does not generally have any decision-making capability with respect to at least the "broad parameters" of the activity (I can't say that I'll join the hockey team: I might, but only if it agrees to play all its games 2-3 Saturday afternoons).

While typically possessing a high degree of fairly "rigid" pre-planning, these activities also may well have a variable/stochastic component that varies from day to day and week to week, and that will therefore need to be accounted for in "dynamically" determining the detailed schedule for the week being simulated.

Finally, note that work and school projects within which the work and school Type I activities are contained are "taken on" and evolved over time through explicit procedures within the overall ILUTE framework. The implication of the above discussion is that "taking on" and evolving other types of projects involving contracts with external agents (e.g. typically group memberships) should also be explicitly part of this longer-term evolutionary simulation process.

Activity Type II: One-Time Contracts

Many activities involve commitments with external agents that are defined within this formulation as being "one-time" in that they do not occur on a regular, weekly basis. Examples include:

- most health/medical activities;
- most personal business (see lawyer, banker, etc.); and
- socialise with friends and relatives (either individually or as a household/family).

Any contract/commitment to interact with an external agent by definition requires pre-planning (book the appointment with the dentist; invite guests to a dinner party; etc.), and this pre-planning may well in a particular instance occur before the week being simulated. In addition, each of these activity types may well have a "routine" element (e.g. visit the dentist every six months). For the purpose of this formulation, however, it is argued that at least one of the following holds true:

1. the interval between episode occurrences is longer than the week, and so there is only a probability that such an episode will occur in the arbitrary week being simulated (e.g. there is presumably approximately a one in twenty-six chance that this will be the week for a person's bi-annual dental check-up);
2. the episode is a specific, unique instance "randomly" generated from a general type of activity (e.g. we may give several dinner parties a year, but the guests will generally be different, and the timing of such parties is idiosyncratic); or
3. the episode is truly a "one-off" event (a friend living in Australia calls up out of the blue and asks if you're free to get together this week).

These cases have been arranged in order of increasing "randomness", or, equivalently, decreasing regularity/fixity. In all cases, despite at least some element of pre-planning, the episodes differ fundamentally from those discussed under Type I activities in that:

1. They do not occur each week with probability one, as the Type I activities (work, school, etc.) are assumed to do. Thus, they probably do not belong on the schedule skeleton that we wish to construct (i.e. we do not "frame" our fundamental weekly activity pattern around the possibility that someone might visit the dentist somewhere along the line).
2. They are inherently more flexible in terms of timing and/or location **at the time of initial scheduling** than Type I activities. As has been noted, for Type I activities, the basic template of the episode attributes is dictated by the external agent to the accepting agent, and so are "fixed and given" as far as the accepting agent is concerned. In Type II activities, the episode attributes are **negotiated** between the two parties ("can you come to dinner Saturday?"; "no, how about the week after"; "sounds good, it's a date"). This flexibility means that these activities will have a greater tendency to be "fit in" around other, pre-existing commitments, as well as the fact that they will appear to be more "random" in their episode attributes, given our inability to observe the detailed considerations/negotiations that actually determine the pre-planned scheduling of a given episode of this type.

For both of these reasons, Type II activity episodes tend to "look like" within-week activities rather than pre-planned activities and are, therefore, treated in this fashion with the model. Three final points to note about this assumption:

1. In speaking about "pre-planned" this makes clear what we are really referring to is the component of pre-planning which deals with the Type 1, weekly/"very regular" activities which are scheduled in advance of the week as part of longer-term decision-making processes (labour market activity, schooling, team sports, etc). This longer-term process deals in "broad brush strokes" with allocations of "generic time blocks" to a longer-term schedule skeleton or template, which serves as the starting point for the "filling in" of the schedule for a specific instance of the week. This, of course, may involve modifying the template, but the template provides an essential means of "getting started".
2. Once initially scheduled, Type II activities become very similar to Type I activities in that a commitment/contract has been made which generally can only be broken or significantly modified at some cost. Thus, while a Type II episode may possess considerable flexibility when being initially being scheduled (and may well not get scheduled at all for the week being simulated if it does not "fit in"), once scheduled, it becomes much more inflexible from a rescheduling/modification point of view.
3. Obviously some cases exist in which Type II activities actually become Type I in nature. People with chronic medical problems requiring weekly (or even more frequently) routinized medical services is but one example of a clearly Type I activity. If such cases can be usefully developed and empirically identified within the model, then there is nothing in the model structure to stop the activity from being included in Type I skeleton-building.

Activity Type III: Within-Household Activities

Many activities do not involve explicit contracts with external agents[12]. Hence, participating in the activity is an internal decision of the household alone, as is the definition of the activity episode attributes, including activity type, frequency of participation, etc. As such, within-household activities are inherently much more flexible in at least some dimensions than the Type I and II activities discussed above.

As with Type II activities, less "firm" information is generally available concerning the specifics of Type III activity episodes for pre-planning purposes. As a result, detailed scheduling of specific episodes is assigned to the dynamic, "within-week" process rather than as part of the pre-planned "skeleton", using exactly the same arguments as for the Type II case.

Note that at least some Type III activities may be very "high priority/utility" for households ("quality time with the kids"; the house desperately needs painting; etc.). Because they do not involve contracts with external agents, however, they may be not only more flexible to schedule initially but also more flexibly/readily rescheduled or even deleted from a given week's schedule if confronted with "harder", less flexible Type II episode requests.

---

[12] "Non-contract" activities may still involve interaction with external objects/agents. For example, shopping involves interaction with stores, and out-of-home activity always involves interaction with the transportation system and fellow travellers. This type of interaction, however, is quite different from that discussed in the previous section, particularly in terms of the lack of any commitment to interact with these agents (i.e. one doesn't have any commitment to visit at a particular store, let alone buy anything there).

Activity Type IV: Serving Dependents

A third type of very high priority "contract" occurs within the family between "independent" and "dependent" family members[13]. A **dependent** is a person who, due to age (very young or very old) and/or physical/mental disability, is unable to undertake certain tasks alone and/or unassisted. The most typical example (and the one we will focus on here[14]) is a young child who cannot be left unattended, and who generally must also be transported to his/her out-of-home activities. Partially because of this and, in most cases, partially simply due to competency concerns, dependents also are not able to unilaterally decide on/commit to their own activities (a child does not "decide" to go to school; permission is required from a parent for a child to visit a friend; etc.). An **independent** family member is (typically an adult) who can make and execute activity decisions unilaterally (not worrying for the moment about intra-household constraint issues).

Serving family dependents is a household project[15]. Activities involved in this project are:

1. **Supervision of the dependent.** Very young children must always be directly supervised by a competent adult. If this is not a family member, then it must be someone contracted with to provide this service (baby-sitter; daycare; neighbour; etc.). The stringency of this requirement decreases as the child grows older and gradually transitions into an independent person. Implications for scheduling include: "someone" must always be home when the child is at home (or, alternatively, the child must "tag along" with the supervising adult if that adult must leave home for some reason); and if work responsibilities, etc. mean that proper supervision cannot be provided, then this must be "contracted out" to a nanny, a daycare, etc.
2. **Chauffeuring the dependent.** Young children cannot travel alone, even if they are physically able to do so. They must be taken, escorted to school, to a friend's house, etc. Again, the need for this gradually reduces as the child ages and the range of their independent activities extends.

---

[13] The shift from household to family in this section is explicit. Dependency as being developed here is fundamentally a family concept. Families "take care of their own"—this is a primal characteristic of the family. A non-family household by definition consists of independent members, as being defined here. This is not say that one cannot extend to notion of "taking care of others" to non-family persons, but it is not something that will be considered at this juncture.

Conversely, family members need not be members of the household, but still may be a dependent that is "served" by a household member. Note that this focus on the family (including non-household family members) as the organising unit for the concept of dependency is a strong rationale for the inclusion of the family class (in addition to the household) within ILUTE, which, indeed, is done in the current ILUTE prototype implementation.

[14] That is, to keep the discussion as simple as possible, the young child example will be discussed throughout. Clearly with appropriate change of language, the concepts apply to an elderly, infirm relative or a person with a significant physical/mentally disability.

[15] That is, the independent persons within a given household collectively have responsibility for serving their collection of family dependents; there should be no ambiguity/confusion in "mixing" between the concepts/objects of family and household in this. It is necessary to define this as a household rather than a family project, since it is households that are processed within ILUTE, not families.

3. **Participation in activities with the dependent**. Going to the doctor with a child, playing with the child, coaching the child's soccer team, etc. all are examples of joint activities with the dependent. Participation often also involves chauffeuring (i.e. getting to/from the activity if is out-of-home) and always implies supervision. But one can chauffeur without participating (one doesn't usually go into the classroom with the child), and one can also supervise without participating (e.g. be home while the child is having a nap or playing alone or with friends).
4. **Attending to the dependent's needs**. Shopping for the dependent, feeding the dependent, etc. are examples of activities which independent family members must undertake on behalf of the dependent.

The generation of serve dependent project activity episodes and episode attributes thus is quite a different process than for other types of projects in that (with the exception of "attending to needs" activities) they are generated in the first instance by the dependent him/herself, who in a sense "requests" service from the serve dependent project ("take me to school"; "supervise my at-home play time"; etc.).

To the extent that these activities being generated are of the external-contract, Type I sort (e.g. school, daycare, team sports), these are included in the pre-planned skeleton, and the associated serve-dependent activity episodes must also be pre-planned. Supervision requirements, even if not "contracted out" generally fall within the pre-planned category as well, given the high priority which must be attached to this very fundamental responsibility.

To put this more formally given the terminology which has been developed above:

1. For every Type I activity in a dependent's schedule, a Type IV activity exists for the household in terms of serving the dependent's activity (take the child to school; attend the child's soccer game; etc.; that is a chauffeur, participate or attend activity exists).
2. For every time period in which the dependent is not otherwise competently supervised (at school, playing at a friend's house, etc.) and requires such supervision, then a household supervise-dependent Type IV activity exists.
3. These serve-dependent Type IV activities must all be pre-planned in terms of being allocated to one or more person's pre-planned schedules (since it is individual persons who execute household-level activities).

An important additional decision issue exists with respect to supervision of dependents that must be resolved as part of the longer-term pre-planning process. This is the question of whether or not to contract with an external agent for all or part of the dependent supervision task, either on an out-of-home (e.g. daycare) or in-home (e.g. nanny) basis[16]. Using an external agent may represent the only way to feasibly pre-plan the primary events (e.g. in the case of two full-time workers who have a pre-school child), or may represent a "utility-maximizing" solution. Note that this very much is a longer-term decision, in that it can tie directly to

[16] Similar decisions exist with respect to the care/supervision of the infirm elderly in terms of out-of-home (placing the person in a "retirement home", etc.) and in-home (nursing, home assistants, etc.) options.

processes such as labour force participation (decide to quit the job and stay home full-time with the child) and even residential location (move to a bigger house to accommodate a room for the nanny). It is also essential that it be part of the pre-planning process, since it may fundamentally affect the feasibility/nature of the schedule skeleton, as well as the travel mode choices that are associated with this skeleton. In particular, if out-of-home daycare is selected, this generates two new out-of-home serve-passenger travel episodes which must be accommodated in the pre-planned skeleton.

Serve-dependent activity episodes that are within-week will also usually generate a serve-dependent episode, but these are all handled by the within-week process.

Finally, otherwise independent adults who do not possess a driver's licence or who do not currently have access to (one of) the household car(s) might be considered to be "mobility dependent" to the extent that household drivers (or the drivers with current access to the household's car(s)) may chauffeur them from one location to another. This form of "dependency" is **not** included in the term as it has been discussed in this section. It is dealt with as mode choice/dynamic scheduling problem (see Miller *et al.* (2005) for a brief discussion of this).

Activity Modes

The set of activities, **A**, in which a person might possibly engage is potentially vast and growing as technology, culture, etc. evolves and "supplies" new opportunities for allocation of our time (e.g. "surfing the web" as an activity did not exist until a few years ago). Further, this activity set varies from one economy/society/culture to another – presumably a Kalahari Bushman and a resident of New York or London are faced with very different activity sets. This observation poses a major challenge to any theory/model of activity scheduling in terms of how to deal with the size and diversity of activity sets.

In the formulation being presented in this paper, it is argued that, in fact, the concept of an activity, *per se*, can be defined in terms of a shorter list of "generic" actions that are relatively universal and invariant over both time and society. What does change over time and from place to place, and what provides the diversity of actions observed over time and place, is the set of **modes** by which a given activity can be executed. As an example: "surfing the web" is one mode of the generic activity "acquire information", which has many other modes, including reading newspapers, visiting a library, attending open houses, going to a store and talking to a salesperson, speaking with friends and neighbours, etc. Similarly, "shopping" is an activity; its modes include in-store, catalogue, internet, etc.

From a modelling perspective, the introduction of the concept of activity mode possesses several potential advantages, including:

1. Similar to the notion of a set of "primary projects", it permits one to identify a finite set of activities which can be taken as "fixed" within the model and programmed as such. This "hard-wiring" of an activity set is almost inevitable in any practical model. What is being argued here is a theoretical/behavioural rationalisation for this assumption, as

well as guidelines for determining this finite, fixed set. That is, as with the primary projects, the activity set should: correspond to major activities of interest within ILUTE; be a fundamental human activity; neither be usefully "aggregated up" into a "meta-activity" nor "split up" into two or more "sub-activities"; and be activities in which all households potentially might engage.

2. The detailed nature of the actions taken by persons in the system being simulated are then defined by the **supply** of activity modes available at a given time within this system. This supply can change over time (thereby introducing evolutionary dynamics into the activity scheduling model) in response to changes in technology, public policy, social norms, etc. As with any supply-side component, the definition and evolution of this supply might either be endogenously generated within the model (if need and capability both exist to model supply endogenously), or, as is more typical in many cases, exogenously presented to the model. That is, the model needs to be "told" that internet shopping exists, in exactly the same way that a conventional travel model needs to be "told" that the commuter rail mode is available for a given trip.
3. Indeed, this concept of activity mode is borrowed directly from and is explicitly analogous to the concept of travel mode. Just as a trip is an episode (in the same way that shopping is engaged in as an episode) which possesses the standard episode attributes of start time, duration and location, so it is argued that all episodes (activity as well as travel) possess an addition attribute: mode. And just as travel as an activity exists in a supply-demand system, in which the supply of alternative modes helps determine the way in which travel demand is realised in terms of an actual trip, so too does the supply of alternative activity modes help determine the way in which activity demand is realised in terms of actual episodes.
4. This notion, that details concerning the activity modes comes from a "supply side", the specification of which may often (at least in the first instance) be exogenously specified again is both a very pragmatic (and largely inevitable) approach to dealing with the complexity of human behaviour, but also, has been argued above, a theoretically justifiable, logical one. In particular:
   (a) it allows one to "get on" with the job of constructing an over-all model without getting "bogged down" in the details of having to construct at the outset a totally inclusive set of detailed activities; and
   (b) it provides a mechanism for introducing new activities (i.e. new activity modes) into the model over time, either through endogenous or exogenous processes (or, for that matter, deleting obsolete modes as they become replaced by new means of doing things).

The Activity Set

Having asserted that a finite set of "universal" activities exists, it remains to define this set. Definition of such a set obviously depends upon the level of detail required in the model, as well as usual caveats concerning data availability, etc. In addition, care is required in differentiating between what is a project, an activity and a mode of activity, and a certain amount of arbitrariness is almost inevitably going to exist in the making of these distinctions. In general, it is expected that the activity set will evolve (and quite possibly grow) over time as model complexity, data base support, etc. grow, and as our experience with models based on these concepts increases. A "first pass" at an activity set sufficient to support an ILUTE-type model, however, is shown in Table 2.2.

In this table, the first column presents a set of fairly "high-level", aggregated activities. In some cases at this level of aggregation, this "activity" corresponds to a previously identified "project" (e.g. "work", cf. Table 2.1). In other cases, the activity "cuts across" two or more projects (e.g. "shopping", "formal group activity"). The second column then breaks these "high-level" activities down into a finer set of activities that might be supported in a more detailed model. These are, of course, illustrative rather than definitive in nature. Note that the "work" activity categorization is typical of "white collar" occupations. A similar categorization can be constructed for "blue collar" occupations. Finally, the third column provides illustrations of alternative "modes" for many of these activities, where these alternatives exist.

The difference between an activity "category" and "mode" is, again, somewhat arbitrarily defined in Table 2.2. In general, "mode" is intended to capture differences in technology/medium of interaction/participation, etc., but often also has locational implications (e.g. in-store versus at-home shopping via catalogue or the internet). In Table 2.2, "in-home" versus "out-of-home" activities are treated as activity categories rather than modes. Certainly one might argue, however, that "out-of-home" may well represent a different "mode" of interaction relative to "in-home". The approach incorporated into Table 2.2 has been adopted based on the assumption that out-of-home options will, in general, exist for most activities, and so may be "hard-wired" as an activity category or type (e.g. out-of-home personal entertainment). The specific types of opportunities for out-of-home activities will then vary with the "supply" of these opportunities provided by a given society/economy/technological state, and these, therefore constitute the modes for this activity (e.g. movies, theatre, etc.).

HOUSEHOLD DECISION-MAKING

Households are a necessary and extremely important behavioural unit in an ILUTE-type model for two reasons. First, they are the fundamental "organising unit" or "container" for persons in society, since persons and families come together within the household to occupy a dwelling unit. Second, critical decisions are made at the household level. These decisions include:

Table 2.2 Suggested List of "Universal" Activities

Activity	Possible Categorization	Example Modes
Entertainment/recreation	In-home, personal	Reading; radio; television; stereo; hobby
	In-home, joint	Radio; television; games; story-telling
	Out-of-home, personal	Movies; theatre; concerts
	Out-of-home, joint	Movies; theatre; concerts; "family outing"
Formal group activity	Team sport game/practice	At sport facility
	Attend religious service	At church, synagogue, etc.; radio; television
	Religious/community group meeting	Face-to-face; teleconference; internet; video conference
	Religious/community group activity	At activity location
Household maintenance	Household dwelling unit	
	Household vehicle(s)	
	Food preparation/clean-up	
Information gathering	By project	Newspapers; radio; television; friends; "window shopping"; internet
Personal business	Doctor	House-call; in-office; clinic; hospital emergency room
	Dentist	In-office
	Bank/Financial	In-office; telephone
Personal maintenance	Sleeping	
	Eating (in-home; out-of-home)	Cook one-self; "frozen meals"; sit-down restaurants; take-out; "fast food"
	Personal grooming (in-home; out-of-home)	
	Exercise (in-home; out-of-home)	Jog in the park; go to the gym; personal trainer
Serve-dependent	Supervise dependent	Household member; non-household family member; "nanny"; "daycare"
	Chauffeur dependent	Household member; school bus; taxi
Shopping	Grocery	In-store; internet
	"Other consumables"	In-store; catalogue; internet
	"Durables"	In-store; catalogue; internet
Socialise with friends/relatives	In-home	Party; dinner/brunch/etc.; "family gathering"; "hanging out"
	At their home	Party; dinner/brunch/etc.; "family gathering"; "hanging out"
	Out-of-home	Restaurant; bar; "hanging out"
Work	Desk-based (working independently)	Usual place of work; at home
	Internal meetings (interaction with co-workers)	Face-to-face; telephone; internet
	External meetings, local clients, conferences, etc.	Face-to-face; teleconference; internet
	External meetings, out-of-town clients, etc.	Face-to-face; teleconference; internet; video conference

1. **The acquisition of durable goods.** Durable goods require the expenditure of significant household resources (money) and significantly affect the resources available to household members in the execution of their daily activities. Thus, decisions concerning the acquisition of durable goods is best conceptualised as occurring at the household level. In ILUTE, durable goods, at a minimum, include the household's dwelling unit (whether rented or owned) and the household's vehicles (again whether owned or leased). Other durable goods which fit into this category but are unlikely to be explicitly dealt with in a typical ILUTE model include: furniture, major appliances, major sports equipment (boats, etc.), vacation home/cottage, etc.
2. **Allocation of household resources to household members.** In order to execute their daily activity schedules, household members must use/expend household resources such as cars and money. These are scarce, limited resources. Resolution of conflicting requests for these resources (two members "need" the car at the same time; money exists to buy a new dish washer or new tires for the car, but not both; etc.) must occur at the household level.
3. **Allocation of household activities to household members.** Households cannot engage in activity episodes *per se*, since a household cannot "act" – only a person can. Households, however, do have projects and activities which must be undertaken. The task of allocating these household responsibilities resides at the household level. Household level activities include: all shelter-related activities; vehicle maintenance; food acquisition and preparation; most family-related activities, notably all serve-dependent (Type IV) activities; family vacation activities; joint social and recreation activities; etc.

Thus, households are responsible for many of the long-term, "big ticket" decisions (housing, cars, daycare, etc.) which define the context for day-to-day activity scheduling. They also affect day-to-day decision-making through the allocation of both household resources and household tasks to individual household members.

SCHEDULING HOUSEHOLD MEMBERS' ACTIVITIES AND TRAVEL

Individuals also make long-term decisions concerning: employment; schooling; acquisition of driver's licences and transit passes; household membership (marriage, divorce, choosing a roommate in a non-family household); and Type I group membership commitments (among, perhaps, others). These decisions all should be included within the longer-term components of ILUTE.

Once these personal and household level decisions are all "fixed" for a given longer-term time step (e.g. for a given year in the simulation), then the daily or weekly activity/travel scheduling decisions for each household member can be considered. As has already been indicated, it is assumed in this paper that a week (Monday through the following Sunday) is the planning period considered in the model. The choice of the weekly period is, of course, arbitrary, but not completely so. Much of life's "routine" (work, school, routine shopping, team sports, etc.) repeats itself on a more or less weekly fashion. The week also permits tradeoffs between weekday and weekend activities and scheduling opportunities to be explicitly considered.

Regardless of planning/analysis time period selected for the model, it must be recognized that considerable "left and right censoring" is occurring within the model. That is, one is trying to model an arbitrary "week in the life" of the household, without knowing what has previously happened (e.g. Person 1 went to the dentist last week and so is very unlikely to go again this week), or what has already been planned for more distant points in time (e.g. the dentist appointment has already been scheduled for next month). This is inevitable in any model which doesn't attempt to model every day in the life of the persons being simulated – clearly an impossible task within the foreseeable future of ILUTE modelling. While this should not stop one from trying to model a "typical week", it does have implications for what such models will and will not include, what modelling assumptions are more or less reasonable, etc.

Given the decision to model a "typical week", as has already been briefly discussed, the scheduling task must decompose into at least two components: determination of a **pre-planned** (i.e. planned prior to the commencement of the week being modelled) "skeleton" or base schedule; and then the "filling in" of this schedule with the **within-week** activities and details that ultimately determine the final executed activity pattern for the week. Note that this "filling in" may well involve modifying the pre-planned skeleton schedule in various ways.

The intent of this paper is not to present a fully detailed activity/travel scheduling model. An operational prototype of such a model is presented elsewhere (Miller *et al.*, 2002, 2005; Miller and Roorda, 2003; Roorda *et al.*, 2005). The following two sub-sections, however, briefly sketch a few elements of such a model. The first sub-section presents several key assumptions/hypotheses which underlie this paper's proposed approach. The second then discusses an issue that is often overlooked in activity scheduling models: the role that costs play within the activity scheduling process.

Hypotheses

Many approaches to modelling activity scheduling are no doubt conceivable. In this paper, the following hypotheses are maintained and provide a basis for model operationalisation.

First, a fundamental assumption is that **scheduling is an event-driven, sequential process**, in which individual episodes are provisionally scheduled as they "arise" out of personal and household projects. That is, at any point in the scheduling process, a partial, provisional (but feasible) schedule exists. When a new episode is "presented" to be scheduled, it is, along with any modifications to the prior provisional schedule required to accommodate the newly inserted episode. It is argued that this is, by and large, how "real life" proceeds. For example, a friend calls up to see if you want to get together Saturday night. More often than not, "on the spot" you will either: (1) agree to meet Saturday; (2) arrange an alternative time slot if Saturday is already committed; or (3) decide it will have to happen some other time in the future, since it isn't going to work out this week.

The first or second outcomes may also involve rescheduling one or more episodes currently in the provisional schedule (e.g. cutting short a shopping trip planned for late Saturday afternoon, or deciding not to go to karate class Friday night so that you can meet your friend instead). A fourth possibility, of course, also exists – deferring the decision until later in the week ("I'd like to, but I may need to babysit my kid sister – I'll get back to you when I know for sure"). In the model being sketched here, this might be handled in at least two ways. First, the episode could be "put back on the stack" to be reconsidered later, which is pretty much what happens in reality. Or, it can be argued that considerable "temporal aggregation" exists within the model. That is, we probably really don't know exactly when within the week a given episode is "presented" to the scheduler – it is at best a random "guess" as to when this occurs. Given this, it may simply be appropriate to deal with each episode once, whenever the model decides (randomly or otherwise) to present the episode to the scheduler and be done with it. In either case, the assumption about "on the spot" decision-making effectively holds.

Given this, a second assumption follows that, in general, **activity/travel scheduling is not an optimizing procedure**. The sequential insertion process briefly described above tends to be myopic and greedy, and therefore may well result in a final schedule which is "sub-optimal" with respect to some appropriate set of criteria (assuming such criteria can be defined and operationalised). In particular, modifications to an existing, provisional schedule tend to be "local" in nature. That is, the person will only rearrange his/her schedule to a certain degree to accommodate a new episode. In particular, "global (re-)optimisation" of a schedule does not generally occur. Over and above practical concerns about computational intensity, etc., this is fundamentally a behavioural argument, which has at least two components. First, as in all aspects of life, it simply isn't clear that people are generally "global optimizers". Second, in the case of activity scheduling, so many activities come with sufficient constraints (start/stop times, minimum feasible duration, fixed locations and hence travel times, etc.) and/or commitments/contracts with others (which we are generally loath to break) that the "degrees of freedom" available for significant (re-)optimisation of an entire schedule are likely to be very minimal.

It can be argued that more opportunity exists for "optimizing" the skeleton portion of the schedule, given its pre-planned, routine nature. This is certainly a possibility, but the constraints/commitments argument posed above certainly holds strongly in this case as well, implying that there is likely only limited opportunity to rearrange the skeleton so as to "optimize" it (again, assuming that an appropriate optimisation criterion exists). To the extent that such modifications are feasible, they can, of course in principle occur.

Perhaps more important (and more typical), it is probably the case that pre-planned (Type I) commitments will only be undertaken (i.e. committed to) if a feasible (and reasonable) skeleton schedule can be constructed (a child can only play a team sport if a parent is able to get him/her to the games and practices on an on-going basis). Thus, feasibility, rather than optimality may be the more important consideration in skeleton schedule construction.

A third fundamental assumption is that travel mode choice (and the associated allocation of household vehicles for individual person travel, as required) is inherent in the activity scheduling process. That is, **travel episodes must be (provisionally) scheduled along with their associated activity episodes**, since activity episode timing and feasibility can not, in general, be determined independent of the travel episodes required to get to and from the activity. Thus, for example, I may only be able to make a shopping trip at a time when a car is available to transport my purchases home, or I may arrange my errands one way if travelling by transit and another if travelling by car (in terms of location(s) visited, timing of the shopping episode(s), etc.).

Fourth, some means must exist for selecting among competing alternatives. These decisions include: allocation of vehicles among competing household members; individual travel mode choices; modification of/choice among activity episodes competing for overlapping time slots; choice of activity location in cases where this not "pre-specified"; allocation of household activities to household members, etc. While many decision criteria/processes are conceivable, it is assumed in this paper that some form of a **utility-based approach** to this problem is virtually inevitable. In making this assumption, the usual arguments against utility and its "rational actor" baggage are recognized. Nevertheless, given the need to determine complex tradeoffs among competing alternatives, the need to relate time, money and activity with one another, and the simple fact that people do receive benefit to varying degrees from alternative courses of action (and that it is not unreasonable to assume that, more often than not, people are at least boundedly rational in their decision-making in that they choose alternatives that are more likely to bring them more benefit than less), all drive one towards a utility-based approach.

In making this assumption, one is not insisting on "perfect information", optimizing, "totally rational" decision-making. As indicated above, "global optimisation" is specifically rejected in this model as an organising principle. Rather, it is simply arguing that people must be able to establish preference orderings over alternatives in order to make choices, both personally and as members of the household collective. And certainly these preference orderings are implicit in these decisions (Person A gets the car; Person B doesn't), even though the persons involved do not explicitly express themselves in these terms[17].

The issue of developing an operational utility-based model will not be explored in detail here. A final, few observations on this issue, however, are as follows.

1. In many scheduling models, the concept of "priority" is used, usually without a tangible definition. In many instances it appears that "utility" and "priority" could be used almost interchangeable terms in any practical model. Both are latent variables, and both seem to be "doing" the same thing in the model. That is, one typically says that "high priority" episodes get scheduled first, or perhaps that if one episode replaces another, then it must be of "higher priority". One can substitute the "utility" for "priority" in the previous sentence without changing its effective meaning. Hence, only one term or the other presumably needs to be used. Given its firm foundations in microeconomic theory, "utility" would appear to be the more sensible, "useful" term to employ. It also would appear to be the broader term, capable of more readily including the various components discussed in the next point, below. One might also

[17] This does not necessarily assume that individual utilities are simply additive within the household, although this is clearly the simplest assumption to make and one that is being pursued in our current modelling work. Other household-level decision rules are obviously conceivable and should at some point be investigated.

note in passing that, in the examples above, "priority" suffers from the same tautological problems that are often attributed to "utility". That is, it is asserted that "high priority" items are scheduled. How do we know that they are high priority? Because they are scheduled first![18]

2. Utility may well derive from many different "components", even within the same activity. These "components" might include:
   - personal pleasure, stimulation, enjoyment, sense of well-being, etc.;
   - the pleasure, stimulation, enjoyment, sense of well-being, etc. of others;
   - commitment to others (i.e. fulfilling contracted obligations);
   - economic reward;
   - sense of security, etc.;
   - status; and
   - urgency; i.e. the need to act now in order to avoid penalties for inactions (finish the term paper tonight or fail the class), the loss of future utility (if I don't mow the lawn today I can't play golf tomorrow), scheduling opportunities, (if I don't go to the store now, I can't fix the screen door tomorrow), etc.
3. Utility, of course, may well have negative components as well. A given activity may be stressful, dangerous, boring, involve physical discomfort, etc.
4. Utility would appear to be largely project-based in its origins. That is, the "benefit" accruing from the shopping episode does not principally lie in the shopping episode *per se*, but in the fact that this episode is a necessary step in holding the dinner party on Saturday evening, from which significant pleasure, etc. will be derived[19].
5. Utility is also likely to be dynamic/context-dependent in nature. That is, the "need" (and hence the utility) of a grocery shopping trip when the kitchen is filled with food is likely to be much less than if the proverbial cupboard is bare.

Finally, **the provisional nature of a schedule must be stressed**. Throughout the scheduling process, the schedule is incomplete, subject to modification – a "work in progress". It is not until the time that "the next episode" is actually executed that the schedule is transformed from a plan into a fixed, realised activity pattern. Thus, while it is assumed that global restructuring of the schedule does not occur, the schedule can be "tinkered with" throughout the scheduling process. This evolutionary (and incremental, "local" rather than "global") building/modifying of the schedule is clearly evident in CHASE type survey data (Litwin *et al.*, 2004), which strongly supports this view of the scheduling process.

---

[18] Having said this, a role for "priority" in scheduling semantics may well still exist. The selection of which activity to schedule first may be based on characteristics of activities other than their utility. For example, it might be rational to schedule activities that are most fixed in time and space (or other dimensions) first, leaving more flexible (and possibly "higher utility" activities) until later. That is, "priority" might refer to the order in which activities are scheduled, and not carry any connotation concerning "utility". The key point here (and, indeed, throughout this paper) is to try to achieve an explicit, consistent, useful semantical representation of the scheduling process.

[19] This is not say that the shopping episode may not have its own, intrinsic utility components, including the stimulation of "getting out and about", of interacting with others, of "hunting for bargains", etc.

Activity and Purchase Costs in Activity Scheduling

It is curious that costs (i.e. the $c_{ie}t_{ie}$ and p_g x_g terms in equation [4]) do not play a significant role in most activity scheduling models, with the exception of travel costs in mode choice calculations. In some cases, this may be appropriate because:

- the activity *per se* involves no cost except possibly for travel (e.g. visit relatives);
- the cost of the activity is trivial within the budget context (e.g. a high-income person stopping at a coffee-shop to pick up a coffee); and/or
- the cost of participating in the activity is a one-time, previously paid, "sunk cost" (e.g. taking a child to her soccer game costs nothing except for travel cost at the time of the activity, but there was a $100 registration fee at the time the child joined the league).

Cost, however, does clearly play a role in activity scheduling whenever this cost is significant relative to one's overall budget (the cup of coffee may be beyond the means of the unemployed person; a family may decide that the $3,000 annual cost of equipment and registration for their son to play "Triple A" hockey is beyond their means; etc.). In addition, even trivial costs add up over the entire activity pattern to yields a significant (or at least discernable) component of the period T budget.

The entire activity "scheduling" process discussed above is fundamentally driven by the need to satisfy the time budget constraint (equation [1] or [2]). Bringing activity and purchase costs into these calculations would clearly add considerably to the process complexity. Nevertheless, this issue should be explored.

Note that in the short-run within which activity scheduling is assumed to occur (i.e. this week in the life), durables are fixed (these are acquired through longer-run processes outside of scheduling *per se*), as are employment income, savings, etc. Thus, most terms are fixed in equations [4] and [5] except the $c_{ie}t_{ie}$ and p_g x_g terms (i.e. the costs of activity participation and consumable goods/services purchases). Perhaps a simple process for dealing with equations [4]-[5] in the scheduling process is:

1. As episodes get scheduled, accumulate costs.
2. If total "out-of-pocket" costs exceed disposable income by more than some threshold, then either no more episodes (which involve out-of-pocket costs) may be scheduled, or some previously scheduled episodes will have to be deleted/modified in favour of other not-yet-scheduled episodes which have higher utility and/or utility/cost "return".

This "knapsack" approach to the problem is basically identical to the approach being used with respect to the time constraint – i.e. "fill up the knapsack" until you run out of room, and then "maybe" take something out of the knapsack to make room for something else (of greater value).

The threshold recognizes the "sloppiness" inherent in both the model calculations as well as in the lack of need for rigorous fiscal accounting within short time periods. That is, the household "books" don't need to exactly balance on a weekly basis – presumably the month (with its monthly cycle of credit card and other bills, etc.) is the shortest time period over which a true balance needs to be struck (at least on a routine basis).

To do this, of course would require simulating purchases as well as activities, as well as in general a much stricter "cost accounting" than is normally undertaken. Whether this is feasible/needed in practice perhaps remains to be seen, but conceptually, the approach is attractive and would appear to warrant further consideration/investigation.

The case of "one-time", up-front costs is an interesting and important case. It is analogous to the purchase price of a durable good – which argues for annualising the cost over the life of the activity. It is also consistent with the overall concepts developed in this paper in that these one-time costs are associated with Type I activities, which are pre-planned and which are "longer-run" in terms of the participate/don't participate decision, in exactly the same way as the "buy car/don't buy car" decision (for example). Thus, it is appropriate to treat these one-time costs as "capital expenditures" rather than as "operating expenses" (i.e. use the same general procedure as for buying houses or cars, although the calculations will generally be somewhat simpler, given the typically smaller amounts and more "informal" financing procedures involved). Note by doing so, there will, indeed, be no $c_{ie}t_{ie}$ term for such an activity on an episode-by-episode basis. This is appropriate, just as the longer run costs of operating a car do not enter into the $c_{ie}t_{ie}$ terms for individual travel episodes, which are only concerned with the out-of-pocket costs of the given trip.

SUMMARY

This paper has sketched a conceptual model of household decision-making that deals with both short-run (daily/weekly activity/travel scheduling) and long-run (auto ownership, housing, etc.) decision processes. It defines individual person-based decision-making as occurring within a household context, within which the household makes long-run decisions about the acquisition and nature of major resources (cars, etc.), and short-run decisions concerning the allocation of these resources for use by household members. Individual persons make decisions concerning the allocation of their time based on this allocation of resources.

The conceptual model discusses the role which both time and money budgets play in the decision-making process. In particular, it discusses the role in which the money budget plays in long-run decision-making concerning the acquisition/replacement of major durable goods, while the role of the time budget is emphasized in shorter-run activity/travel decisions.

The concept of the project is introduced and discussed in considerable detail as a fundamental organising principle for modelling household decision-making. A project is a coherent set of logically related activities. It is argued that all of one's activities can be represented by a set of projects, and that it is projects which "generate" the daily activities in which we engage.

Activities, in turn, are grouped into four fundamental types, characterized by the extent to which they tend to be preplanned, their flexibility with respect to scheduling/rescheduling, as well as the extent to which they involve commitments with other agents, either external to or within the household. The concept of commitment is viewed as an extremely important one in terms of defining priority, flexibility, etc. in the scheduling process.

Although a fully elaborated model of activity scheduling is not presented in the paper, some key features of the activity scheduling process discussed include:

1. It is a dynamic, event-driven process in which activity episodes are provisionally scheduled as they "arise" out of their generating projects.
2. It is a largely rule-based procedure, in which "global optimal" schedules are not sought. Rather, "localised", limited adjustments to the provisional schedule are made in order to accommodate (where possible) new activity episodes as they are generated.
3. At the same time, it is argued (although not pursued at great length within this paper) that "utility/priority" assessments of alternative activities, allocations of household vehicles to household members, etc. occur as part of the process as a means of deciding among competing demands (two episodes competing for the same time slot; two persons competing for the same vehicle; etc.).
4. Travel is viewed as an integral part of the activity scheduling process. Indeed, a trip is defined a travel episode, which must be (provisionally) scheduled along with the associated activity episode being "served" by the trip. Mode choices for these travel episodes must also be determined as an integral part of the scheduling process, since modal travel times affect the timing and even feasibility of a given activity episode.
5. Just as travel episodes have mode, it is argued that so too do many activities. It is argued that the concept of an "activity mode" is an attractive one in that it permits a relatively stable set of activities to be defined, with variations of and extensions to these activities being defined by the modes available for engaging in these activities. Activity modes are largely defined by technology, culture, the economy, etc. As such, they vary over time and from one society to another, and, from a modelling point of view may need to be largely exogenously specified to the model (in the same way that travel modes are).

The conceptual model discussed in this paper has been implemented in a very simplified form in terms of a prototype one-day activity/travel model (i.e. with longer-run decisions assumed to be known and fixed) for the Greater Toronto Area (Miller *et al.*, 2002, 2005; Miller and Roorda, 2003; Roorda *et al.*, 2005). Work on extending the model to a full week and to include longer-term decision-making within a full ILUTE microsimulation framework is on-going.

REFERENCES

Axhausen, K. (1998). Can We Ever Obtain the Data We Would Like to Have? In: *Theoretical Foundations of Travel Choice Modelling* (T. Gårling, T. Laitila and K. Westin, eds.), pp. 305-324. Pergamon Press, Oxford.

Doherty, S.T., E.J. Miller, K.W. Axhausen and T. Gårling (2001). A Conceptual Model of the Weekly Household Activity-Travel Scheduling Process. In: *Travel Behaviour:*

Patterns, Implications and Modelling, (E. Stern, I. Salomon and P. Bovy, eds.), pp. 148-165. Elgar Publishing Ltd., Cheltenham U.K.

Litwin, M., E.J. Miller and S.T. Doherty (2004). "Investigating Competition Patterns in Household Activity Scheduling Processes". Presented at the 83rd Annual Meeting of the Transportation Research Board, Washington, D.C., January.

McCann, B. (2000). *Driven to Spend.* Surface Transportation Policy Project, Washington, D.C.

Miller, E.J. and A. Haroun (2000). "A Microsimulation Model of Residential Housing Markets". Presented at the 9th International Association for Travel Behaviour Research Conference, Gold Coast, Queensland, Australia, July 2-5.

Miller, E.J. (1996). Microsimulation and Activity-Based Forecasting. In: *Activity-Based Travel Forecasting Conference, June 2-5, 1996 Summary, Recommendations, and Compendium of Papers* (Texas Transportation Institute, eds.), pp. 151-172. Travel Model Improvement Program, U.S. Department of Transportation and U.S. Environmental Protection Agency, Washington, D.C.

Miller, E.J. and M.J. Roorda (2003). A Prototype Model of Household Activity/Travel Scheduling. *Journal of the Transportation Research Record,* No. 1831, pp. 114-121.

Miller, E.J., M.J. Roorda and J.A. Carrasco (2005). A Tour-Based Model of Travel Mode Choice. *Transportation,* **32**(4).

Miller, E.J., M.J. Roorda, L. Eberhard and M. Litwin (2002). "A Microsimulation Model of Daily Household Activity and Travel". Presented at 98th Annual Meeting of the Association of American Geographers, Los Angeles, March.

Miller, E.J. and P.A. Salvini (2001). The Integrated Land Use, Transportation, Environment (ILUTE) Microsimulation Modelling system: Description & Current Status. In: *The Leading Edge in Travel Behaviour Research* (D. Hensher, ed.), Chapter 41. Elsevier Science, Oxford.

Mohammadian, A. (2001). *Dynamic Model of Household Automobile Transactions within a Microsimulation Framework.* Ph.D. thesis, Department of Civil Engineering, University of Toronto, Toronto.

Roorda, M.J., S. Doherty and E.J. Miller (2005). Operationalising Household Activity Scheduling Models: Addressing Assumptions and the Use of New Sources of Behavioural Data. *In this volume.*

ACKNOWLEDGEMENTS

This paper was supported by the Social Sciences and Humanities Research Council of Canada, through a Major Collaborative Research Initiative, 2000-2005: *Access to activities and services in urban Canada: behavioural processes that condition equity and sustainability,* and GEOIDE, the Network of Centres of Excellence in Geomatics, Canada.

Integrated Land-Use and Transportation Models: Behavioural Foundations
M. Lee-Gosselin and S. Doherty (Editors)

3

Operationalising Household Activity Scheduling Models: Addressing Assumptions and the Use of New Sources of Behavioural Data

Matthew J. Roorda, University of Toronto, Toronto, Canada
Sean T. Doherty, Wilfrid Laurier University, Waterloo, Canada
Eric J. Miller, University of Toronto, Toronto, Canada

Introduction

It is only recently that operational models of the household activity scheduling process have begun to emerge. These models share a desire to realistically replicate the sequence of decisions that leads to observed patterns of human activities and travel - including the decision of what activities to conduct, by whom, for how long, at what time and location, and by what mode. Understanding the sequence of these decisions is crucial in developing a dynamic model of scheduling behaviour that can represent the way people reschedule over time in response to changing opportunities and constraints.

While a fairly well-developed theoretical basis exists for such models, one of the biggest stumbling blocks for their development has been the lack of observed data on the actual process of decision-making as it occurs within the household. Traditional trip-diary data provides limited insight into this process, since only the executed schedules, which are a product of a large number of scheduling decisions, are actually observed. Attempts to use trip-diary data as a basis for developing operational scheduling models require that many assumptions be made concerning the decision process. These assumptions include how decisions are ordered, how conflicts are resolved, how schedules are modified, and how/when group decisions are made.

At this stage in development, it is important that researchers attempting to operationalise such models communicate their experiences, especially with respect to the underlying assumptions they have had to make. Such information would provide a more definite objective for future data collection efforts and help set priorities for empirical analysis.

This chapter begins with a review of the theoretical background for activity scheduling models and a brief review of emerging methods. The chapter then describes a recent attempt to operationalise an activity scheduling model with household agents for the Toronto area using trip diary data from the 1996 Transportation Tomorrow Survey. A description of CHASE©, one of several survey instruments used in a pair of complementary panel surveys in Toronto and Quebec City, are described in the next section, followed by a description of assumptions made concerning the scheduling decision process and how they could be improved. Finally, key areas for future development of the operational scheduling model are discussed, considering expected availability of more in-depth scheduling behaviour survey data currently being collected in Toronto.

ACTIVITY SCHEDULING MODELS

It has long been recognized that improving our understanding and ability to forecast travel behaviour requires a focus on the activities that give rise to the need for travel. This realisation has led to the development of an activity-based approach in an effort to improve travel demand forecasting and better assess the impacts of emerging transportation policies and trends (Jones *et al.*, 1983; Kitamura 1988; Ettema and Timmermans 1997). Travel is considered to form part of a continuous pattern of daily behaviour, depicted as a single sequence of activities in time and space. The roots of the activity-based approach can be traced to the space-time geographic concepts pioneered by Hägerstand (1970) and as further explored by Chapin (1974, 1978) and Cullen and Godson (1975) in the 1970's. These widely cited authors emphasized the importance of opportunities and constraints in the formation of activity patterns, building upon the concept of space-time "prisms", widely viewed as most complete approach to understanding activity patterns. However, the majority of activity-based research to follow focused on attempting to collect information on revealed patterns of activities and travel behaviour with increasing accuracy.

It is only relatively recently that travel behaviour researchers have been calling for more in-depth research into the underlying activity scheduling process as a means to improve our understanding and ability to model observed activity schedules (e.g. Pas 1985; Jones *et al.*, 1990; Axhausen and Gärling, 1992; Lee-Gosselin, 1996; Axhausen, 1998; Bhat and Lawton, 2000). However, developing models of activity scheduling has met with considerable challenge. Two basic approaches have been adopted. The first approach focuses on the simultaneous choice among a set of activity-travel patterns, normally under a utility maximization framework (e.g. Jones *et al.*, 1983; Recker *et al.*, 1986a, 1986b; Kawakami and Isobe, 1990). In light of behavioural criticisms of this approach, (Gärling 1994), a second approach emerged that focuses more explicitly on the modelling of the sequencing of decisions made during the scheduling process. Some of the first attempts focused on the sequential choice of activities and locations (e.g. Kitamura 1983; van der Hoorn 1983). Building on this approach, Kitamura *et al.* (1997) proposed a simulation model in which the next activity type to be engaged in is first chosen, followed by its duration, then location, assuming that activities are planned and executed in sequence.

A series of additional sequential models have evolved from the SCHEDULER conceptual framework as proposed by Gärling *et al.* (1994) including SMASH (Ettema *et al.*, 1993), and models by Gärling *et al.* (1998) and Kwan and Golledge (1995). Similar in construct is the ALBATROSS model developed more recently by Arentze and Timmermans (2000). The shared theme in these models is the notion of activity priority acting as a primary determinant in the sequencing of activities. However, in practise, simplifying assumptions are most often adopted in the specification of priority - for example, by assigning priority by activity type.

THE TRAVEL ACTIVITY SCHEDULING MODEL WITH HOUSEHOLD AGENTS

General Description of the Operational Model

In Toronto, Canada, there has been a push to develop an operational microsimulation model of the activity scheduling process. The Travel Activity Scheduling Model with Household Agents (TASHA) is currently operational and is based on trip diary data from the 1996 Transportation Tomorrow Survey. The scheduling model microsimulates a 24-hour schedule formation process for residents of the Greater Toronto Area while attempting to maintain the underlying concept of the "project", proposed by Axhausen (1998) as an organizational theme for activities in the schedule. The major features of the operational model are as follows:

The model makes use of the concept of the project to organise activity episodes into the schedules of persons in a household. Activity episodes are the basic unit of analysis in the formation of a schedule. Each activity episode has a start time, a duration, a location and an activity type. A project is defined as a group of logically connected set of activity episodes that contribute towards a common objective. Projects may involve one or more household members as well as differing activity types. Projects have project agendas, which are a list of activity episodes that are associated with the project. The concept of the project, as discussed in depth by Miller (2005), can be used to combine short-term decisions (such as activity scheduling) with long-term decisions (such as residential location choice) in a unified conceptual framework. One of the other key advantages of using the concept of the project in an operational model is that it allows household interactions between household members to be handled cleanly and realistically, particularly with activities that involve joint participation (e.g. going to a movie together) or joint responsibility (e.g. taking care of children).

The model features interactive household agents. The schedules of the persons within the household are generated simultaneously to allow for interaction between family members that normally occurs within a household. The key way that household members interact in the current model is through the generation of joint activity episodes as part of household level projects. Joint activity episodes are activities where more than one person in the household takes part in an activity. The participation in joint activities requires that, in general, the activity takes place at the same time, for the same length of time at the same location. Therefore a "window" of opportunity must exist or be created in the schedules of all of the household members taking part in that activity for it to be a feasible joint activity. There are many other ways in which household members interact to "coordinate" their schedules. Obvious examples include the allocation of household vehicles, the coordination necessary for the care of children and the household decisions of which household member should do the necessary tasks for the maintenance of the household such as grocery shopping. While for the time being the only form of interaction between household agents is that of generating and scheduling joint activities, several additional interactions are intended to be captured in future versions of the model.

The model is a microsimulation of a 5% sample of households in the Greater Toronto Area. A total of approximately 89,000 households and 243,000 persons are represented as individual entities in the computer model. Activity/travel schedules are generated for each person individually.

The model was designed using an object oriented programming technique. Object orientation is a modelling paradigm that attempts to mirror real life objects relevant to the scheduling process directly as "classes" in the program code. The class design for the Travel activity scheduling model is shown in Figure 3.1. Households, persons, projects, activity episodes and travel episodes are each represented as explicit entities in the model. Each household has persons, and each person has a person schedule which contains all activity episodes and travel undertaken by that person. Both households and persons have projects, each of which has an agenda consisting of all activity episodes relevant to that project. The model also contains a spatial representation of the Greater Toronto Area, as well as a series of probability distributions for activity episode frequency, start time and duration.

The model assumes broad project and episode types. Projects are assigned as either person-level or as household-level projects. Broad project types are assigned to each household and to each person as shown in Figure 3.2. The serve-dependent household-level project has not been incorporated into the working version of the operational model but represents an important point of interaction among household agents (i.e. adults must coordinate with each other over the care of their children) and will incorporated into the next version of the operational model.

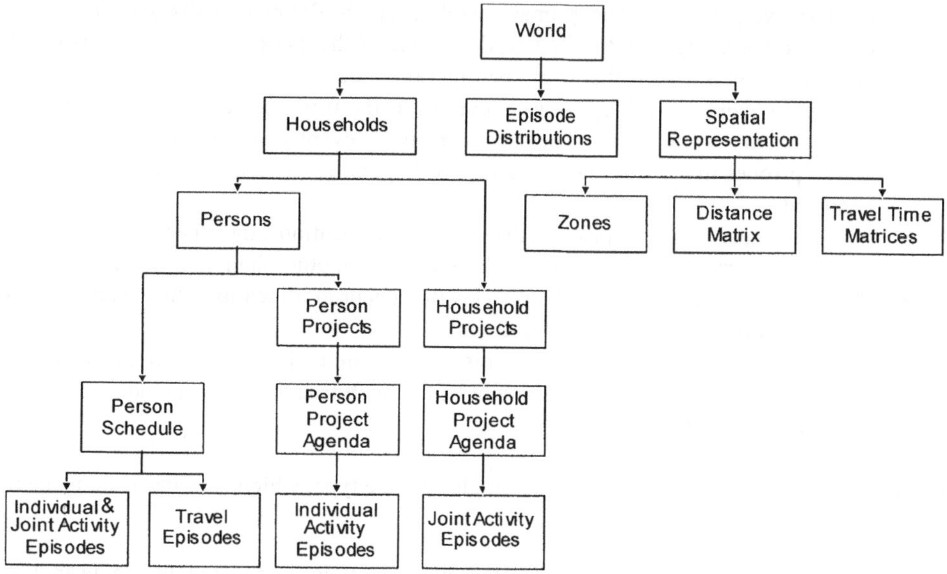

Figure 3.1 Object Oriented Class Structure

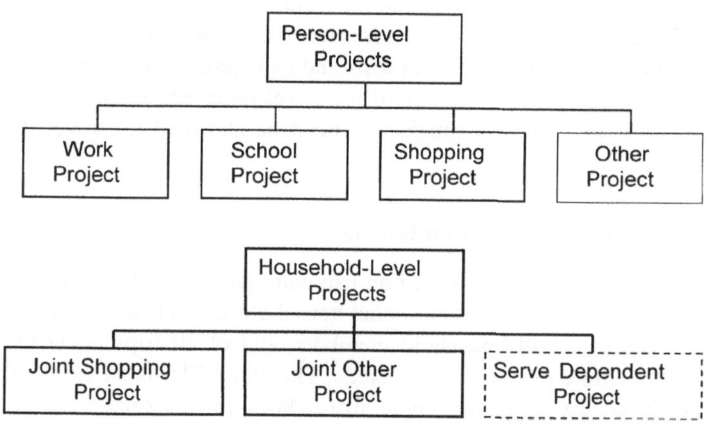

Figure 3.2 Project Types

Within each project, one or more episodes types are incorporated. The work project is the most complex project as it houses several activity episode types. These episode types include:

- *Primary work* – work episodes occurring at the usual place of work that are part of the primary work event. The primary work event is defined as the sequence of work episodes beginning with the first work episode of the day plus any work episodes from subsequent work chains that begin before 3:00 p.m.,
- *Secondary work* – work episodes occurring at the usual place of work that are part of the secondary work event. The secondary work event is defined as the a sequence of work episodes in a chain that starts after 3:00 p.m., given that a primary work event has occurred,
- *Work business* – work episodes that occur at a location other than the usual place of work, for a person that normally works at a location other than their home,
- *Work-at-home business* - work episodes that occur at a location other than home, for a person that normally works at home, and
- *Return home from work* – at-home episodes that are embedded within the primary work event. These episodes can be thought of as lunch trips but may include other at-home activities.

Each of the other projects contains only a single episode type, which is of the same name as the project.

The model assumes sequential household decisions. It is recognized that households and household members make decisions simultaneously on many different aspects of their lives. Mode choice, for example, is a decision that is very strongly integrated with decisions about where and when to undertake activities. In the current operational model, however, the numerous household decisions are assumed to be made sequentially. That is, residential location, work location and auto ownership are assumed to be exogenous inputs into the model. The schedule formation process then provides trips and trip chains, which are the inputs for a mode choice model. Currently, modelled auto drive travel times are assumed in the schedule formation procedure to determine how much time must be allocated for travel in a person's schedule[1].

The Current Data Source – Uses and Limitations

The Travel activity scheduling model is based on trip diary data from the 1996 Transportation Tomorrow Survey (TTS). The TTS is a traditional household-based trip diary survey in which attributes of the household, of all household members, and of all trips made by household members over a 24 hour time period are collected. The 1996 TTS is an extremely valuable data source because it has a very large sample size (5%), allowing for a very good level of confidence that it is representative of the entire population.

The 1996 TTS data are used for the generation and validation of activity schedules in three ways. First, the database provides the base population on which the schedule model is run.

[1] A future improvement to the model will be to incorporate a mode choice/vehicle allocation module within the scheduling model so that travel times reflect the chosen mode.

Attributes of the households (e.g. number of vehicles, residence location) and person attributes (age, sex, employment status and location, etc.) from the TTS are considered to be exogenous inputs into the scheduling model. For a base year model run, schedules for each person in the base population are regenerated.

Second, the TTS trip data are used as the basis for generating activity episode attributes including their frequency, start times, durations, the number of people involved and the location of the activity. Some manipulation is necessary to extract activity attributes from the trip database. Activity durations, for example, are determined by comparing the start times of two consecutive trips and subtracting an estimated travel time for the first trip.

For the current model a set of observed probability distributions for each activity episode attribute are developed based on TTS data. Joint probability distributions for frequency-start time and for start time-duration are used to reflect the correlation between these inter-related activity attributes. The probability distributions are developed for different classifications of household and person attributes. For example, distributions for individual shopping activity episodes are developed using classifications of age group, gender and employment status. The classifications for each episode type are shown in Table 3.1.

Table 3.1 Classifications for Activity Episode Generation

Activity Type	Explanatory Variables used for Classification
Primary work	Age, occupation, employment status
Secondary work	Occupation, employment status
Work-business	Age, occupation employment status
Work-at-home business	Age, occupation employment status
Return home from work	Occupation, employment status
School	Age, student status
Independent other	Age, possession of drivers license, employment status
Joint other	Presence of children, number of adults, number of vehicles
Independent shopping	Age, employment status, gender
Joint market	Presence of children, number of adults, number of vehicles

The third use of TTS data is for validation of modelled activity scheduling outcomes against observed schedules. Preliminary model validation results have been provided in Miller and Roorda (2003), and have shown that the model replicates observed trip-making within reasonable limits for a prototype model, but that there is significant room for model refinement.

Three major limitations of the 1996 Transportation Tomorrow Survey for the purpose of modelling activity scheduling are as follows:

First, only the executed trips are observed, from which executed activity schedules are derived. Since the executed activity schedule is the product of a large number of scheduling decisions prior to execution, it only provides a limited understanding of how a person's schedule is derived. Thus, creating a fully dynamic scheduling model is near to impossible based on TTS data alone.

Second, the survey is limited to a 24-hour window on each person's schedule. The decisions that are made about the frequency with which a person conducts certain types of activities, however, are closely intertwined with the decisions made for prior and subsequent days. It is clear that many activities are planned on a weekly basis. For example, events such as grocery shopping, organised sports, and university courses, most often occur at the same time each week. The scheduling implications of such events cannot be adequately examined with a 24-hour survey.

Third, activities can only be classified to a very aggregate set of activity types. The activity classifications are limited to home, work, school, shopping, and other. Since the attributes of activities within each of these classifications can vary widely, it is difficult to determine scheduling rules that are appropriate for each classification. For example, social visits and doctor's appointments are both considered to be in the "other" activity category. However, doctor's appointments are usually planned well before the day, have relatively inflexible start times and durations, usually occur during the day and are high priority compared with most other activities. Social visits, however, can be much more loosely planned, flexible and are much more likely to occur in the evening. An aggregate activity classification forces very heterogeneous activities to be scheduled in the same way.

Finally, TTS data does not provide any information about how individual episodes are organised into projects. Activity episode attributes including location, start time, duration, mode, activity type and the number of household members involved are observed. However, no additional information is available to determine whether activities of different types are linked to each other logically to fulfill a common objective. Therefore, it is necessary to implement the project concept based only on activity type.

The Scheduling Model Approach for the Current Operational Model

In the current operational model, person schedules are constructed from scratch based on the following steps:

1) Activity episodes are generated for insertion into each project agenda based on 1996 TTS distributions of activity attributes;
2) These activity episodes are inserted into project agendas where they are placed into a preliminary time sequence with other activity episodes that are connected by a common purpose;
3) Once the project agendas have been formed, person schedules are constructed by taking activity episodes from the project agendas and adding them to the person schedule. Activity attributes are modified and travel is added as necessary to result in a coherent consistent schedule;

4) A "clean up" algorithm is applied to reflect final scheduling / fine tuning just before or during execution of the schedule.

As such, the procedure is a "bottom up" approach to activity scheduling, which is in contrast to the "top-down" approach used in a variety of other scheduling models in which "patterns of activities" are chosen from a large, but finite set of observed activity patterns, (Jones *et al.*, 1983; Recker *et al.*, 1986a, 1986b; Kawakami and Isobe, 1990). It is felt that the "bottom up approach" is more conducive to dynamic scheduling in which schedules are constantly changing due to new opportunities and constraints that a person encounters prior to the execution of their schedule.

In order to develop realistic activity patterns, a fairly large number of assumptions are made throughout the process of generating schedules. The following section outlines the procedure for generating schedules for members of a household and the underlying assumptions for each of the schedule formation steps that are necessary to ensure reasonable schedules.

1) Activity episodes are generated for insertion into each project agenda. At the outset, the project agenda is blank for each project. Activity episodes are then generated based on 1996 TTS probability distributions for frequency, start time and duration along with reasonable rules to ensure that the resulting agendas are logical. Joint probability distributions for frequency-start time and for start time-duration are used to reflect the correlation between these inter-related activity attributes.

The generation of activity episodes for the work project is particularly complex given the number of activity types that are included in this project. In general, the following principles are adhered to when generating work project episodes:

- No work episodes are generated for children under 11 years old, or for people whose employment status is "not employed". No work-at-home episodes are generated for people under 19 years of age.
- Episodes are generated and inserted in the following order: primary work episode (W), work business episodes for people with a usual place of work (B), secondary work episodes (R), return home from work (L), work business for people with no usual place of work (B), work-at-home business (A)
- Frequency of episodes of each type is randomly chosen from the appropriate frequency distribution for that person/episode type (a maximum of 1 episode is allowed for episode types W, R, and L).
- Given the frequency, start time is then chosen randomly from the joint frequency-start time distribution so that:
    - Work business episodes fall within the primary work event;
    - Secondary work episodes must start at least one hour after the end of the primary work event and after 3:00 p.m.;
    - Return home from work (i.e. lunch) must conclude before 3:00 p.m. and must result in at least 30 minutes of work before and after the returning

home. Return home from work episodes are not generated if the start time of the primary work event is later than 12:00 p.m.

- Given start time, duration is then chosen randomly from the joint start time-duration distribution within appropriate constraints, as follows:
  - Work-business episodes must be less than the duration of the primary work event;
  - Secondary work episodes must start at least one hour after the end of the primary work event and must conclude by the end of the day;
  - Return home from work (i.e. lunch) episodes must be at least one hour less than the duration of the primary work event.
- For primary work and return to work episodes for people with a usual place of work, the employment zone is assigned to the primary work episode. For all other episodes, other than the return home from work, a location choice model based on 1996 TTS data is used to generate a random zone for that episode.

For other person-level activity episodes, the generation of episodes is done simply by assessing the frequency, duration and start time from the appropriate probability distributions for that person/episode type. The complexities of the work project do not apply because only a single episode type is assumed to be included in each of these projects.

2) Activity episodes are inserted into person-level and household-level project agendas. Once activity episodes are generated, they are added to project agendas along with other activity episodes with a common purpose. Activity episodes are inserted into the appropriate project one by one such that each project is internally consistent, or in other words, there are no activities within a project agenda that overlap in time. As such, construction of the project agendas represents the first step that a person takes to begin to organise their desired activities into a preliminary sequence. A project agenda does contain "gaps" with no planned activities. It is also noted that a project agenda does not include travel, which is only accounted for when a person constructs their actual schedule.

There are four different cases that occur when a "new" activity episode is being inserted into a project agenda that already contains "existing" activity episodes that were previously inserted. The cases include splitting an episode (Case 1), inserting an episode between "prior" and "posterior" existing episodes (Cases 2 and 3) and overlapping an existing episode completely (Case 4). These cases are described in Figure 3.3.

The process of inserting episodes into project agendas involves the application of the following rules/assumptions:

- When the new episode being inserted overlaps with part of at least one existing activity episode, the following steps are followed to attempt to create an appropriately sized "gap" for the new activity.
  - If either of the prior or posterior existing episodes is a "gap" in the project agenda then the new episode is shifted to replace all or part of the "gap",

- The prior episode is shifted if a "gap" exists in the project agenda immediately before the prior activity,
- The posterior episode is shifted if a "gap" exists in the project agenda immediately after the posterior activity,
- The durations of the new episode and the existing episodes are reduced in proportion to their existing durations to a minimum of 50% of their original duration, and
- If all of the above fail, the insertion is considered to be infeasible and the new episode is rejected.

- The insertion of work-business episodes into a primary work event is a special case in which the primary work episode is not shifted, but rather is partially replaced by the work business activity episode. This is allowed because work business episodes are included in the duration of the primary work duration distribution.
- If activities are rejected due to scheduling conflicts, they are regenerated with a new randomly generated start time and duration up to a maximum of 10 attempts.

3) Person schedules are constructed by taking activity episodes from the project agendas and adding them to the person schedule. While projects are used to organise partially elaborated activities into sequence with other activities with a common purpose, the final timing of the activities must be coordinated with activities from other projects. The process of generating a person's schedule, therefore, involves taking episodes one-by-one from the project agendas and adding them to the person schedule in order of priority.

The order of priority is chosen to reflect to the greatest extent possible the degree of preplanning and commitment to other parties (such as employers or other household members) associated with each episode type. The order used in the current operational model is as follows:

- Work-business episodes,
- Primary work episodes,
- All other work episodes,
- School episodes,
- Joint other episodes,
- Joint shopping episodes,
- Individual other episodes, and
- Individual shopping episodes.

The construction of the person schedule proceeds in a manner similar to that of the project agendas. The major differences are that in the person schedule, travel episodes are added to account for the time necessary for trips between activities with different locations, and that the person schedule does not include "gaps" or areas with no planned activities. It is noted that at the beginning of the scheduling process, a person's schedule consists of a single "at-home" activity, which is the default if no other activities are added to the schedule.

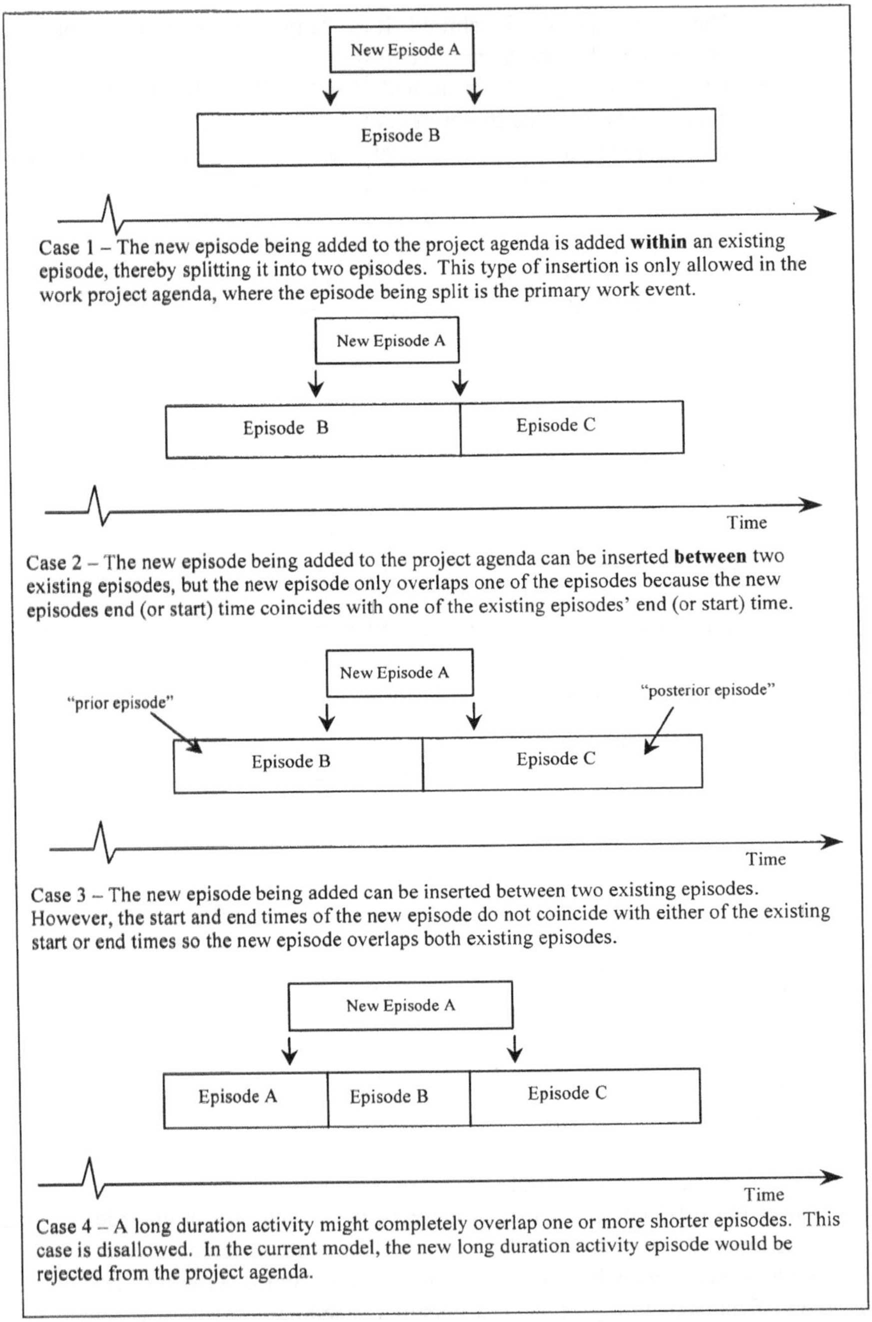

Figure 3.3 Inserting Activity Episodes into Project Agendas

An example of an episode insertion into a person schedule is shown in Figure 3.4. As shown in this diagram, a number of steps are followed to properly insert an episode into a person schedule.

- The travel episode from episode 1 to episode 2 is deleted,
- New travel episodes are defined from existing episode 1 to the new episode and from the new episode to existing episode 2.
- Episodes 1 and 2 are shifted forward and backward, respectively, to allow for the necessary room to insert the new episode and the two new travel episodes.

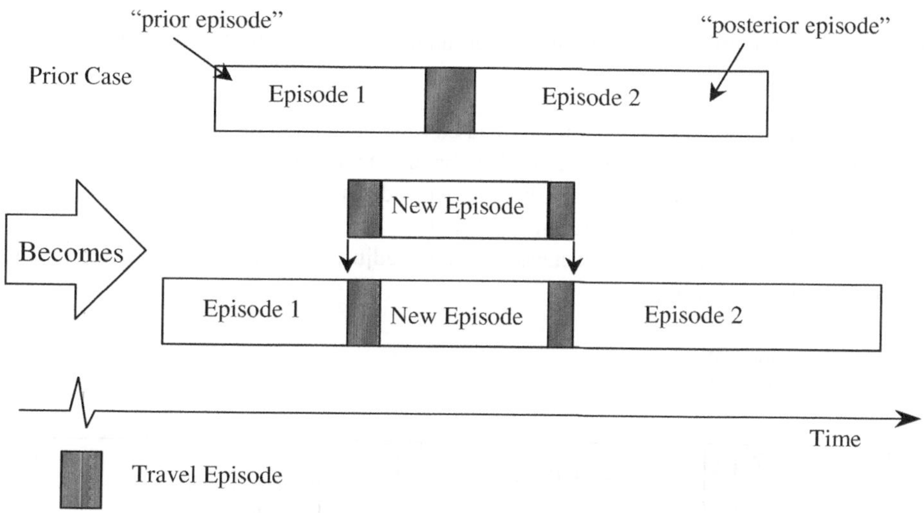

Figure 3.4 Inserting an Activity into a Person Schedule *with Travel*

- If "non-home" episodes exist directly before episode 1 and directly after episode 2, then there is assumed to be no room for shifting of episodes. In this case, episodes 1, 2 and the new episode are truncated in proportion to their durations to a maximum of a 50% reduction in duration.
- If all of the above steps are unsuccessful then the insertion is considered to be infeasible.

4) *A "clean up" algorithm is applied to reflect final scheduling / fine tuning just before or during execution of the schedule.* Once a skeleton or preliminary 24-hour schedule is derived based on steps 1, 2 and 3 above, further scheduling changes can be made to reflect decisions made just before or during the execution of a person's schedule. These may include cleanup or optimisation algorithms, but also may allow for the introduction of random events, impulsive changes, and further modification, revisions, re-sorting and planning.

At present, this stage of the scheduling process is limited to a single cleanup algorithm that is applied to rearrange the schedule to remove work episodes with duration less than or equal to a 30 minute duration.

An unrealistically high number of "short" work episodes are generated by the model because start times are generated randomly. We believe that schedules are constructed according to a rational process, although it is recognized that not all schedules are arranged optimally. It is felt that using a 30 minute threshold is an appropriate compromise that improves the schedules without fully optimizing them. An example of the schedule clean-up procedure is shown in Figure 3.5.

In this example a number of steps are undertaken to "clean" short work episodes in the schedule.

- The "short" work episode and its associated travel episodes are deleted;
- The primary work episode is extended forward to include the duration of the "short" work episode, and thereby retain the total work duration;
- Travel episodes are regenerated appropriately;
- The start time of the work business episode is adjusted to account for the reduced total amount of travel time.

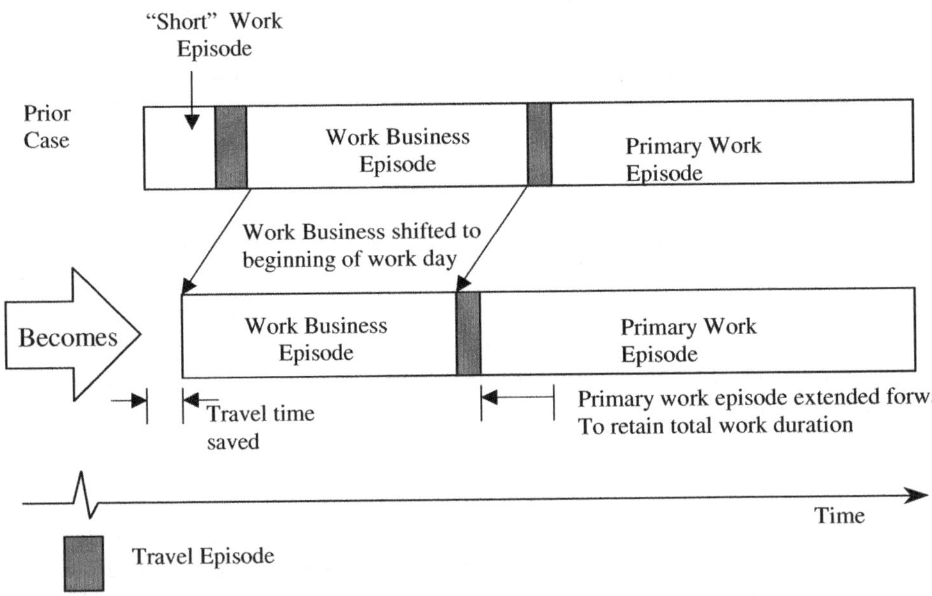

Figure 3.5 Example of Cleaning a Schedule that has a Short Work Episode

Opportunities for Addressing Assumptions Using CHASE© Data

There exists an opportunity for a set of fundamental improvements to the current operational model. Complementary panel surveys are currently underway in Toronto and Quebec City, in which information about the activity scheduling and planning *process* is collected in addition to the household's executed activities and travel. The first wave of the Quebec City panel survey is done using the OPFAST survey (Lee-Gosselin, 2004), a paper and pencil instrument in which respondents update their weekly activity plan every day, fax the changes to an interview centre and proceed to answer qualitative and quantitative questions about their activity scheduling process. The first wave of the Toronto panel uses the CHASE (Computerized Household Activity Scheduling Elicitor) survey instrument (Doherty and Miller, 2000). CHASE builds upon previous laboratory-based methods (e.g. Hayes-Roth and Hayes-Roth 1979; Ettema *et al.*, 1994) by providing a means to observe the scheduling process as it occurs in reality in a household setting over a week-long period. In this way it is able to capture both routine and complex scheduling processes as well as observe those scheduling decisions made during the actual execution of the schedule – which have turned out to be frequent.

The use of computers was necessitated by the desire to automatically trace the timing and sequencing of data entry, and seamlessly prompt for additional information when and where needed - tasks that are quite manageable via computer programming, but can be quite complex via paper-and-pencil surveys. The provision of an electronic scheduling style interface seemed to be a natural interface for the entry of activity-travel information.

Following an upfront interview, household members are instructed to begin the week-long CHASE survey by adding activities "that they have already at least tentatively planned" to their schedule for the following week-long period, leaving any unplanned time blank. A dialog box prompts for all activity information, and activities show up as boxes on screen, as shown in Figure 3.6. After the initial data entry, household members are asked to login daily to the program for the rest of the in order to make further additions, modifications and deletions to their activities and travel as the decisions are made in reality. Additional information about when and why each decision was made is automatically prompted for. All add/modify/delete entries made to arrive at the final observed schedule are recorded in sequence along with observed patterns - the key difference from traditional diary surveys.

Several enhancements to the program were made available for the Toronto Panel survey. Most notably, the program was modified to start on any day of the week, locations are coded by address or nearest intersection, and more detail sought on joint activities with children. A detailed "End-of-week-review" was also implemented to capture attributes of activities that go unobserved during the week, such as spatial, temporal and interpersonal flexibility/fixity, and usually frequencies and durations.

Based on the original a sample of 40 households in Hamilton (Canada), Doherty and Miller (2000) demonstrated that CHASE was associated with a reasonably low respondent burden - an average of about 16 minutes per adult per day and 10 minutes per child was required to complete their daily schedule (calculation based on time stamps of first and last data entries on computer). Further analysis of the data revealed a very dynamic scheduling process – including a large portion of activities planned impulsively (~30%) or close to execution (planned day-of or day before ~30%), with continuous modification during execution, often jumping out several days to make sub-plans (Doherty and Lee-Gosselin, 2000; Doherty *et al.*, 2001). Many of these decisions would be difficult to capture in a laboratory-based or single day experiment, further validating the approach adopted. Additionally, observing scheduling over several days also allows the sequence, and more importantly the situational factors of each decision in context of past and future plans, to be examined in much more detail as they actually evolved.

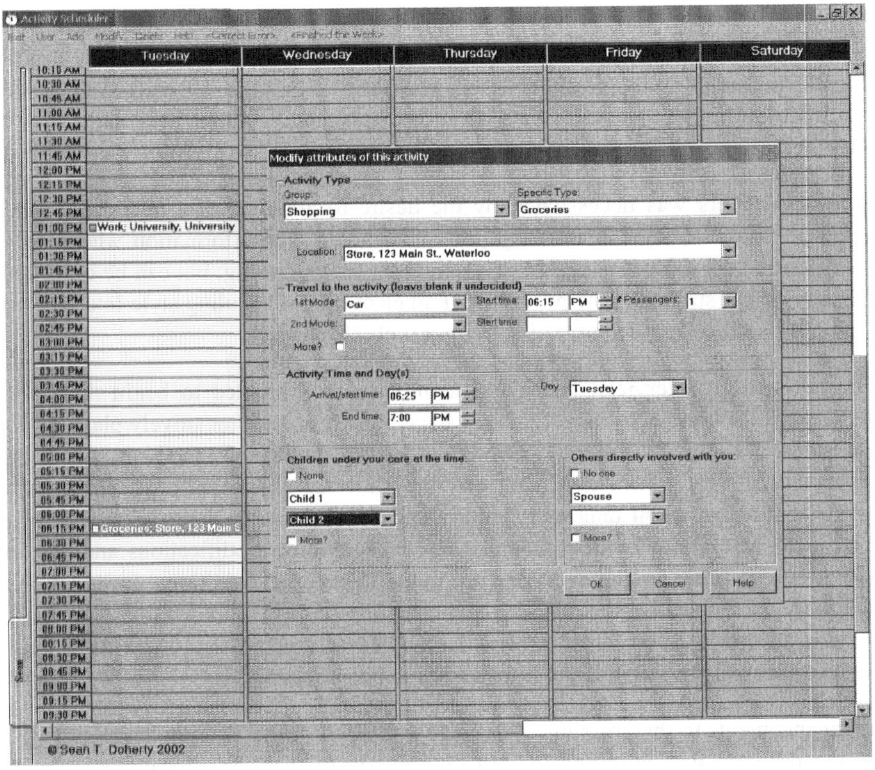

Figure 3.6 CHASE Main Scheduling Interface, Showing Add/Modify Dialog BoxUsed to Capture Attributes of Activities

The scheduling process data that will become available from even the first wave of the CHASE panel survey has the potential to allow for the development of a fully dynamic schedule process model. It will also provide data to enhance the current scheduling model in a number of different ways, as is discussed in the following sections.

BEHAVIOURAL ASSUMPTION SUMMARY AND ASSESSMENT

At this stage in development of scheduling models, it is important that the underlying assumptions embedded in current models be fully understood, and that strategies be developed to improve the behavioural validity of the models. Such information provides a more definite objective for future data collection efforts and helps to set priorities for empirical analysis and model development. The major assumptions that have been made in the current operational model are summarized in Table 3.2.

For each assumption, an attempt has been made to outline alternative modelling approach that would result in a more behaviourally appropriate model. The data sources available or required for the new developments have been outlined and some preliminary qualitative judgements have been made regarding the ease of analysis or development of the model formulation and the ease with which the modification could be integrated into the current modelling framework. It is noted that the improvements to the behavioural assumptions in Table 3.2 have been ordered (roughly) in order of their priority for implementation in future versions of the activity scheduling model, as will be discussed in the next section.

Table 3.2 Behavioural Assumption Summary and Assessment

Behavioural Assumption Made in the Current Model	Alternative Modelling Approach to Improve Behavioural Validity	Existing or Potential Data Source to Support the Alternative Modelling Approach	Ease of Development/ Analysis of the Alternative Approach	Ease of Integration of the Alternative Approach into the Current Modelling Framework
Mode choice is a decision that is assumed to occur in sequence after the schedule is formed.	Mode choice decisions should be incorporated as part of the schedule formation procedure since mode has an impact on travel time and therefore affects the schedule. Scheduling should be sensitive to vehicle availability in particular, which requires interaction between household members when there is competition for vehicles	TTS data can be used to assess vehicle availability. CHASE can provide only limited insights on the sequence of mode choice in context of other activity attributes.	Moderate (Work is currently underway to develop a mode choice model that is sensitive to vehicle availability. Adequate data exists.)	Moderate (Requires addition of a new mode choice model within the scheduling framework)

Behavioural Assumption Made in the Current Model	Alternative Modelling Approach to Improve Behavioural Validity	Existing or Potential Data Source to Support the Alternative Modelling Approach	Ease of Development/ Analysis of the Alternative Approach	Ease of Integration of the Alternative Approach into the Current Modelling Framework
Activity location is generated based on household, person and zone characteristics only, and may not be modified during the scheduling procedure	Activity location should be sensitive to the activity's position in the person's schedule and the locations of activities before and after the activity.	Trip-based Transportation Tomorrow Survey data could be further analyzed to provide this linkage. CHASE data will provide insight into location changes as the schedule develops over time	Moderate (Data exists and a more elaborate location choice model could be readily re-specified)	Less Difficult (Little program restructuring, relatively simple function modification)
Joint household decisions are limited to the generation of joint activities (i.e. activities that include more than 1 household member)	Additional joint household decisions / interaction should be incorporated including vehicle allocation, the care of children and who does other tasks for the household such as grocery shopping.	CHASE data from households can provide detailed information on joint activities, such as what activities in the household are applicable to household members, how spouses may differentially plan and conduct joint activities, and the difference between adult activities that involve children vs. serve the children vs. take place when children are present. Some household interactions, such as vehicle allocation, can be captured in TTS data. Others, such as the decision of who should do household tasks requires new survey data that targets the differentiation between household and person activities.	Difficult (Household interactions are complex and are likely to vary greatly from household to household)	Moderate (The current model is structured to allow for household interactions if clear rules/ methodology are discovered
A single set of schedule formation decision rules apply to all people	Different strategies for scheduling behaviour could apply to different people	CHASE data from the Toronto Area Panel Survey will provide much more information on how and when people plan different activities. Other smaller focussed surveys have been developed to capture the planning of different activity attributes. Some combination of such data sources may provide guidance in attempts to derive different sets of rules for different types of people.	Difficult (Rules are difficult to derive since each person is unique and many act in idiosyncratic ways)	Difficult (If drastically different rules apply to different people, significant model restructuring may be required)

Behavioural Assumption Made in the Current Model	Alternative Modelling Approach to Improve Behavioural Validity	Existing or Potential Data Source to Support the Alternative Modelling Approach	Ease of Development/ Analysis of the Alternative Approach	Ease of Integration of the Alternative Approach into the Current Modelling Framework
Schedules are planned over a 24 hour period	Schedules of at least 7 days (and possibly longer) in length would capture more fully the dynamic nature of the scheduling process	CHASE data from the first wave of the Toronto Area Panel Survey will provide 7 day schedule data.	Moderate (Similar method to that used currently could be addressed using 7 day activity generation)	Moderate (Straightforward model extension, provided that computational burden can be overcome)
Projects are based on very broad activity types (work, school, shopping and other) and only include activities of that type	More detailed project definitions could be developed that more closely reflect the true concept of the project, which allows for different activity type, provided that there is a common overall objective for those activities	In CHASE, user defined definitions of activity type are used that allow for a highly detailed assessment of activity types. Therefore, the potential exists to more precisely redefine activity and project classifications. Neither CHASE nor TTS have information about whether activities are logically linked as part of a project. Results of the OPFAST survey in Quebec City could provide such insights.	Difficult (New data may be required. Necessary to merge insights from several data sources. There exists no proof that the project is a valid concept in all cases)	Less Difficult (Given clear conclusions regarding the classification of projects and activities, the model could be readily extended to incorporate them)
The "execution" stage of schedule formation includes only a single algorithm to "clean-up" short work episodes	The "execution" stage of the scheduling process could be extended to include "impulsive decisions" to add/modify/delete activities and "random event" such as unexpected delays, cancellations, changes in duration, etc.	CHASE data provides a detailed tracing of information on impulsive additions and modifications to schedules as they occur leading up to and during execution. The reasons for these changes are also probed (for instance, to distinguish modifications due to random events versus user-initiated changes. New or follow-up surveys to CHASE could also be considered in order to examined re-scheduling processes in more depth.	Difficult (The "execution stage" is not well-researched and new data may be required)	Moderate (If clear rules can be deduced, they could be relatively straightforward to implement)

Behavioural Assumption Made in the Current Model	Alternative Modelling Approach to Improve Behavioural Validity	Existing or Potential Data Source to Support the Alternative Modelling Approach	Ease of Development/ Analysis of the Alternative Approach	Ease of Integration of the Alternative Approach into the Current Modelling Framework
Activities are scheduled in order of priority based only on activity type. There is no explicit time dimension describing when an activity is planned.	Activities could be added to a person's schedule in a fully dynamic way, whereby some activities of different types are planned well in advance to form a "skeleton" schedule, while others are planned spontaneously, very close to the time of execution.	CHASE data from the first wave of the Toronto Area Panel Survey captures the point in time when an activity is first planned and inserted into the schedule, which serves as a strong indicator of its priority. Data is also systematically collected on a wide range of attributes of activities (spatial fixity, temporal fixity, normal frequency and duration, interpersonal fixity). Therefore, it would be possible to develop a priority "function" that is sensitive to the attributes of activities as well as key situational factors observed during scheduling (e.g. time to next event; time available on schedule) making it possible to dynamically predict the priority of an activity.	Moderate (The CHASE survey is specifically designed to capture such dynamic planning)	Moderate (Straightforward replacement of "hard-wired" priority to a "priority function")
The model generates a single schedule at one point in time, without regard to a person's history of schedule formation	We would expect some resemblance between a person's schedule from one year to the next. Therefore, schedule formation should be placed in the context of a person's scheduling history.	The Toronto Area Panel Survey allows for the observation of the schedule formation procedure for the same households at one-year intervals.	Difficult (Significant research required to understand long term scheduling dynamics)	Difficult (Would require storage/retrieval of scheduling history)
No learning or habit formation is incorporated	Learning could be applied by implementing modifications to the episode generation routine, such that attributes of generated activities reflect past experiences	CHASE provides some information on the historical formulation of habitual activities, but no information on new habit formation over the course of the weekly data collection. The Toronto Area Panel Survey data will be needed at minimum to observe changes in scheduling behaviour over time. An in-depth survey would probably be needed to fully explore learning and habit formation	Difficult (Learning/habit formation is not well-understood and would require an innovative new approach and data collection effort)	Difficult (Significant changes would be required to existing activity generation procedures)

Behavioural Assumption Made in the Current Model	Alternative Modelling Approach to Improve Behavioural Validity	Existing or Potential Data Source to Support the Alternative Modelling Approach	Ease of Development/ Analysis of the Alternative Approach	Ease of Integration of the Alternative Approach into the Current Modelling Framework
Activity attributes are initially generated based on the final executed schedules observed in the TTS data	Activity episodes should be initially generated based on their "desired" attributes and then modified to reflect the time and resource constraints a person is faced with in the scheduling procedure	CHASE data will provide some insight into "normal" activity attributes prior to execution, such as normal frequency and durations. A new in-depth survey would be useful to fully understand how and why "desired" activities differ from "executed" activities.	Difficult (New data required. Difficult to define and capture "desired attributes" reliably)	Less Difficult (Would require modification to model inputs only)
A large number of detailed assumptions have been made to handle conflicts arise during the insertion of activities into project agendas and person schedules	Although no outstanding problems are currently seen with these assumptions, opportunities for verification of the details of the process exist.	CHASE data provide a large number of observations of activity insertions into schedules. These data could be examined to verify or invalidate the rules currently being used in the model. Details such as which episode is appropriate to truncate or which direction to shift episodes when scheduling conflicts arise would be assessed.	Moderate (The CHASE survey is specifically designed to capture modifications to the schedule over time)	Unknown

Planned Improvements to the Operational Model and the Need for New Survey Data

The priorities for further development of the Travel Activity Scheduling model for Household Agents (TASHA) rely heavily on the availability of data. Given that the appropriate data are available, improvements to the model would logically be done by addressing those assumptions that would yield the greatest improvement in behavioural validity for the least effort in both development / analysis of the new approach and in the integration of the new approach into the current operational modelling framework.

The highest priority model improvements for the time being, therefore, are those that require no additional data (i.e. can be done using TTS data). These include:

- Incorporation of the mode choice model into the scheduling procedure
- Improving the activity location choice model
- Incorporation of additional joint household decisions such as the care of children and vehicle allocation

Once the first wave of the Toronto Area Panel Survey (based on the CHASE survey instrument) becomes available, a new set of improvements will be possible, including:

- Applying different strategies for scheduling to different people
- The movement from a 24 hour model to a week-long model
- A much more detailed representation of episodes and projects
- A more detailed representation of the "execution stage" of schedule formation
- Scheduling of episodes that is based on observed information about those episodes regarding their priority, flexibility and when they were planned. This is in contrast to the simplified and rather ad hoc priority ratings currently being used in the operational model.

We expect that future versions of the TASHA model will continue to be estimated using data from travel diary surveys such as the TTS, however the model structure would be informed by the more in-depth information provided by CHASE. Therefore a "CHASE-like" survey would not be required for the implementation of TASHA in other urban areas.

Further improvements will require new data, as follows:

Clearly there is a need for a Toronto based survey that attempts to capture directly how a person's activities of different types are linked together with a common purpose or objective. This is crucial since the scheduling model is built upon the concept of the project and it would be much better to fully elaborate the concept with observed data. One possibility for collecting such information would be through an "add-on" to the core CHASE survey instrument being used in the Toronto Area Panel Survey. A series of probing questions could be asked of respondents to establish links between activities that they have entered into the week-long schedule diary. Notably, questions of this nature have been asked in the OPFAST instrument in the Quebec City panel survey, as discussed by Lee-Gosselin (2004). Adding such questions in Toronto could be effective, but risks overly burdening CHASE participants, which would also have implications for data quality. Another possibility would be to design a "targeted" face-to-face survey asking respondents about a general activity or event that they are planning for the near future (such as a home maintenance project, a family outing or a dinner party). Respondents would then be asked probing questions to determine all of the activities that they have already done or will do that are related to that event.

The ways in which attributes of activities may be differentially decided is an important concern – for example, when planning activities, people often only partially plan (or do not plan at all) certain activity attributes such as location, exact timing, the travel mode, or involved persons. Assuming that all attributes are decided at once, in the same sequence, will certainly not hold all the time. Investigating these sequences is key to future model development.

Another area of important concern is that of "decision rules" utilized throughout the scheduling process. Although the application of rule-based approach to travel behaviour modelling has been taken up in the literature (e.g. Gärling *et al.*, 1986; Lundberg 1988), very little direct empirical evidence has been published on the nature of the decision rules utilized during the activity scheduling process. One technique for investigating the actual rules used to is to ask people to "think aloud" during a problem solving exercise (see also Ericsson and Simon 1993).

An alternative would be to provide some means to query people live during a scheduling survey, perhaps via telephone or other recording device. Such data could lead to the development of alternative rule-based modelling structures as opposed to adopting traditional utility maximizing frameworks.

A new in-depth survey would be useful to fully understand how and why "desired" activity attributes (before any scheduling has taken place) differ from "executed" activity attributes (after time and resource constraints are realised in the scheduling process). The design of such a survey would not be straightforward. Nonetheless, the concept should be further explored because the current model implementation, which uses "executed" activity data from TTS to generate "desired" activity attributes is not entirely satisfactory in a behavioural sense.

CONCLUSION

The Travel Activity Scheduling Model for Household Agents (TASHA) represents a very successful attempt to microsimulate the formation of 24 hour activity schedules for a large sample of people, based on trip diary data provided in the 1996 Transportation Tomorrow Survey. However, it is clear that the further model improvements to effect the desired behavioural basis of the model will require data sources beyond the Transportation Tomorrow Survey. The CHASE survey, which is a high quality data source soon to be available for modelling purposes, has been designed to provide much of the necessary information to further improve the model in a number of key ways. However, there is also a need for continued development of new survey techniques to analyze several of the important concepts underlying the current modelling effort and to address and improve on the assumptions made in the model.

REFERENCES

Arentze, T.A. and H.J.P. Timmermans (2000). *Albatross: A Learning Based Transportation Oriented Simulation System*. The European Institute of Retailing and Services Studies, Eindhoven, The Netherlands.

Axhausen, K. (1998). Can we ever obtain the data we would like to have? In: *Theoretical Foundations of Travel Choice Modeling* (T. Gärling, T. Laitila and K. Westin, eds.), pp. 305-323. Pergamon, Elsevier Science, Oxford, UK.

Axhausen, K.W. and T. Gärling (1992). Activity-based approaches to travel analysis: Conceptual frameworks, models, and research problems. *Transportation Reviews*, **12**(4), 323-341.

Bhat, C.R. and K.T. Lawton (2000). *Passenger Travel Demand Forecasting*. Transportation Research Board, National Research Council (USA). Millennium Paper.

Chapin, F.S. (1978). Human time allocation in the city. In: *Human Activity and Time Geography* (T. Carlstein, D. Parkes and N. Thrift, eds.), Vol. 2, pp. 13-26. John Wiley and Sons, New York.

Chapin, F.S.J. (1974). *Human Activity Patterns in the City*. Wiley-Interscience, New York.

Cullen, I. and V. Godson (1975). Urban networks: The structure of activity patterns. *Progress in Planning*, **4**(1), 1-96.

Doherty, S.T. and M.E.H. Lee-Gosselin (2000). Activity scheduling adaptation experiments under vehicle reduction scenarios. Paper presented at the *9th International Association for Travel Behaviour Research Conference*. Gold Coast, Australia, July 2-7.

Doherty, S.T. and E.J. Miller (2000). A computerized household activity scheduling survey. *Transportation*, **27**(1), 75-97.

Doherty, S.T., E.J. Miller, K.W. Axhausen and T. Gärling (2001). A conceptual model of the weekly household activity-travel scheduling process. In: *Travel Behaviour: Patterns, Implications and Modelling* (E. Stern, I. Salomon and P. Bovy, eds.), pp. 148-165. Elgar Publishing Limited, Cheltenham U.K.

Ericsson, K.A. and H.A. Simon (1993). *Protocol Analysis: Verbal Reports as Data*. MIT Press, Cambridge.

Ettema, D., A. Borgers and H. Timmermans (1993). Simulation model of activity scheduling behaviour. *Transportation Research Record 1413*, 1-11.

Ettema, D., A. Borgers and H. Timmermans (1994). Using interactive computer experiments for identifying activity scheduling heuristics. Paper presented at the *Seventh International Conference on Travel Behaviour*, Satiago, Chile, June 13-16.

Ettema, D. and H. Timmermans (1997). Theories and models of activity patterns. In: *Activity-Based Approaches to Travel Analysis* (D. Ettema and H. Timmermans, eds.), pp. 1-36. Pergamon, Oxford, UK.

Gärling, T. (1994). Behavioural assumptions overlooked in travel choice modelling. Paper presented at the *Seventh International Conference on Travel Behaviour*, Santiago, Chile, June 13-16.

Gärling, T., T. Kalén, J. Romanus and M. Selart (1998). Computer simulation of household activity scheduling. *Environment and Planning A*, **30**, 665-679.

Gärling, T., M.P. Kwan and R.G. Golledge (1994). Computational-process modelling of household activity scheduling. *Transportation Research B*, **28**(5), 355-364.

Gärling, T., J. Saisa, A. Book, and E. Lindberg (1986). The spatiotemporal sequencing of everyday activities in the large-scale environment. *Journal of Environmental Psychology*, **6**, 261-280.

Hägerstrand, T. (1970). What about people in regional science? *Papers of the Regional Science Association*, **24**, 7-21.

Hayes-Roth, B. and F. Hayes-Roth (1979). A cognitive model of planning. *Cognitive Science*, **3**, 275-310.

Jones, P.M., M.C. Dix, M.I. Clarke and I G. Heggie (1983). *Understanding Travel Behaviour*. Gower, Aldershot.

Jones, P., F. Koppelman and J.-P. Orfeuil (1990). Activity analysis: State of the art and future directions. In: *Developments in Dynamic and Activity Based Approaches to Travel Analysis* (P. Jones, ed.), pp. 34-55. Oxford Studies in Transport. Avebury.

Kawakami, S. and T. Isobe (1990). Development of a one-day travel-activity scheduling model for workers. In: *Developments in Dynamic and Activity-Based Approaches to Travel Analysis* (P. Jones, ed.), pp. 184-205. Avebury, Aldershot.

Kitamura, R. (1983). Sequential, history-dependent approach to trip-chaining behavior. *Transportation Research Record 944*, 13-22.

Kitamura, R. (1988). An evaluation of activity-based travel analysis. *Transportation*, **15**(1/2), 9-34.

Kitamura, R., C. Chen and R. Pendyala (1997). Generation of synthetic daily activity-travel patterns. *Transportation Research Record 1607*, 154-162.

Kwan, M.P. and R. Golledge (1995). Integration of GIS with activity-based model in ATIS. Paper presented at the *Transportation Research Board 74th Annual Meeting*, Washington, DC.

Lee-Gosselin, M.E.H. (2005). A data collection strategy for perceived and observed flexibility in the spatio-temporal organisation of household activities and associated travel. In: *Progress in Activity-Based Approches: Models, Data and Applications* (H.J.P. Timmermans, ed.). Elsevier, forthcoming.

Lee-Gosselin, M. (1996). Scope and potential of interactive stated response data collection methods. In the proceedings: *Household Travel Surveys: New Concepts and Research Needs*, Irvine, California, March 12-15, 1995, pp. 115-133. Transportation Research Board Conference Proceedings 10.

Lundberg, C.G. (1988). On the structuration of multiactivity task-environments. *Environment and Planning*, **20A**, 1603-1621.

Miller, E. J. (2005). Propositions for modelling household decision-making. *In this volume*.

Miller, E. J. and M.J. Roorda (2003). A prototype model of household activity/travel scheduling. *Transportation Research Record 1831*, 114-121.

Pas, E. (1985). State of the art and research opportunities in travel demand: another perspective. *Transportation Research Part A*, **19A**(5/6), 460-464.

Recker, W.W., M.G. McNally and G.S. Root (1986a). A model of complex travel behavior: Part I - Theoretical development. *Transportation Research*, **20A**(4), 307-318.

Recker, W.W., M.G. McNally and G.S. Root (1986b). A model of complex travel behavior: Part II - an Operational model. *Transportation Research*, **20A**(4), 319-330.

Van Der Hoorn, T. (1983). Development of an activity model using a one-week activity-diary data base. In: *Recent Advances in Travel Demand Analysis* (S. Carpenter and P. Jones, eds.), pp. 335-349. Gower, Aldershot, Gower.

ACKNOWLEDGEMENTS

This paper was supported by the Social Sciences and Humanities Research Council of Canada, through a Major Collaborative Research Initiative, 2000-2005: *Access to activities and services in urban Canada: behavioural processes that condition equity and sustainability*.

Integrated Land-Use and Transportation Models: Behavioural Foundations
M. Lee-Gosselin and S. Doherty (Editors)

4

Conceptual Perspectives for Explaining Spatio-Temporal Behaviour in Urban Areas

Thierry Ramadier, Université Louis Pasteur/CNRS, Strasbourg, France[1]
Martin E.H. Lee-Gosselin, Université Laval, Québec, Canada
Alexandre Frenette, Graduate Center, City University of New York, USA[1]

This chapter resulted from a theoretical and methodological discourse about daily mobility in urban areas that took place during the design of a longitudinal survey for the Quebec City region[2]. This chapter has two objectives: (a) to propose a systematic conceptual framework for analysing and explaining the travel behaviour of urban dwellers; (b) to underline the multidimensionality of decision-making processes, and of rules of behaviour and their evolution, arguing that the economic and temporal dimensions should not be as dominant as they have been in conventional frameworks. In particular, we propose that the spatial and temporal dimensions of behaviour should be considered simultaneously and given equal weight.

Our aim in this discourse was to elaborate tools for understanding, rather than just describing, trips in urban areas, using an activity-based theoretical approach as a starting point. Although

---

[1] Work reported here was carried out while Thierry Ramadier was a Post-Doctoral Fellow, and Alexandre Frenette was a Graduate Research Trainee, both at Université Laval.

[2] This survey, which became known as the Quebec City Panel Survey, was one of pair of longitudinal surveys undertaken by the PROCESSUS Network. This chapter explains the theoretical background for the survey design. A description of the adopted survey methods will be found in Chapter 5 of this volume.

the proposed conception of daily mobility is strongly functionalist in nature, it is important, in conceptual and methodological terms, that these tools used may open possibilities for analysis from other perspectives. Thus, as proposed by Kaufmann (2000), to explain the complexity of mobility behaviour, it is important to question the weight given to theories purporting that the individual makes rational choices primarily based on economics. It is impossible to study the range of rationales underlying daily mobility if, *a priori,* they are supposed to be subsumed by economic motivations such as the personal gain of time or money. But how can we approach "alternative rationalities," whether their origin be social (norms, social representations) or from the individual's cognitive activity (spatial representations, the spatio-temporal basis of decisions)? From a theoretical point of view, we seek to show the relevance of the systemic transactional approach, initially developed by Dewey and Bentley (1949). From a methodological point of view, we assert that a longitudinal data-collection protocol, inspired by this approach and reflexive in character, will open interesting possibilities for measurement.

TWO INSEPARABLE ENTITIES: TRANSPORT INFRASTRUCTURE AND THE USER

Presently, the principal mission of the planners of transport infrastructure consists of optimizing the relationship between the transport system and the demand for mobility instead of, as was previously the case, investing in the construction of new elements to be added to the system (Andan *et al.*, 1995). To meet this mission, two strategies are generally envisaged: efforts to improve the infrastructure; and efforts to improve the use of the transport system. Note that "improving the use" in this context does not usually imply a reduction in motorised travel, which can become a goal in the face of impacts such as atmospheric pollution, greenhouse gas emissions, noise, unsafe traffic, or urban sprawl: the interaction between users and infrastructure is so important that a whole process of regulation has grown up without incorporating any such goal. Indeed, these two entities form a system of mutually defined parts. The characteristics of the infrastructure originate in part from the demand patterns of its users (rush hours, frequency of public transport services, fares, etc.), whereas the users partly define their lifestyle as a function of the characteristics of the transport infrastructure to which they have access[3].

These two strategies are not contradictory and can be operated in parallel. Indeed, the planning of the transport infrastructure may be oriented towards peak demand while trying to meet types of unmet demand: an example would be providing access to some classes of off-peak employment that, historically, have been ill-served by the system. Thus, the levers of change that we wish to identify in our research depend equally on the behaviours of the user and the characteristics of the infrastructure. This vision of the management of transport infrastructures assumes that we must renounce the idea of a stable and permanent equilibrium of the

[3] The notion of "infrastructure to which they have access" includes whether or not the users have the use of private motor vehicles: the substantial part of transport infrastructure that is dedicated to private vehicles holds meanings for a user's physical environment that are very different for motorists and non-motorists.

user/infrastructure system, as the evolution of an entity may disturb the system as a whole. Instead of trying to anticipate the development of the system while taking for granted the notion of steady-state equilibrium, which has been the main thrust of the notion of development in the past, should we not focus on explaining the evolution of the system by attempting to predict foreseeable futures as a function of emerging phenomena? Here the notion of evolution assumes that the seeking of equilibrium is a process that rarely has a stable outcome. To understand this process of equilibration, it is revealing to relate observable (albeit transitory) states of equilibrium to individuals' activity/mobility patterns. An example illustrates this idea. During many daily work commutes, time spent in traffic congestion, because is it predictable, is used more and more by certain drivers for activities generally associated with home: listening to music, making phone calls, learning a language, etc. The idea of the automobile as an extension of one's living space has long been associated with this mode of transportation: arguably, "non travel" behaviours, allowing drivers and passengers to add value to traffic congestion, rapidly appeared as an adaptation to the systematic and predictable character of traffic jams. This adaptive evolution in the perception of daily travel allows the user/infrastructure system to come into balance, but it is a balance from a broader perspective than that of traffic flow. The timing of congestion on certain arteries becomes a characteristic of the infrastructure, because it is well-known and predictable, whereas the individual reduces the cognitive dissonance imposed by the situation by finding value in the "free time" with an array of activities. However, if we judge that this situation is not desirable from energy, environment, or social point of view, and our goal is to develop and propose realistic alternatives, we have no choice but to make explicit the decision-making processes and rules of behaviour that lead the individual to renounce or maintain a car as a mode of transportation under given circumstances.

Opting to move from a logic of description and prediction to one of explanation and dynamics opens the possibility to consider spatial behaviours as the product of a transaction within the individual/urban infrastructure system.

DAILY MOBILITY FROM A TRANSACTIONAL PERSPECTIVE

The user/infrastructure system does not form a system of mutually defining entities unless the activities that motivate urban travel are considered. Indeed, as travel has long been seen as a demand derived from the activity (Raux & Chapleau, 1995), we can distance ourselves from simply analysing traffic flow without asking why people travel, or from seeing the individual as an encumbrance to the efficiency of the infrastructure. The individual can no longer be subordinate to the characteristics, and more particularly the constraints, of the transport infrastructure. Indeed, more than three decades of time geography and activity-based methods in travel behaviour research (e.g., Jones *et al.*, 1983) have established the importance of the individual activity/mobility patterns.

This theoretical position seems supported by Gärling *et al.* (1986), who suggested that individuals first decide the sequence of their activities when they must travel to many places: only after that do they seek to minimise costs and travel. In other words, they deal first with their priorities and constraints, and then travel. Unlike predictive models of traffic flow, activity-based models focus on the flexibility of behaviours. The study of decision-making processes that precede or accompany urban travel therefore become central when the objective of inquiry consists of understanding rather than describing daily mobility. This allows the formulation of a hypothesis that now has to be validated: *the primacy of the nature of the activity on the planning of travel suggests that, for a given individual, the rules of behaviour related to a trip differ according to the activity*[4]. It remains to define *a relationship between rules of behaviour and the execution of activities*. From the point of view of daily mobility, we can see this relationship in terms of the *lifestyle* of the individual. By lifestyle, we refer to what Zapf (1987) defined as "a relatively stable schema for the organisation of daily life within a framework comprising life circumstances, the available resources and the orientation that each individual gives to her life"[5]; in addition, we recognise the influence of social mores, and the close association between lifestyle and observed patterns of consumption. The relationship between rules and activities has much to do with the individual's orientation: with the accumulation of experience, she must assess whether she wishes her lifestyle to remain constant, or to see it to evolve in ways that her life circumstances and resources permit. This is the cognitive setting for the relationships we may observe between an individual's activity pattern and his/her set of constraints and opportunities.

Thus, a further step now orients the research: this consists of integrating individual lifestyle into the analysis of daily travel and its evolution. In other words, analyses should seek to take into account the representations, values and social norms and, more generally, the relationship with urban space, around which daily mobility is built, and progressively rebuilt. Up to now, these global approaches have often been descriptive, providing typologies of individuals as a function of the characteristics of their day-to-day trips. To go beyond, we would broaden our hypothesis thus: *each lifestyle generates behavioural rules that are specific to its set of associated activities, and these rules are orchestrated by decision processes that pertain to the situation in which the individual is engaged.*

Considering daily mobility to be a function of lifestyle supposes that the individual is engaged in a particular relationship with the urban environment. Consequently, as proposed by the field of ideological cognition (Beauvois *et al.*, 1987), an individual's knowledge is structured as much by the possibilities of doing something with some physical objects as by the intrinsic properties of those objects. This knowledge is more evaluative than descriptive: is produces more information about what these objects are worth than about what they are. In other words, taken separately, neither a person's characteristics, nor the physical context in which she lives, can explain each of her actions, because it is at the conjunction of these two dimensions that they are generated. Together, they reveal as much about specific configurations as about urban life styles. It is in this sense that that the relationship between the individual and the

[4] Note that this refers to an activity as it is experienced and perceived by the individual : it does not imply that all activities belonging to some externally defined category, such as "work" or "shopping ", will be the subject to the same rules.

[5] Approximate translation

infrastructure constitutes a system in which these entities define each other, which constitutes the principal characteristic of the systemic transactional approach. Ittelson (1973) picked up on this major characteristic of the transaction, as he declared it in the course of research into individual-environment relationships: "...all aspects of the situation are included as active participants (...) and these aspects do not appear as entities that are already constituted and that interact only such that their identities are unaffected"[6]. The transactional approach puts the emphasis on the temporal dimension and the evolution of the system. However, the system is not oriented towards a state of equilibrium that can be defined in advance (Altman and Rogoff, 1987). It is an open system: any equilibrium that is achieved will be different from that which prevailed before any disruptions that the system may encounter. Thus, the system is not predetermined or fixed to a particular origin, direction or form (Altman, 1992). Indeed, the transactional perspective postulates that disruptions are a given, an element of the system. In consequence, the situation, the changes that intervene, and the ongoing processes of the system are all inseparable elements.

This theoretical approach puts the emphasis on the principle that the individual is active in her relationship with physical objects, because she makes choices on the basis of her evaluations of these objects. The temporal dimension of the situation is even more important given that the individual acts on those choices in order to travel. Thus, motivations are of primary importance and, given our central research question, so are the activities that motivate trips. This is why the systemic transactional perspective resonates well with activity-based approaches to understanding travel behaviour.

Activity Patterns and Choice Sets

Analyses of activity patterns mostly focus on constraints and on individual's freedom of movement in space, resulting from: individual resources (time, money, etc.); self-imposed rules (routines); sociological characteristics (norms, social networks, household structure, etc.); and urban form (supply of urban amenities and their locations). However, the understanding of trips is mostly limited to simple descriptions: pairs of trip-ends, time-of-day and perhaps duration. Decision processes and behavioural rules are difficult to discern from these factors, because they are revealed much less by the pattern of spatial behaviour and the choices or strategies exercised by individuals, than by the spatial and temporal articulation of activities. However, if the spatial and temporal dimensions are given equal importance and taken into account together, a particularly detailed description of activity attributes is possible. We can see that research into decision processes only rarely use a spatio-temporal framework, and analyses tend to favour the temporal dimension to the detriment of the spatial dimension.

The factors that govern activity patterns should not be envisaged simply as inhibitors of action. They also structure activities and, having done so, can create opportunities for individual actions. The resulting room for manoeuvre constitutes the individual's choice set (Andan and Raux, 1995). The existence of a choice set presupposes that the individual is active, develops

[6] Paraphrase

intentions (Gärling, 1995), and that she acts as her own "urban planner" as she locates all the functions and amenities that exist in her conurbation. On this last phenomenon, Nagy and Baird (1978) proposed a model for the way in which children go about planning the location of their activities.

The existence of a choice set also presupposes that there is an impulsion to locate activities in space, as pointed out by Bourdin (1996), when he stipulates that, over and above mobility, there is a "choice of anchors". Bourdin defines four rules for locating an activity. The first is the constraint that an activity needs a place in which it can be carried out, a constraint that is conditioned by the role of norms and power. This first rule provides the articulation between the choice set and the activity pattern. The second results from the interest that the place evokes, its unique resources, notably: its "native appeal" (found uniquely at that place by people with local knowledge, and very difficult to transmit to another place); a sociability that is more or less transportable to other places than the one in which it started up; and the facilities that make possible either or both of these. The third rule comes from the interest to undertake an activity in this particular location, not because of the special characteristics of the place itself but because of its relative position – for example choosing a cinema or a shop because it is close to a motorway. The last rule consists of locating an activity so as to avoid the constraints of alternative places that otherwise offer the same resources, such as being willing to go to a more distant restaurant because the ones closer to home are too noisy. The second and third rules make reference to utility theory, where the main process is the optimisation of gain, but the last rule fits better the principle of satisficing, in the sense of psychological comfort.

This classification has three important theoretical consequences. First, in postulating that locating activities is a necessity, this amounts to considering the choice set as a set of intentions rather than as a set of preferences. Thus, the implication is to put the emphasis on the motivations that are present and the on the projects[7] that are associated with the choice, whereas preferences refer mostly to inherited social norms. Secondly, it envisages an articulation, on one hand between choices and constraints, and on the other hand between utility and satisficing. Thirdly, in this classification, choices and constraints coexist, but above all satisfaction is not seen exclusively as an outcome of utility maximisation. Satisfaction is derived also from psychological comfort that is based on an optimisation between a way of living that is being pursued, and the experiences that are afforded by the envisaged places and/or projects.

This effort to classify activity location choices contributes to current efforts by researchers to explore the spatial strategies of individuals, households and even social groups, through an understanding of cognitive processes and behavioural rules. Although spatial strategies are often considered independently of temporal strategies, some research shows the importance of considering these two dimensions simultaneously. For example, Chalmé *et al.* (2000) showed

[7] We use "projects" here in the sense of sets of related activities with a common goal. See the related discussion in Miller, Chapter 2 of this volume

that, when planning successive trips under given spatial and temporal constraints, some people focus on the possible timing of tasks that they must undertake. But under the same given constraints, others develop essentially spatial strategies, for example by limiting themselves to a group of activities that can be carried out in the same part of town: for them, the urban area is divided into zones that they evaluate in this way. Yet others respond to the given constraints with a spatio-temporal strategy, incorporating both the possible timing of tasks and a set of perceived zones.

These decision processes reveal something about behavioural rules. For example, Gärling *et al.* (1986) show that that individuals do not necessarily seek to minimise distance between pairs of local destinations: rather, they seek to minimise their overall distance of travel taking into account whole sequences of planned activities and the places involved. This rule is even more used when the number of activities and their spatial dispersion increases. In fact, when there are many activities, an individual may use a coarse division of space that does not allow the minimisation of local distances because the planned trips will be categorised using these large pieces of geography as reference points.

A challenge to this research is that the temporal dimension tends to be invoked only with regard to the planning of trip sequences. However, not all our activities are arranged in advance. Among other things, Gärling *et al.* (1995) note that many errors in the prediction of behaviour come about because intentions are studied only in relation to planned behaviour. Thus, the predictive models leave out habits, intentionless acts, and impulsive actions for which intentions have changed between the recording of the intention and the execution of the behaviour. Gärling *et al.* thus propose to classify behaviours according to their degree of spontaneity: habitual, pre-arranged or impulsive. However, in improving thus the temporal aspects of the decision processes, spatial aspects may become secondary, perhaps based only on conventional notions such as trip distance. Arguably, distance is perceived and reflected upon less and less in terms of kilometres, but rather in terms of distance-time, hence reinforcing once again the predominance of time in decision processes. Is there not a *prima facie* case for an approach to analysing decision processes and behavioural rules in which both the temporal and spatial dimensions are considered simultaneously, without giving more weight to one than the other?

Behavioural Rules and Decision Processes Underlying Spatio-Temporal Behaviour

A more detailed conceptual framework for decisions about activities and travel, seen as an interface between the individual user and the physical context, is given in Figure 4.1. As discussed above, it is contended that the individual's characteristics alone cannot explain travel behaviour: these are shown on the left as the combined roles of biological determinants and of five components of an individual's lifestyle, expressed through needs and desires. Note that one of the five lifestyle components is the *repertoire of possible activities* that is essential to the specification of certain types of activity-based model. Similarly, the physical context alone cannot provide a sufficient explanation: this is shown on the right of Figure 4.1 terms of the organisation, in the urban environment, of different types of activity location, and of the transport infrastructure, including roads, public transport, and facilities for pedestrians and cyclists. The potential for mobility, according to this framework, comes about by the interaction of four "situations" that are conditioned both by an individual's lifestyle and by the physical context. These are:

> *The cognitive situation:* the individual's knowledge and interpretation of the urban environment -- what it has to offer, how it is organised and how quickly different places can be reached;
>
> *The spatio-temporal situation:* where the individual is at the moment and where she expects to be in the short term, the location(s) of her residence, work and/or study places, and her temporal and spatial constraints;
>
> *The social situation:* the spatial structure of the individual's social networks, availability of people to accompany her in activities, and her social status;
>
> *The material situation:* the availability of personal vehicles and telecommunications equipment, the size of living accommodation, and the availability of home facilities (such as working space).

The cognitive situation by itself gives rise to behavioural intentions.[8] The remaining three contexts – spatio-temporal, social and material, act together to define a set of behavioural proposals, but for which intentions have not yet been formed. These two cognitive dimensions lead to a first level of negotiation between, on one hand, desires and needs, and on the other hand, intended and proposed behaviours. The negotiation is a mental process that may, or may not, involve exchanges with others: it functions according to the behavioural rules that an individual wishes to enforce. If the four situations are stable, notably the spatio-temporal, social and material situations, then this "global" negotiation leads directly to travel and activity behaviour. But if the situations are unstable, or if there are transitory problems with the physical context, such as exceptional traffic congestion, the global negotiation initiates a detailed negotiation to reconcile the possible behavioural outcomes with the current state of the spatio-temporal, social and material situations and, if applicable, the transitory constraints of the physical context. Such a detailed negotiation results in a modified activity programme, upstream from the executed behaviours. Although we have referred mostly to the individual in this section, the same mechanisms apply to household-level decisions about daily mobility, but with increased chances for instability. Finally, a number of short-term and long-term feedback loops for learning, adaptation and adjustment are shown in the figure. Note that among these is the short-term adaptation of the rules themselves, a mechanism for making exceptions and "bending rules" if the negotiation processes become blocked.

To summarise, if lifestyle is a stable part of an individual's characteristics, this conceptual framework postulates a mechanism for the behavioural rules it generates to strongly influence activity and travel behaviour. However, the dynamics of an evolving lifestyle are much

---

[8] Strictly speaking, the individual's social status, which is a component of her social situation, also influences behavioural intentions rather directly.

affected by the individual's confrontation with her physical environment, which itself feeds the negotiation process with revised intentions and proposals for new behaviours. To use this framework as a guide for survey design, a further step was required: to incorporate the notion of the degree of spontaneity of behaviour in a way that balances the temporal and spatial dimensions.

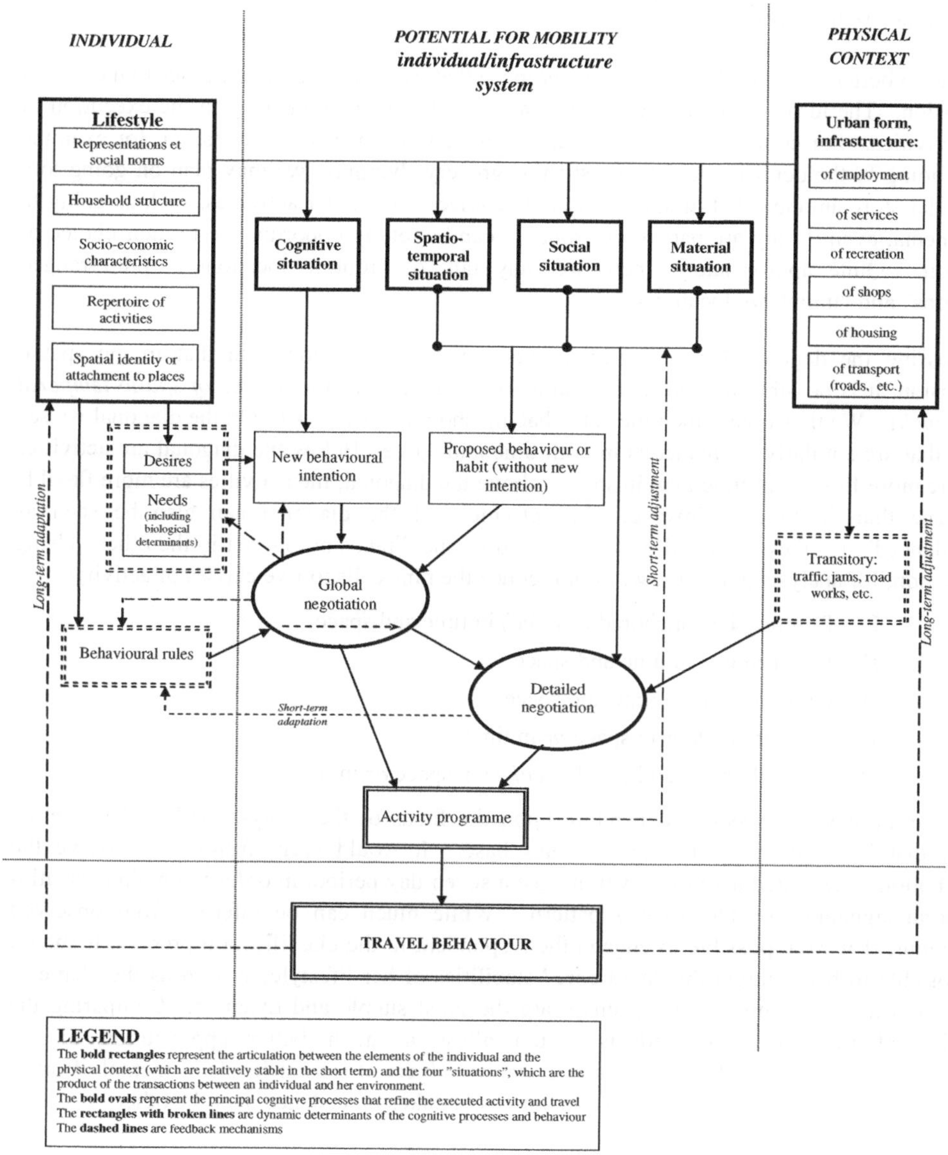

Figure 4.1 Conceptual Framework for Spatio-Temporal Behaviour

A SPATIO-TEMPORAL ANALYSIS MATRIX FOR ACTIVITIES

Coming back to Gärling *et al.* (1995) classification of activities into routine, planned and spontaneous, this is usually thought of in terms of the temporal dimension. Indeed, the observation of the amount of routine, planning and spontaneity in executed activities has been extensively discussed in recent years in connection with activity scheduling surveys (e.g. Doherty and Miller, 2002; 2003).

In order to better model daily mobility, we propose that we apply the same classes to the spatial dimension. The idea is simple: there is also a level of spatial spontaneity in the execution of our activities. For certain activities, we do not have to even pose the question, for example when going to the dentist. For others, such as grocery shopping, we may plan on going to a particular store among a list of several feasible choices. Yet other activities, such as going to the cinema, even if they are part of our routine, can be totally opportunistic in their choice of location. Quite appropriately, then, we may talk of "routine locations", "pre-arranged locations" and "impulsive locations".

In crossing the degree of spontaneity in the temporal dimension with that in the spatial dimension, it is possible to characterise an activity in terms of the *degree of spatio-temporal spontaneity*. We thus obtain the nine cells that are shown in Table 4.1. On the diagonal we see those that are similarly spontaneous in both time and space. Below the diagonal are activities that are more flexible in time than in space: above the diagonal, the activities are more flexible in space than in time. However, the activities off the diagonal can also be seen as "unbalanced", i.e. they are fixed or pre-arranged in one dimension but not in the other. Using this notion, for analysis purposes, we could reduce the nine cells to five classes of activity:

- I Routine (i.e., anchored or fixed) in time and space
- II Pre-arranged in time and space
- III Opportunistic in time and space
- IV Anchored *either* in space *or* in time
- V No anchorage, and flexible *either* in space *or* in time

Other groupings are possible, depending upon the focus of the analysis and/or the level of detail that is sought in elaborating a rules base. It would seem fruitful to observe the distribution of executed activities over at least a seven-day period, in order to obtain a "spatio-temporal signature" of the activity pattern. While much can be inferred from observed behaviour, if it were possible to engage the respondent in the classification process, it should be possible to better target the "hard core" stabilities of her lifestyle, as well as the elements that, from a spatio-temporal perspective, are the most supple and dynamic. Comparing the matrix patterns across respondents would allow a much better appreciation of the circumstances under which time and space are the subject of similar levels of mental effort in the organisation of activities, or when one dimension takes precedence over the other.

Table 4.1: Proposed classification of activities by their degree of spatial and temporal spontaneity, with examples

Degree of spontaneity		**in space**		
		Habitual	**Non-habitual, pre-arranged**	**Impulsive**
in time	**Habitual**	**I** ROUTINE *"I am always at my office from about 8am until 4.15pm"*	**IVa** Time fixed, place arranged *"This week, let's have our regular Friday lunch meeting at O'Grady's"*	**IVb** Time fixed, place flexible *"I really must have a coffee every morning close to 10am: I just pick one up at the closest place I can get one"*
	Non-habitual, pre-arranged	**IVc** Time arranged, place fixed *"I'll meet you at our tennis club tomorrow at 7.30 pm"*	**II** PRE-ARRANGED *"I want to see that new film just released all over town: let's go next Tuesday at the Odeon, the 6.30pm showing"*	**Va** Time arranged, place flexible *"I must call the baby sitter between 6pm and 6.15pm – I'll stop at the first pay phone I see"*
	Impulsive	**IVd** Time flexible, place fixed *"I must drop off my father-in-law's mower at his house sometime this week"*	**Vb** Time flexible, place arranged *"OK, I'll stop when I am going by the Ste-Foy branch of Sears and pick up a catalogue"*	**III** OPPORTUNISTIC *"Peaches at $1.99! I'll stop there and buy some"*

METHODOLOGICAL IMPLICATIONS FOR THE PROCESSUS SURVEYS

The discourse summarised above led to a number of design principles for some new kinds of survey for the PROCESSUS programme. It was intended from the outset that these surveys would be research tools, focussing on the observation of decision processes as well as on behavioural outcomes[9]. The main preoccupation of PROCESSUS is the behavioural foundations of Integrated Land-Use, Transport and Environment (ILUTE) models. These models attempt to encapsulate interactions between the decisions of urban actors – including individuals, households, firms, and public agencies – over both the long and short term. In the methodological framework of ILUTE, multi-agent microsimulation plays a predominant role. Among its most significant challenges is the specification of submodels of daily mobility, for which conventional data on daily activities and/or travel are inadequate. The main objectives

[9] Surveys of decision processes was an approach already well established in our empirical research programme through Activity-Scheduling and Stated Response surveys

of the new surveys were to explore in depth the relationships between daily activity patterns and travel and to identify behavioural mechanisms that could be built into ILUTE submodels.

Among the design principles that resulted from the discourse, the most important were:

- The use of longitudinal panel methods, to allow the observation of the evolution of decision processes through repeated measurement of the same sample
- The simultaneous observation of multiple adults (and children if possible) within the same functional household, for multi-day periods
- The use of a "reflexive" approach, in which respondents would participate in observing both their own behaviour and the extent to which participation in earlier waves of the panel survey had changed their appreciation – and possibly the functioning – of their decision processes. Accordingly, instrumentation was changed from wave to wave.
- A strong focus on questioning the "room for manoeuvre" in the temporal and spatial attributes of executed activities, and the mechanisms of planning and modification of activities, with a view to eliciting behavioural rules. A key feature of this was the inclusion of *perceptions* of both temporal and spatial flexibility to compare with the actual *exploitation* of such flexibility.

These principles were followed with varying emphasis in the PROCESSUS Panel Surveys in Toronto and Quebec City[10].

CONCLUSION

Until the middle of the 1980s, cognitive approaches to mobility focussed more on the ways that individuals record places in memory than on the ways in which they chose destinations and how to travel between them. Currently, the tendency is rather the inverse. We have tried to build upon the accumulated findings from both of these approaches to spatial behaviour, as well as on findings from research on activity planning in which the temporal dimension has dominated. This discourse led to the design of new data sources for research and model specification purposes.

Recalling the broad hypothesis discussed above: *each lifestyle generates behavioural rules that are specific to its set of associated activities, and these rules are orchestrated by decision processes that pertain to the situation in which the individual is engaged,* our conceptual framework (Figure 4.1) suggests that the rules act primarily through the global and detailed negotiation processes – which we noted includes the possibility of "negotiating" with oneself. We have thus put a substantial emphasis in the panel surveys on decisions that are associated with the mental effort of pre-arrangement, and on opportunistic decisions. By appreciating these decisions in both time and space, and exploring their interpersonal implications, these in-depth surveys allow us to build a base of rules and heuristics that should make microsimulation

[10] The methods for the Toronto Panel Survey are described in Doherty et al. (2004) and Roorda et al. (2005 – Chapter 3 of this book). Again, the survey method adopted for the Quebec City Panel Survey' *Observed and Perceived Flexibility of Activities in Space and Time* (OPFAST), is outlined in Chapter 5

models more realistic. In this regard, we hope to learn how to encapsulate both decision-making and mediation, as discussed by Miller in Chapter 2.

The survey work is also expected to contribute to methodological research. Partly as a result of initial comparisons of the lessons learnt from the Toronto and Quebec City Panel Surveys, we see *four priority areas for further development of such tools*. The first is a better integration of social norms in the survey methods: it may not be reasonable to try to measure adherence to social norms directly in an activity/travel study, but we would like to know more about the circumstances in which their influence is strong. Secondly, we would ask a similar question about the workings of the influence of social status. The third priority is to know more about the reassessment of behavioural rules by an individual based on accumulated experience. Finally, although we believe that the mechanisms in the conceptual framework are realistic for both individual and conjoint decisions, we find the framework weak in the interpersonal domain: as noted above, we are conscious that many decisions about daily mobility are made at the household level.

This work was predicated on the relevance of the systemic transactional approach, which makes the case that changes in either the physical environment or in human behaviour bring about adaptations and adjustments in the other. Our starting point for the empirical work has put strong emphasis on the activities and projects of individuals and households, notably by measuring perceptions of the degree of associated temporal and spatial flexibility. It is hoped that this will throw new light on the mechanisms and the levers for those adaptations and adjustments.

REFERENCES

Altman, I. (1992). A transactional perspective on transitions to new environments. *Environment and Behavior*, **24**, 268-280.

Altman, I. and B. Rogoff (1987). World views in psychology: trait, interactional, organismic, and transactional perspectives. In: *Handbook of environmental psychology* (D. Stokols & I. Altman, eds.), pp. 7-40. New York: Wiley.

Andan, O. and C. Raux (1995). L'univers de choix et les leviers de changement des comportements: l'apport d'une méthode interactive de simulation. In: *Les enquêtes de déplacements urbain: mesurer le présent, simuler le futur* (P. Bonnel, R. Chapleau, M.E.H. Lee-Gosselin, C. Raux, eds.), pp. 418-435. Lyon: Centre Jacques Cartier.

Beauvois, J.-L., R.-V. Joule and J.-M. Monteil (1987). *Perspectives cognitives et conduites sociales.* Cousset (Fribourg) Suisse: Editions Delval.

Bourdin, A. (1996). L'ancrage comme choix. In: *Mobilités et ancrages, vers un nouveau mode de spatialisation?* (M. Hirschhorn & J.-M. Berthelot, eds.), pp. 37-56. Paris: L'Harmattan, coll. Villes et Entreprises.

Chalmé, S., W. Visser and M. Denis (2000). Cognitive aspect of urban route planning. Communication in *International Conference on Traffic and Transport Psychology*, Berne, Switzerland.

Dewey, J. and A.F. Bentley (1949). *Knowing and the Known*. Boston: Beacon.

Doherty, S. T. (2002). Interactive Methods for Activity Scheduling Processes. In: *Transportation systems planning: methods and applications* (Goulias, K., ed.). New York: CRC Press.

Doherty, S. T. and Mohammadian, A. (2003). Application of Artificial Neural Network Models to Activity Scheduling Time Horizon. *Transportation Research Record* 1854: 43-49.

Doherty, S. T., Nemeth, E., Roorda, M., and Miller, E. J. 2004. Design and Assessment of the Toronto Area Computerized Household Activity Scheduling Survey. *Transportation Research Record* 1894, 140-149.

Gärling, T., R. Gilholm and A. Gärling (1995). Reintroducing attitude in travel behavior research: the validity of an interactive interview procedure to predict car use. In: *Les enquêtes de déplacements urbain : mesurer le présent, simuler le futur* (P. Bonnel, R. Chapleau, M.E.H. Lee-Gosselin, C. Raux, eds.), pp. 337-353. Lyon: Centre Jacques Cartier.

Gärling, T., J. Saïsä, A. Böök and E. Lindberg (1986). The spatiotemporal sequencing of everyday activities in the large-scale environment. *Journal of Environmental Psychology*, **6**, 261-280.

Ittelson, W. H. (1973). Environment perception and contemporary perceptual theory. In: *Environment and cognition* (W. H. Ittelson, ed.), pp. 1-19. New York: Seminar Press.

Jones, P. M., M. C. Dix, M. I. Clarke and I. G. Heggie (1983). *Understanding travel behaviour*. Gower: Aldershot.

Nagy, J.N. and J.C. Baird (1978). Children as environmental planners. In: *Human behavior and environment* (I. Altman and J.F. Wohlwill, eds.),vol.3, pp. 259-292. New York: Plenum Press.

Raux C. and R. Chapleau (1995). Conclusion. In: *Les enquêtes de déplacements urbain: mesurer le présent, simuler le futur* (P. Bonnel, R. Chapleau, M.E.H. Lee-Gosselin, C. Raux, eds.), pp. 459-468. Lyon: Centre Jacques Cartier.

Roorda, M., S. T. Doherty and E. Miller (2005). Operationalising household activity scheduling models: Addressing assumptions and the use of new sources of behavioural data. *In this volume.*

Simon, H.A. (1957). *Model of man.* New York: Wiley.

Stokols, D. (1978). Environmental psychology. *Annual Review of Psychology*, **29**, 253-295.

Zapf, W. with others (1987). *Individualisierung und Sicherheit. Untersuchung zur Lebensqualität in der Bundesrepublik Deutschland.* Verlagsbuchhandlung C.H.Beck, München, 157 p.

ACKNOWLEDGEMENTS

This paper was primarily supported by the Social Sciences and Humanities Research Council of Canada, through a Major Collaborative Research Initiative, 2000-2005: *Access to activities and services in urban Canada: behavioural processes that condition equity and sustainability,* with additional support from the Québec Ministry of Transport and the PICS programme for scientific exchanges, CNRS, France. The authors are grateful for the constructive suggestions of three referees.

Integrated Land-Use and Transportation Models: Behavioural Foundations
M. Lee-Gosselin and S. Doherty (Editors)

5

MEASURING ACTIVITY AND ACTION SPACE/TIME: ARE OUR METHODS KEEPING PACE WITH EVOLVING BEHAVIOUR PATTERNS?

Talia M. McCray, University of Rhode Island, USA[1]
Martin E.H. Lee-Gosselin, Université Laval, Québec, Canada
Mei-Po Kwan, The Ohio State University, USA

ABSTRACT

With the evolution of in-depth methods, researchers are now increasingly equipped to capture and analyse spatio-temporal behaviour and underlying decision processes, using a variety of computer-aided and manual methods. In this paper, several theoretical frameworks for emergent spatio-temporal behaviour are briefly reviewed as background to a discussion of leading-edge issues, including the links between personal accessibility and gender, health and land-use, and the social and equity impacts of new information technologies. The focus of the paper is on the adequacy of advances in data collection to deal with the complexity of behaviour in time and space, including cyberspace. It is argued that much remains to be done to ensure that such data collection and analysis techniques are adaptable to a full range of subpopulations, including some that are sometimes difficult to recruit for in-depth travel

[1] Work reported here was conducted while Talia McCray was a Ford Foundation Postdoctoral Fellow at Université Laval, Quebec City, Canada.

behaviour studies, such as low-income or immigrant groups. It is concluded that the current advances in modelling require, at least at the specification stage, multi-method data designs that yield relatively holistic descriptions of how people perceive the temporal, spatial and cyberspatial attributes of potential activities, and how they interact with others to organise them. Three examples of such multi-method approaches are described from recent and ongoing methodological and empirical research in Quebec City and Portland, Oregon.

INTRODUCTION

The central focus of this chapter is "moving beyond the obvious" to better understand what people do (their activity "space") and how it differs from what they might do (their action or knowledge "space"), with increased attention to the temporal dimension. Of course, such a concern with liberty of action and not just with inferred preferences is far from new: see for example Hanson and Burnett (1981). In the 1970s, notions such as spatio-temporal "prisms", activity space, and action space led many travel behaviour researchers to question the adequacy of travel-oriented data collection and analysis methods, a debate that continues to this day. At first, gaming-simulation and other in-depth techniques responded to a need for a more holistic observation of interdependencies within and beyond the household. However, the connection with travel modelling was initially tenuous. Many useful lessons led to the growth of the activity-based tradition, which by the 1990s had reached out to mainstream household travel surveys and models. More recently, the merger of new methods from time-use and spatial behaviour research has rekindled the debate.

Our purpose here is to bring together, from current research, new insights about how to gather data on spatio-temporal behaviour and its underlying decision processes, in order to address issues of social equity and accessibility, to examine the role of household interdependencies and social networks, and to begin to understand how perceptions of space and time affect individual travel behaviour.

Spatial and temporal activity patterns are arguably more variable and less anchored than they were 30 years ago. In the North American context this is often related to employment – there is usually more than one employee in a two-headed household; employment is becoming less stable and more flexible; and increasing numbers of individuals hold second jobs. This reality encourages researchers to reinforce the temporal dimension of action and activity spaces. This is particularly relevant to our understanding of the perceived realistic opportunity set, incorporating such issues as interpersonal commitments or the self-imposed limits that may result from perceptions about personal security. Also relevant to broadening our understanding of activity and action space/time is incorporating the ways in which technology, notably the use of cell phones and the Internet, increases opportunity sets, by aiding in activity scheduling or by offering new kinds of human interaction in wider cybernet communities. Drawing examples from their work and others, the authors argue the value of "going the extra mile" in data collection on activities, travel and underlying decision processes, notably by including interdependencies and respondent perception and cognition. We see this as indispensable to developing models that better address changing behaviour and emerging issues, and to represent the behaviour of women and a full range of subpopulations, including those who are difficult to engage in research, such as some low-income and immigrant groups.

The chapter begins with conceptual frameworks that address activity space/time, and then focuses on the extent to which our data collection (and modelling) techniques are keeping pace with measuring accessibility and equity issues for the purposes of guiding policy makers in their decision making. The main perspective of this paper is the measurement and representation of changing behaviour in time and space, including cyberspace. It is argued that much remains to be done to ensure that such data collection techniques are adaptable to a full range of subpopulations, and that an "ecumenical" approach to data collection is required to meet the requirements of emerging models and analysis approaches. We discuss the need for multi-method data designs that yield a fuller explanation of how activity engagement comes about.

Three examples of such multi-method approaches are described from recent and ongoing methodological and empirical research. The first of these is a package developed for a panel study of 270 Quebec City households: using in-depth interviews and diary instruments, it traces approaches to organising household activities and travel and then engages respondents in interpreting revealed patterns, including perceived levels of spatial and temporal fixity of each activity undertaken during seven days. The second consists of a quantitative and qualitative analysis of non-work related activities of low-income women, also in Quebec City: data collection methodologies include revealed behaviour questionnaires, focus-group discussions, map exercises to follow up the focus group, GIS-based analysis, and the use of large OD data sets to benchmark focus group data. The third example is drawn from GIS-based 3D geovisualisation and geocomputation to analyse activity-travel diary data collected in Portland, Oregon, and from research on individual activities and "travel" in cyberspace.

We conclude with a summary of the prerequisites for working through the complexities of modelling social equity and accessibility issues, and some priorities for the work ahead.

Conceptual Frameworks that Address Activity Space/Time

Evolving Notions of Travel Behaviour

Transportation planners have traditionally observed people's daily movement patterns within cities, developing calibrated models using data from observed behaviour, and using the models to forecast future travel. These models were designed to estimate travel demand within urban centres by analysing the impacts of travel congestion, transportation infrastructure changes, changing transit fares, and land-use changes on travel behaviour. Because the maximum demand for transportation typically occurs during the time people travel to and from work, attention was typically given to the work trip, and the location of jobs and workers. However, as changes took place in the transportation network and concern was raised about energy use, emissions, safety, and transportation demand management, more attention was given to the behaviour of the individual traveler within the transportation system. Land-use was mostly treated exogenously.

Early attempts to model travel behaviour were based on an approach to urban travel demand forecasting that conceptualised trips within the urban area by a model system, (Pas, 1984; Stopher & Meyburg, 1975). The urban area was divided into traffic zones, and the models sought first to replicate existing vehicular volumes on the road system and to use these patterns to forecast future travel. Data for the calibration of the models came from household travel surveys, observations of traffic volumes, and measurements of existing capacity.

Focusing on the need to measure the effects of land use development on accessibility, Hansen (Hansen, 1959) pioneered the gravity model. Gravity based accessibility models have a long history (Handy, 1992; Handy & Niemeier, 1997; Ingram, 1971a; Ingram, 1971b; Morris, Dumble, & Wigan, 1979). Accessibility at a location, defined by a zone or single point, is assumed to be directly proportional to the attractiveness or magnitude of the activity and inversely proportional to some measure of distance. Although there are a variety of different forms, the gravity model is valuable as a measure of place accessibility, but has several limitations when measuring individual accessibility (Kwan, 1998). The model does not support the analysis of trips that are sequential in nature (multipurpose, trip-chaining), nor does its mathematical integral form allow for the analysis of spatio-temporal constraints that can render activities unreachable.

The paradigm shift from a trip-based framework to an activity-based framework, starting during the mid-1970s, opened up new avenues for researchers to better understand and address the human behaviour elements that drive travel. This change was essential to creating a research environment where issues of social equity and accessibility could be better understood. Accessibility issues were not a focus during the trip-oriented era, and only recently gained attention during the activity-based era. A need to better understand the complex and multiple-faceted factors affecting accessibility within an urban environment has now emerged. This focuses on the needs and desires of individuals to engage in activities at various locations at different times, including needs and desires that do not get satisfied. Designing data collection that keeps pace with to evolving activity/travel behaviour and accessibility issues under new social and technological influences would seem to require, at the least, detailed measures of both space and time. We look next at the theoretical foundations of spatio-temporal behaviour and disaggregate space-time measures.

Theoretical Foundations of Space-Time Measures

Activity-based framework

The foundation of the activity-based framework was developed in North America through the research of Chapin (Chapin, 1974), an urban planner who focused on human behaviour patterns with a transportation focus, and earlier in Europe by Hägerstrand and the Lund University group, who used a time geographic approach (Hägerstrand, 1970). Chapin elaborated the theory that travel behaviour patterns are a result of individuals traversing space and time to satisfy needs and wants. Hägerstrand dealt with human behaviour patterns by focusing on the constraints that conditioned the traversing of space and time.

Chapin (1974) identified two fundamental components that must be in place in order for an individual to engage in an activity. First, the individual must desire to engage in the activity, and second, the opportunity or means to engage in the activity must exist. The desire to engage in an activity reflects the level of importance the individual assigns to the activity. The opportunity to engage in the activity encompasses the quality of the activity and the spatial/temporal context, such as distance and opportunity costs. Equally important are the means to engage in the activity, reflecting the availability of affordable transportation and entitlement, the right to use the service.

Hägerstrand (1970) conceptualised the human-activity space by applying the concept of a life path. From birth to death an individual follows a path that is shaped by constraints that serve to promote or inhibit human interactions. The set of potential human interactions are constrained by the individual (*capability constraints*) due to biological needs and/or the tools available to the individual, e.g. a private vehicle, and by the need to coordinate joint activities (*coupling constraints*), e.g. childcare drop-off. Constraints are also imposed by the culture or society, which limits access to locations in space and time (*authority constraints*), e.g. the opening hours of the shopping centre.

To date, the majority of activity-based data collection has reflected the requirements of the utility maximization framework in general, and discrete models in particular (McFadden, 1974; (Ben-Akiva & Lerman, 1985; Jones, Dix, Clarke, & Heggie, 1983; Kitamura & Kermanshah, 1983; Wen & Koppelman, 2000). However, the utility maximization framework has proven to be limited in its ability to represent underlying behavioural processes that shape accessibility and influence travel (Axhausen & Gärling, 1992; Garling, 1998) (Doherty & Miller, 2000). Gärling (1998) argues that the framework is not a fully adequate description of how people make decisions, because people may use principles other than utility, and may use different principles in different circumstances. Moreover, preferences can be inconsistent, change over time, and be influenced by other factors. In addition, critics point out that activity decisions may be in conflict, and in particular that utility theory has not been satisfactorily applied to activity choices that need to be reconciled as part of a larger plan (Miller, 2005), for example activities and travel needed to prepare for a dinner party. All of these concerns highlight the limitations of conventional travel data. Nevertheless, a number of recent activity-based models, such as FAMOS (Pendyala *et al.*, 2004), MORPC (Vovsha *et al.*, 2004), have helped clarify data needs, particularly improved space-time measures.

Disaggregate views of space-time accessibility

Recent advances in Geographic Information Systems (GIS) are now allowing researchers to better process microdata and to address some of the complexities of individuals and household decision processes that affect accessibility. Disaggregate space-time accessibility measures were developed on the above mentioned time-geographic framework of Hägerstrand (1970). The mobility allowed by a particular set of constraints, facing an individual, can be described in the form of a space-time prism, which can be represented as a potential path area (PPA) through projecting the three-dimensional prism onto two-dimensional geographic space (Lenntorp 1976, 1978; Dijst and Vidakovic 2000; Dijst *et al.*, 2002;). A potential path area is the geographic area that can be reached within the space-time constraints established by an individual's fixed activities. It is the area that an individual can physically reach after one fixed activity ends, while still arriving in time for the next fixed activity.

The approach offered by these space-time measures - coupled with the geocomputational ability of GIS and individual-level activity-travel data - has potential for overcoming many limitations of travel demand models. This is not just because the use of GIS may greatly enhance our ability to represent real-world complexities (e.g. complexities of the structure of urban human activity-travel behaviour). Rather, with GIS and increasingly available disaggregate data, highly refined space-time measures of individual accessibility can be operationalised, and the conceptualisation and modelling framework are no longer conditioned by *a priori* schema of areal or zonal units. The latter is particularly significant because analytical results in studies involving areal units such as census tracts or traffic analysis zones are affected by the modifiable areal unit problem (MAUP) (Openshaw and Taylor, 1979). The disaggregate spatial framework of space-time measures is a useful way for overcoming its effect.

Concepts of space-time accessibility based on individual-level data have been widely implemented within vector and raster GIS (Forer and Huisman 2000; Kim and Kwan 2003; Kwan 1998, 1999a; Miller 1999; O'Sullivan *et al.*, 2000; Weber 2001, 2003; Weber and Kwan 2002, 2003), creating a resurgence of interest in these concepts. Recent studies using disaggregate space-time accessibility measures have yielded important results about individual differences in accessibility and the effect of the temporal attributes of the urban environment. For example, Kwan (1998) has shown that space-time measures produced different spatial patterns when compared to conventional measures based on gravity models. Gender differences become more evident when applying space-time measures than conventional methods. This suggests that space-time measures have the ability to capture certain types of differences among individuals, such as the effect of space-time constraints on individual accessibility.

Recent research on space-time measures has also shown that individual accessibility is affected by facility opening hours and variable travel speeds at different times of the day and in different parts of the city. Weber and Kwan (2002), for example, examine the effect of travel time variations and facility opening hours on individual accessibility using a variety of data. The results show that link-specific travel times produce very uneven accessibility patterns, with access to services and employment varying considerably within the study area. The time of day that activities were carried out has also been shown to have an effect on accessibility. Evening congestion sharply reduced individuals' access throughout the city. The effect of this congestion on mobility is highly uneven spatially. Further, the use of business hours to limit access to opportunities at certain times of the day illustrates a problem that non-temporally restricted accessibility measures produce inflated values when these opportunities are treated as if they were available at all times of the day. Incorporating these time variables reveals not only that accessibility may be lower than would otherwise be inferred, but that it also produces a very different, and perhaps unexpected, geography of accessibility (Weber and Kwan 2002).

Social psychological perspectives on spatio-temporal behaviour[2]

Other research traditions have approached spatio-temporal behaviour in terms of the relationship that an individual builds with space-time as s/he perceives it. A social psychological understanding of an individual's relation to space may permit a more accurate appreciation of the dynamics of behavioural change in a given day or week (e.g. Gärling *et al.*, 1986). This recognises dynamic, decision-making processes that depend on their specific and limited spatio-temporal context.

Studies in environmental psychology suggest that attachment to places or spatial representations of cities, the most stable factors in space/individual relationships, strongly affect the definition of "lifestyle", a term which in common use refers mostly to patterns of consumption, but here includes adherence to sub-cultural and social norms. However, some theoretical perspectives (e.g. Dewey and Bentley, 1949) have long seen the individual and its environment as mutually defined, placing more emphasis on the environmental situation. According to this perspective, the decision-making processes related to daily mobility are a product of the relation between the individual, defined by his/her lifestyle, and the spatio-temporal constraints of the moment. It is thus important to study both temporal and spatial fixity as viewed by individuals, households and other personal networks.

For their part, behavioural rules are closely linked to the value system of the individual. Whereas some research has attempted to analyse behavioural rules from the perspective of behavioural intentions and attitudes related to mobility, Ramadier *et al.* (2005) propose eliciting these rules by studying *individuals' representations* of daily mobility. Nevertheless, there is no such thing as a global representation of mobility. The significance of a trip, in almost all cases, depends on the activity served. For this reason it is important to distinguish each representation according to categories of activity that are related to particular spatial behaviours, in order to compare lifestyles on the same basis.

These theoretical perspectives from social psychology suggest that the relationship between the individual and space/time is undoubtedly a powerful way to reveal *both* behavioural rules and decision processes linked to spatial mobility. This relationship allows us to move beyond the mere description of spatial behaviour and attempt to understand the dynamic aspects of the observed phenomena.

Summary

There is clearly much overlap in the understanding of spatio-temporal behaviour represented by the three theoretical perspectives discussed. Each of them addresses the notion that an

[2] This subsection is a précis of part of a broader conceptual framework that is presented in this volume by Ramadier *et al.* (2005)

individual acts within bounds that are feasible and acceptable. They differ in the extent to which behaviour is understood by the observation of behavioural outcomes, versus the investigation of how the set of opportunities in time and space are perceived and how decisions are made.

However, the overriding question becomes: What are the data requirements that would allow researchers to better understand human behaviour? We are collecting data, but are we collecting the right type of data? Our conceptual models for activity-based research are well developed. We understand that an individual's access to urban opportunities is affected by the individual herself/himself and the reality of the environment in which she/he lives. The individual traveller acts out of needs and desires, which change over time. Gathering data to understand an individual's needs and desires by exploring the effects of gender, culture, income, race, and lifestyle on activity space/time is challenging at best. The answer appears to lie not in seeking generalisations about needs and desires, but in building data collection around specific executed activities, and related sets of activities, over multi-day periods and then seeking to understand their antecedents. Just as important is to examine the role that perceptions played as respondents committed themselves to those activities: what were their expectations at the time about what was available to them? Using these microdata on these specifics, differences attributable to gender, culture, income, race, etc., can be meaningfully explored. In the past, such data have been seen as purely descriptive. By contrast, data describing the temporal and physical attributes of the urban travel environment, such as volume-capacity characteristics or travel time, have proven easier to obtain and represent in our models. But advances in computing and GIS promise a more holistic approach to utilising both individual behavioural and environmental microdata, perhaps providing a platform for representing the decision processes that affect accessibility.

The challenge lies in the inclusion of "hard to reach" groups (e.g. low income, immigrants), and in gathering meaningful microdata on how individuals within a household organise and prioritise their activities, taking into account at least the perceived importance, the availability of transport, and the management of time constraints. The following section examines a broad range of accessibility issues and the efforts of researchers to understand them.

KEEPING PACE WITH WHAT'S IMPORTANT

Understanding how factors of policy, land-use form, and human behaviour interact and are interconnected is key to defining activity and action space/time. Challenges arise when researchers must determine which societal elements should be measured, and which accessibility issues are most assistive to policy makers in their decision making. To date, there has been an exhaustive amount of research on the effects of modal choice and its implications for accessibility. However, how well do we understand the effects of gender, socio-economic status and culture on the decision to travel at all? Do technologies, the quality of the destination and travel time carry the same implications for all population groups? This section highlights some of the more important issues we believe should be addressed by developing creative data gathering and modelling techniques.

Most of our current preoccupations concern the equity of accessibility and the inclusiveness of data collection. Often, when transportation sustainability studies are undertaken to evaluate equity issues, the focus is on the use of the roadway system, congestion, and parking policies. In other words, who pays and how much do they pay? To a much lesser degree do these studies directly address social equity issues of the transportation disadvantaged. Who are the transportation disadvantaged? Denmark (Denmark, 1998) characterizes this group as people who "due to poverty, disability, frailty, or other conditions have found their mobility increasingly restricted as the shift to automobile dependencies continues unabated." His definition points to individuals who have become socially excluded due to transportation policies that favour wealthier travellers, i.e. auto users who have the ability to pay, through land use and housing policies that have systematically isolated the urban poor from regional labour markets. This concentration of urban poverty in the US has created a culture of "not working" (Wilson, 1987) where youth neither have access to jobs, nor do they see esteemed family members and friends taking part in the labour market. Policy interventions to improve the quality of life for socially excluded groups must consist of transportation, land use, housing, social welfare, and technological solutions. In this section, we focus on a number of issue areas in which the adequacy of data collection is of particular concern.

Social and Equity Impacts of ICT

What are some of the promising impacts of ICT (information and communication technologies) on addressing social and equity issues of travel behaviour? Mokhtarian (Mokhtarian, 2003) provides a literature review of a number of relationships that can exist between physical travel and telecommunications. Her review suggests that even though incomplete, the empirical evidence suggests that complementarity is more prevalent than substitution. With the telecommunications boom of mobile cell phone use, email, internet, and fax machines, which allows individuals to interact without physically moving in space, the thought was travel would decrease or be eliminated in some cases (US DOT, 1993). Researchers looked to the increasing trends of businesses embracing new interactive modes of tele-working and teleconferencing, more shopping via the internet, and educational institutions using the internet as a research tool and as a course provider. However, several studies have shown the ICT impact to have stimulated more travel due to an increasing need for face-to-face interactions and for the delivery of internet purchased merchandise (Salomon, 1986; Couclelis, 1999). At another level ICT, mostly in the form of mobile cell phone use, assists individual travellers to rearrange, change, or add to their schedules in a real-time fashion.

Focusing on the effects of urban space on employment, Shen (1998) developed an accessibility measure based on a gravity model to measure both travel and tele-working opportunities in the Boston Metropolitan area. Levels of tele-working were predicted for the year 2005 based on 1990 data. Results of his simulation give little hope of decreasing the negative effects of distance for low-income households, but showed a tremendous benefit to wealthy suburban communities. The study is valuable in that it sought to address the impacts of ICT for the transportation disadvantaged, by incorporating a measure of future tele-working opportunities. However, the study has several limitations including the method used to forecast tele-working, the emphasis on one mode (auto), and the use of an aggregate model to address individual accessibility.

In another study, which uses time-use data within an activity-setting framework, the impacts of social contact on the travel behaviour of a representative sample of individuals from Canada, Norway, and Sweden are analysed (Harvey & Taylor, 2000). They found that [pg. 53] "there is a tendency for persons with low social interaction to travel more. It is argued that individuals need, or want, social contact and if they cannot find it at the workplace, they will seek it elsewhere thus generating travel." Even though this study does not directly deal with issues of social exclusion, their methodology could help establish the links between a lack of social interaction and the negative impacts of high levels of joblessness and urban decay in many U.S. urban centres. The effects of social exclusion have a negative impact on sustaining the social and economic well-being of a community. An interesting future study might address the degree to which vehicle-less low-income households can maintain healthy and meaningful networks within the community and outside the community.

Gender and Accessibility

Recent studies have shown that there are considerable differences in travel behaviour and activity patterns between men and women (e.g. Hanson and Pratt 1995). Ignoring these gender differences may lead to policy measures that not only ignore women's travel needs but also aggravate the difficulties they face in their daily lives (e.g. Rosenbloom and Burns 1993). An important issue is that a substantial proportion of women's travel consists of multi-purpose and multi-stop trip chains, and this characteristic of women's travel often renders travel demand models problematic. An example is the extension of the work trip in order to access essential facilities such as childcare centres. Since many women with children need the automobile in order to be able to undertake linked trips and to meet specific travel needs of their children, sustainable transport measures or travel demand management policies that aim at reducing car use in general may seriously affect the possibility for them to fulfil their household responsibilities.

Further, conventional models often ignore the role of individual time-budget and space-time constraints in women's lives. As women often face specific sets of space-time constraints imposed by the space-time fixity of their obligatory activities, their accessibility is constituted in the sequential unfolding of these activities (Pirie 1979). In other words, women's individual accessibility is determined more by the number of opportunities that are within reach given their particular life situation and adaptive capacity, instead of by the number of opportunities that are located close to their home or workplace. Women's access to jobs and urban opportunities is determined more by their space-time constraints in everyday life than by factors such as travel mobility or relative location to opportunities (Palm 1981; Hanson and Pratt 1990). Further, those constraints associated with socially ascribed gender roles tend to be spatially and temporally more binding for women than for men (Pickup 1984, 1985; Tivers 1985; Kwan 1999a,b).

Health and Accessibility

Understanding how transportation and land-use patterns facilitate or inhibit access to health services and opportunities for exercise is vital to addressing the societal needs of today. Unlike public health experts who have had a long history in analysing the effects of distance on healthcare accessibility, transportation researchers are just beginning to seriously address the link between health and urban land-use form. According to (Moore, 2003), this recent focus in the U.S. is partially due to the heightened attention the media has given to blaming urban sprawl and the lack of pedestrian friendly neighbourhoods for contributing to the increase in the number of Americans who are physical inactive, resulting in an epidemic of obesity. Moore's USA Today article points out that even though health professionals are calling for Americans to exercise more by simply walking, the reality in many neighbourhoods is that people cannot walk owing to the lack of sidewalks, the impermeability of many suburban layouts, and the location mismatch between households and daily activity locations, such as shopping centres, employment centres, and schools. With the spreading out of neighbourhoods, zoning constraints, and the dominance of the automobile, some research foundations are funding new initiatives that attempt to bring together public health and urban planning researchers to identify ways of changing transportation and land-use policies that negatively impact walking and bicycling – these last now increasingly labelled "active-transportation".

Another important health link with transportation is the study of the spatial distribution of healthcare facilities in developing and developed countries. Low-income women are often disproportionately affected by the distance between home and healthcare facilities because they do not have access to private vehicles, and must walk or rely on public transport, which may not meet their needs. In two studies on prenatal care accessibility, McCray (2000, 2004) found travel-related constraints exist when analysing the travel patterns of low-income women in Detroit, Michigan, and rural South Africa. Using focus groups and geovisualisation of public transit routes, McCray (McCray, 2000) found that low-income women's transportation needs for prenatal care could not be fully met by Detroit Department of Transportation (DDOT) bus routes, and therefore supported the use of specialised healthcare paratransit services. The monetary cost of the fixed-route system, unreliable scheduling of DDOT buses, attitudes of other DDOT riders, personal safety concerns, and the shortage of direct bus routes to healthcare facilities prevented or discouraged women from obtaining healthcare. The majority of the participants had children under the age of six, and would have needed to take at least two buses to reach their chosen healthcare facility. Within this environment, the specialised paratransit service, the Healthy Baby Service, was created to lessen the travel burden and encouraged low-income women to keep their prenatal care appointments by utilising drivers who have been trained to address some of the social and health needs of their clients while en route.

In a study conducted in northern KwaZulu Natal, South Africa, McCray (McCray, 2003) addresses the environmental factors and socio-economical factors that motivate or discourage women from utilizing prenatal care in the Ubombo Magisterial District. A multinomial logit model is used to explore the many factors that inhibit or support a woman's use of prenatal care services. Geovisualisation is also used to explore the effects of travel mode, and to relate the levels of prenatal care utilization to the spatial distribution of healthcare facilities. In order to address these factors, one must understand how the environment and culture play a role in shaping women's lives and their travel patterns. Travel is understandably difficult in this rural low-income area where transport options are limited. However, many of the documented factors in the public health literature, as affecting prenatal care utilization, do not apply in this region. Fetching water as a daily activity, which is usually not associated with prenatal care utilization, has a significant negative impact on utilization. An important piece of this exploration is addressing time constraints and the cultural context. After time has been spent attending to essential tasks upon which the entire family is dependent, do rural women have "*remaining time*" left to visit a healthcare facility?

Land-use Issues and Accessibility

Over the past few decades, researchers have sought to understand the linkage between land-use patterns and travel behaviour, mostly related to the journey to work. A variety of factors or combination of factors have been used to characterize urban form, including distance from the urban centre, average distance from commercial facilities, distance from the nearest transit stop, population density, housing stock, the degree of land-use mix, and a measure of neighbourhood walk-ability, to name a few. Characterization of travel can be average trip length, trip frequency, modal split, or total vehicle miles travelled. Researchers who support the land-use and travel behaviour linkage argue that individuals who reside in transit/pedestrian-oriented developments tend to walk or use transit more and make shorter trips. The studies are many, ranging from aggregate models (Friedman, Gordon, & Peers, 1994; Holtzclaw, 1990) to disaggregate models (Cervero & Kockelman, 1997; Handy, 1996). Still, other researchers apply caution in interpreting these results, and suggest that people living in different communities cannot be adequately compared due to sample biases (Crane & Crepeau, 1998), and socio-economics factors may have a stronger effect on travel patterns than simply land-use factors (Hanson, 1982a; Stead, 2001). A number of analytical techniques, including multiple regression analysis, were used to consider how the variations in socio-economic factors explain travel distance. Added to the differences caused by socio-economic levels, individual attitudes, perceptions, and values (Gärling, Gillholm, & Gärling, 1998) have also been shown to significantly affect travel behaviour and account for a larger proportion of the travel variance than land-use (Kitamura, Mokhtarian, & Laidet, 1997).

These studies support the inclusion of socio-economic and personal factors in land-use/accessibility studies, especially those studies that seek to better understand, model, and address the accessibility needs of the transportation disadvantaged.

Summary

The four issue areas in this section of the paper highlight topics that merit increasing attention in our efforts to address some of the more important accessibility issues. The question becomes: what should we be measuring? Who determines what is most important? The authors believe that our work should be policy driven. Our work should attempt to answer some of the societal needs of families from diverse economic, racial, cultural, and lifestyle backgrounds. Pressing topics in the urban centres of high-income countries include long commutes to places of employment that have a negative impact on our environment, the inaccessibility of active recreation and healthcare services, and the inaccessibility of affordable and quality shopping. Land-use patterns are linked to travel behaviour. However, relying heavily on a few summary land-use variables, such as population or housing density, in models used to analyse levels of accessibility can lead to oversimplified conclusions about policy. Similar concerns can be raised about specific features and "fixes". For example, the presence of sidewalks in a middle/upper income community may promote walking; however, sidewalks in a lower income community may not promote walking. This is because the underlying issue is not just about physical facilities, but also about perceptions, socio-economics, culture, and values. Data collection methods should be geared towards understanding land-use variables in relationship to a populations' socio-economic environment, and to the number of opportunities that are within reach given their adaptive capacity.

In order to change our focus, we must go beyond the "conventional" data collection methods. Addressing the relationship of accessibility to issues of social equity, the use of ICT, gender differences, health and land-use requires us to conduct our studies differently. We cannot rely on a single data collection technique such as traditional travel diaries, especially if periods of observation longer than one or two days are indicated. These may work for "mainstream" groups, but may be regarded with indifference, suspicion or even fear by members of marginalised groups. Depending on the focus of the study and the population of interest, the most effective approach may involve mixes of qualitative and quantitative techniques. Among these may be techniques as varied as focus group discussions, games, "self mapping" of individual space/time, or participant observation of both activity patterns and underlying decision processes in their social contexts, perhaps enhanced with monitoring technologies. Equally important, as noted above, is the inclusiveness of our recruitment of study respondents. In order to reach and involve those populations whose accessibility needs are not well understood, such as low-income and immigrant populations, transportation researchers must be prepared to leave the academic environment and form partnerships with community leaders. In exchange, they must offer to develop studies that are beneficial to the communities, and expand the knowledge base required if activity modelling is to be relevant to social equity. The evolution of data collection tools to meet these challenges is the subject of the next two sections.

TRAVEL BEHAVIOUR DATA COLLECTION METHODS: DO THEY CAPTURE INFORMATION THAT MATTERS?

Travel behaviour data collection can be divided into two main categories: the observation of actual or "revealed" behaviour and its underlying factors and processes, versus the elicitation of possible future behaviour or "stated" responses. We discuss mostly the expansion of the former in the light of the needs discussed above, with again a particular focus on space-time measures. Of course, these two categories are interdependent: an adequate understanding of respondents' actual spatio-temporal behaviour is a prerequisite for collecting data on their responses to future conditions; and an appreciation of stated responses can alert us to problems of completeness and balance in revealed behaviour instruments.

Revealed Behaviour Methods

The collection of data on revealed household travel behaviour (RB) is a mature art that has served well the needs of travel demand forecasting – insofar as the focus (as noted above) is on the capacity of road and public transport networks, notably for weekday peak-hours journeys to and from work. Even within these limits it is not, however, without problems. Successive international conferences on transport survey methods have drawn attention to issues of quality, and in particular to the representativeness of residual samples, especially in countries that rely heavily on landline telephone methods. The content of trip-oriented surveys, by definition, is incomplete for the study of dynamic and evolving behaviours in both work and non-work domains, even if data are collected on all days of the week. It was in this context that a move to activity-based methods became widely accepted in the 1990s.

Revealed Behaviour data collection has been affected in stages by the activity-based paradigm. Initially, the focus was on survey instruments that asked respondents to record daily sequences of activities, and the travel needed to undertake them was observed by asking "how did you get there?" only after responding to "what did you do next". This approach was implemented in a number of metropolitan travel surveys in the USA. Among the best known is described in Stopher (1992). This was shown to produce a fuller picture of the travel day, notably with regard to short trips that were often missed by trip diaries. In subsequent stages, the data requirements of activity-based models led to a variety of instruments that focussed primarily on patterns of activities and related tours. By the mid 1990s, activity-based modellers sought to extend data collected to include activity generation/participation, activity scheduling and duration, dynamics over horizons from within-day to long term, cognitive processes, and the policy sensitivity of a wide range of measures that affect activities (Lee-Gosselin and Polak, 1997).

In the most recent stage, attention has shifted to the observation (and not just the inferring) of *decision processes* alongside the observation of behavioural outcomes — the activity/travel patterns themselves — a preoccupation that has a longer history with some Stated Response data collection methods, as discussed below. Much of the attention has been given to observing scheduling processes, incorporating the notions of: a "repertoire" of possible activities on which a respondent draws to fill their schedule; the progressive refinement of schedules involving combinations of routinised, individually planned and impulsive activities; different levels of advanced planning, coordination and negotiation within households and social networks; and the simultaneous planning of multiple activities. This implies a level of detail in data collection that is feasible only for small and medium-sized samples.

Doherty and Miller (2000) developed a computer-aided self-interview (CASI) package, CHASE[3], that keeps an automatic log of all entries, modifications and deletions made by a respondent in the course of seven days' use of an electronic time planner. Respondents are encouraged to plan and re-plan the days ahead and are queried automatically about the reasons and timing each time something is changed. They are also queried about associated travel, and other data are collected through interviews at the beginning and the end of the observed week. The theoretical bases of an alternative approach with related objectives is described by Ramadier *et al.* (2005). Like CHASE, this approach traces the evolution of decisions about activities and travel throughout a seven day period, but with less detail on each change in the schedule, and with a much more substantial follow-up interview seeking to understand holistically both spatial and temporal activity planning that was revealed during the observed week. This led to OPFAST[4], a manual survey package that is a major part of the methodology for the Quebec Panel Survey, which is described in some detail below.

A concern with these recent developments is the substantial response burden that is associated with obtaining detailed data on activity location and timing over multi-day periods, alongside detailed data on temporal and spatial scheduling. Technology is playing an increasing role. Current experiments by the authors and others involve augmenting scheduling surveys with monitoring technologies, notably vehicle- and person-based Global Positioning System (GPS) loggers. Sometimes these are linked to portable CASI devices, sometimes as passive devices in combination with prompted recall techniques through which respondents provide information such as mode, accompanying persons or activities at stops that cannot easily be observed or inferred electronically[5] (Stopher *et al.*, 2003). In addition, for some purposes there is a compelling case for observing revealed behaviour over multi-week periods, and this is increasingly feasible with such technologies as monitoring devices that run automatically, transmitting data using telemetry, such as was recently introduced in the Atlanta region on a sample of 1200 private vehicles to monitor vehicle use and conditions immediately preceding crashes. However, even diaries are sometimes tolerated for long observation periods, such as six weeks in a European study (Axhausen *et al.*, 2002). At the same time, retrospective techniques covering decades, inspired by biographical and ethnographic approaches, are starting to find their place in transport studies. Examples include car-use (Lee-Gosselin and Turrentine, 1997), car ownership (Mohammadian and Miller, 2003) and residential mobility (Séguin and Thériault, 2000); in the first example, information from a retrospective instrument was used to build customised scenarios for a Stated Response survey.

There has been considerable effort in recent years to collect new types of data on the relationships between travel and information and communications technologies (ICTs). At first this focussed on telecommuting (see Golob and Regan, 2001; Kwan 2002). Useful conceptual frameworks have also been proposed for dealing with the complex relationships between

---

[3] CHASE: Computerised Household Activity Scheduling Elicitor

[4] OPFAST: Observed and Perceived Flexibility of Activities in Space and Time

[5] In principle, other passive techniques such as monitoring electronic financial transactions could be employed, but these are usually infeasible for travel behaviour researchers.

people's activity-travel behaviour and their use of ICTs (e.g. Mokhtarian 1990). For example, Kwan (2001) has formulated a cognitive model that conceives the individual as an extensible agent and conceptualises individual access in the information age through notions such as cyberspatial cognition and computer-built long term memory. Dijst (2003) has developed an action space model for studying the impact of ICTs on individual access to opportunities based on different types of constraints on ICTs. However, collecting data on the interactions between individual activities or travel in the physical world and in cyberspace poses a serious challenge to the researcher, especially with respect to people's Internet activities. Some researchers have used an adapted version of conventional activity-travel diary as instrument for collecting people's activities in cyberspace (e.g. Mokhtarian and Meenakshisundaram 1999), while others have attempted to develop and use new data collection methods.

Stated Response Methods

Arguably, Stated Response (SR) data collection methods – those that elicit behavioural responses, relevant constraints, or both, under hypothetical situations – have made a seminal contribution to the methods now used to research spatial and temporal activity, whether revealed or stated. Much of the SR data collected are rather narrow in scope. For example, the most widely used class of SR surveys, Stated Preference (SP), are confined by design to tradeoffs between limited numbers of attributes and pre-defined options, although some "adaptive" SP techniques do take respondents through a filtering stage to identify relevant choice sets among a wide range of possibilities. However, other classes of SR involve more complex problem-solving in time and space. In particular, gaming-simulation methods such as HATS (Jones, 1979) or CUPIG (Lee-Gosselin, 1989) illustrated long ago the value of exploring the decisions underlying activity patterns, and engage respondents in observing and interpreting their own decisions "reflexively".

Broad reviews of SR are given in Lee-Gosselin, (1996; 2003), but some particular developments are relevant here. First, the more complex methods are generally built around a "revealed base", that is, the framing of hypothetical situations is tailored to the context of recent activities, and to the influences on the decisions to execute those activities. Thanks to the developments in "revealed" observation techniques, discussed above, the revealed base may be richer in space-cyberspace-time measures. Moreover it may contain information on the set of options that were perceived to be possible at the time it was decided to engage in the activity, and sometimes on options that were rejected as infeasible – that is, on preferred activities that were suppressed or delayed. Secondly, SR methods are ideally suited to explorations of perceived spatio-temporal flexibility. Using a rich revealed base, Stated Adaptation methods – those that elicit hypothetical behavioural outcomes under given constraints – can induce to respondents to "relive" how they would have coped in the specific circumstances of one or more recently executed activities in the face of a perturbing factor, such as the illness of a child or a blocked motorway. Thirdly, SR methods can provide structured protocols for investigating the behaviours evoked by new technologies and policies that alter familiar time-space constraints, such as advanced location-based services search software on "smart" mobile phones, the advanced reservation of a parking space by Internet, or the opening of a pedestrianised town centre to motor traffic in the evenings.

Increasingly, Stated Response methods are being combined with Revealed Behaviour methods in the same instrument package to provide a fuller explanation of how choices about activity engagement come about, and to what extent people are at liberty to maximise their preferences.

Where next with travel behaviour data collection?

It can be said that the data collection toolkit has evolved considerably in recent years. With the consent and the collaboration of respondents, it is possible to collect microdata on sequences of action in time and space, and on the personal and inter-personal decision processes that led to those actions, in considerable detail and over multi-day periods (and sometimes longer). These are the data that are sought by several emerging modelling and analysis approaches, such as multi-agent microsimulation.

However, to what extent have packages using this toolkit succeeded in addressing the full range of accessibility and other issues on which surveys are expected to inform public policy? Thus far, success has been limited, but there is reason for optimism. Households and individual decisions have been successfully modelled using time-space measures, but mostly in the domain of consciously weighed, longer-horizon decisions such as residence location or vehicle ownership. Micro-decisions about daily activity and travel patterns are only partly conscious. Policymakers need to know about the macroscopic consequences of myriads of such decisions, and the sensitivity of the consequences to policies and technologies. This requires analysis and modelling methods that can simulate observed micro-decisions, giving weight to those classes of decision for which survey respondents themselves identify the potential for leverage, through conscious processes such as information search or negotiation. Nevertheless, the specification of rule bases and other components of simulation is not only the only use of the rich spatio-temporal datasets that are starting to become available. Explanatory analyses can reveal much about respondents' room for manoeuvre and about the circumstances under which behaviour may change and hence why some policies succeed better than others. Unfortunately, rare are those applications of either RB or SR that effectively investigate needs and desires that do not get satisfied, or the difference between activity and action spaces (e.g., Djist, 2003; McCray 2004), despite the noted potential for the tools to do this.

Once again, it is clear that much remains to be done to ensure that such data collection and analysis techniques are adaptable to a full range of subpopulations. It is also evident that demographic, economic and technological developments are leading to more complexity in behavioural dynamics. *It is our view that the current advances in modelling require, at least at the specification stage, multi-method data designs that yield relatively holistic descriptions of how people perceive the temporal, spatial and cyberspatial attributes of potential activities, and of how they interact with others to organise them.* In the following section, we present three case studies of data collection packages that have attempted to respond to parts of this challenge.

EXAMPLES OF MULTI-METHOD SPACE-TIME DATA COLLECTION

The three case studies that are outlined below are examples of a large number of possible multi-method approaches to data collection that can broaden the scope of data on activities in time and space, and their underlying decision processes. Each data collection package was undertaken primarily to aid the design and specification of behavioural models, but each also yields empirical results that are of interest for exploring trends and evaluating the potential impacts of policies and technologies.

The Quebec City and Toronto Panel Surveys

As part of a large programme on the behavioural foundations of long-term simulations of land-use and transport interactions[6], two longitudinal surveys were initiated in 2002. One was in Toronto and the other was in Quebec City. Each of these was designed as a three-wave panels survey of 270 households. The objectives included both the generation of new empirical information for the broader research programme, notably to aid in the specification of microsimulation models, and the testing and comparison of new survey methods. These surveys were specifically designed to broaden the scope of time-space measures in ways that address, in particular, three of the concerns discussed above. First, they were to be in-depth, multi-day designs, allowing the spatio-temporal organisation of activities and travel to be appreciated in context. Secondly, decision processes underlying activity engagements were to be observed rather than inferred, using both prospective and retrospective techniques. Thirdly, stable panels were to be employed primarily to observe how the dynamics of household and individual behaviours responded to changes within the household, such as employment, income, the arrival and departure of household members and associated shifts in time commitments: any shifts in external factors such as transport supply, fuel price, retail destinations or economic activity would be studied opportunistically if they occurred during the three year life of the survey, but these were not essential to the main objectives.

Unlike most panel surveys in transport, the intention here was not to keep the instruments constant from wave to wave and produce the equivalent of a repeated cross-section on the same sample. In both of these two surveys the instruments were varied so that each wave built on the experience of respondents in the previous wave. Different but complementary designs were adopted in the two regions, in part to meet the methodological objectives. The Toronto Panel Survey is centred on activity scheduling using the CHASE package (see above, Section *Revealed Behaviour methods*) for the first wave, and is described in Roorda *et al.* (2005). The Québec Panel Survey shared with Toronto the parallel collection of data on both behavioural outcomes and underlying decisions, but placed greater emphasis than Toronto on the perceptions of respondents about the ways they organise their activities in time and space. As mentioned above, the Québec survey sought less detail than Toronto in the systematic collection of activity scheduling attributes, and more detail on a holistic interpretation of the

---

[6] SSHRC Major Collaborative Research Initiative, 2000-2005: Access to activities and services in urban Canada: behavioural processes that condition equity and sustainability

activity and travel patterns that, in the first waves of both surveys, were recorded by respondents over a seven-day observation period.

The Quebec Panel Survey instrument package for the first wave, which was by far the most detailed of the waves, is known as OPFAST[7], which consists of:

1. *A startup in-home interview of the household, during which:*
   - Basic information is gathered on the structure of the household, household vehicles, residential mobility history, activity repertoire with travel modes, etc.
   - All adults over 16 are instructed in the use of the paper instruments
   - A fax machine is lent to the household and installed

2. *Paper instruments:*
   - A single Weekly Planning Sheet, organised into 7 columns with a time scale similar to a desk diary, on which each adult is asked to record all known activities for the coming week during the startup interview; on the evening of each day, they are asked to add, delete and modify entries (by crossing out, erasing, etc.) for any of the days remaining (see Figure 5.1, left)
   - A seven daily Activity and Travel Log on which executed behaviour is recorded, one day at a time (see Figure 5.1, right)
   - Once per day, the current state of the Weekly Planning Sheet, and that day's Activity and Travel Log, are faxed to the university; the date and hour of transmission is automatically recorded on each sheet
   - Support is provided by telephone based on material received

3. *In-depth home interview soon after the 7 day observation period, including:*
   - Validation of paper instruments
   - A sorting game that addresses the perceived spatial/temporal fixity of activities
   - Retrospective recording of details on planning horizons and interdependence
   - Holistic interpretation by each adult of the way they approached the spatial and temporal organisation of their activities
   - Semi-structured group interview about "projects" (sets of jointly-planned activities), activity negotiation, use of telecommunications, unsatisfied demand, and expectations about the future

4. *Data entry programme:*
   - Entry screens and relational database in Microsoft Access™
   - Designed to align with CHASE data format, where possible
   - Usable for direct entry by interviewer in startup interview

[7] Observation and Perceived Flexibility of Activities in Space and Time

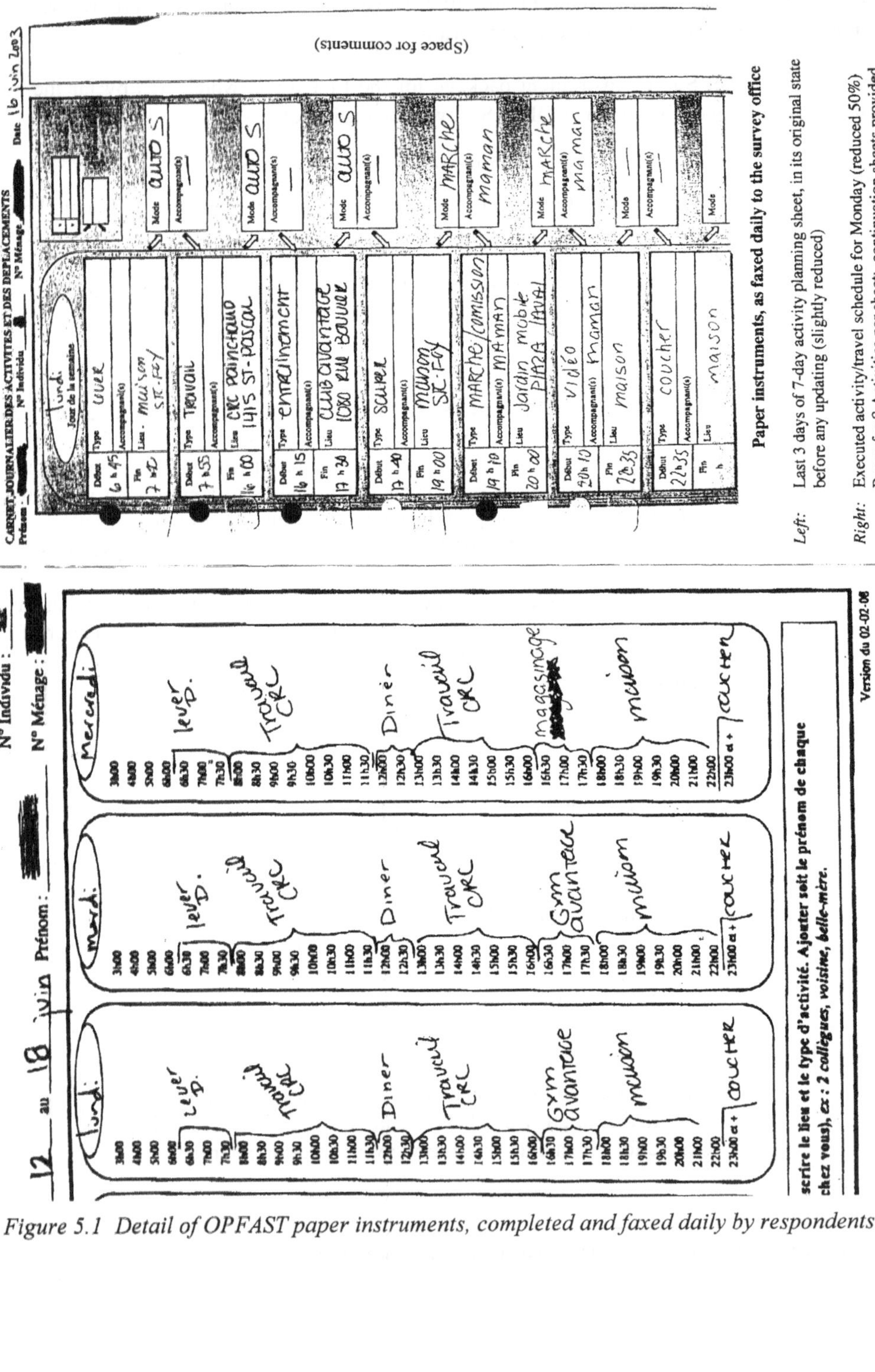

Figure 5.1 Detail of OPFAST paper instruments, completed and faxed daily by respondents

Extensions of the OPFAST package under test and development include the use of Global Positioning System (GPS) personal loggers to reduce respondent burden in the tracking of daily activity patterns, and the integration of GIS support to the geocoding of activity locations.

The use of a reflexive approach, changing the instrumentation in the second and third waves to recognise and exploit the cumulative effects of participation in the earlier wave(s), has a particular consequence: no sample refreshment is possible. Our view is that panel "conditioning" is normal and substantial given the extent of participation required, and that the researcher and respondents are in a joint learning process.

To start up Wave 1, a representative range of households with different life-stage, residential location and vehicle ownership characteristics were drawn from published directories using screening and quota sampling. Recruitment was effected by letter with telephone follow-up, and by approaching disadvantaged subpopulations through intermediaries (see below). It was a requirement of the Québec panel that all adults under the same roof who were involved in each other's activities participate for a household to be eligible: as a result, about 60% of the sampled households yielded information on the interactions between two or more adults. A choice of financial incentives was offered worth between Cdn$20 and Cdn$30.

OPFAST is a complex package that draws on a number of the theoretical frameworks and on developments in both Revealed Behaviour and Stated Response survey methods discussed above. In particular, it is designed to strengthen the understanding of spatio-temporal activity organisation, building on respondents' own perceptions about personal and household activity. The resulting data are used both for exploratory analysis and to generate a base of rules and meta-rules needed in multi-agent simulation. Initial results from the early stages of the Wave 1 were presented by Joud *et al.* (2003). After setting aside the first 20 households (whose instruments had not reached full maturity), these were based on 64 households, 115 adults, 2556 activities and 800 trips. These indicated, for example, that spatio-temporal routinisation is highest in among the over 75s and among 25-54 year olds, and they already showed in detail the extent to which access to services, shopping and social interaction requires advanced planning, negotiation and coordination with others. The full dataset on the mature instruments contains 250 households and 381 adults who executed 26,267 in-home[8] and out-of-home activities and 12,840 trips.

In their second waves, Toronto and Quebec City used very similar instrument packages that were designed primarily by Matthew Roorda of the University of Toronto. As respondent retention was of utmost importance in a sample that could not be refreshed, these packages had much lower respondent burden than CHASE or OPFAST. A Computer-Aided Telephone Interview (CATI), filled in details of two days of executed activities and travel that were recorded by the individual respondents on a mailed memory jogger. In most cases, the memory jogger was mailed back and activities are entered into the CATI to reduce the duration of the interview. The CATI included a limited Stated Adaptation component that selects one

[8] It should be noted that many in-home activities were grouped into classes such as "morning preparations" to avoid tedious detail.

single activity and one pair of related activities for which hypothetical changes in constraints (for example a traffic delay) are posed. Respondents are questioned on how they would have coped with this situation, and this was used to build an understanding of how flexibility is used to deal with different classes of activity. Quebec City used the same content but interviewing was completed using paper forms, and an adaptation of the CATI programme was used for later data entry. In part, this was because Quebec City extended the procedures to include the perceived spatio-temporal flexibility classification for each activity in the jogger, and to repeat the Wave 1 questions on activity negotiation, use of telecommunications, unsatisfied demand, and expectations about the future.

The data collection strategy for Wave 3, to be completed in both Toronto and Quebec City during 2005, was the same as that for Wave 2 except that the Stated Adaptation component was replaced by a new component that focused on a seven-day overview of activities that are routine in time. During the interview, additional attributes were sought for each routine activity: Quebec's more detailed routine activity form means that this part of their telephone interview was shorter.

On the critical issue of panel retention, much attention was given to keeping in touch with panel households, for example using newsletters and pre-wave motivational letters. Both Toronto and Quebec City achieved close to 84% retention between Wave 1 and Wave 2. A fuller description of the Quebec City Panel Survey methodology will be found on Lee-Gosselin (2004).

Quebec Low-Income Women's Project

Much research has focused on the differences in travel behaviour between women and men, and many studies agree that gender, household responsibilities, and life-cycle are significant factors that influence travel behaviour (Hanson, 1982a; Hanson, 1982b; Lee-Gosselin, 1989; Niemeier & Morita, 1996; Rosenbloom, 1995a; Rosenbloom, 1995b; Wilson, 2001). However, little attention has been given to how transportation and land-use patterns affect the activity choices of low-income women. Women, due to household and employment responsibilities, face high levels of space-time constraints (Kwan, 1999). These constraints inhibit their abilities to choose beneficial activities that affect the quality of life.

The Quebec low-income women's travel behaviour study (McCray, Brais, & James,) was a multi-method stand-alone study that complements the Quebec Panel Study. It was designed both to address, in detail, issues that are concerns for low income women, and thereby to meet one of the shortcomings discussed in the foregoing section: measuring needs and desires that do not get satisfied. It was also to test ways that the transportation disadvantaged could be effectively engaged in the research. This population is often invisible or underrepresented in travel behaviour studies that involve the use of a daily activity diary over multiple days. Part of the challenge in recruiting this population is that historically, low-income individuals have had a mistrust of government and university research. Key to developing a trust relationship is working directly with community leaders and showing early on, conceivable benefits to the community, in return for their input.

Thirty-four participants, with an average age of 40 years, were recruited from 5 women centres, 3 in the city and 2 in the suburbs. The centres are designed to address the needs of low-income women, the majority of which are unemployed. Twenty-one of our study participants were unemployed and 8 were students. Approximately 50% (N=17) of the women had children living with them in the household. Only 7 of the women had access to a car in the household. Of the 34 women participating in the focus group discussions, 8 volunteered to enter the Quebec Panel Study.

The value added in conducting this study goes beyond a simple recruitment tool. Using an innovative activity-based methodology, the travel patterns of study participants were analysed using focus-group discussions, map exercises, and questionnaires. This triangular approach allowed one to examine activities undertaken on a regular and occasional basis, to explore issues of personal safety dependent on the time of day, and to understand the challenges of using public transportation. Information was also gathered on places women would like to access, including leisure activities and discount shopping, but cannot for various reasons, such as using the bus. The focus group results were then confronted in the analysis with quantitative data on the travel patterns of women in vehicle-less households, derived in GIS using data from a regional Travel Survey.

The FG discussions served as a means of assessing accessibility needs related to health care, school, employment, shopping (grocery and malls), social activities (visiting family or friends), and entertainment. Women shared their experiences of using public transit and the challenges that exist if one has small children, or is disabled or elderly. Also during this time researchers gained an understanding of the extent to which the women were knowledgeable of the public transit system and available fare saver programs, such as the one-day pass that allows unlimited rides. These discussions often reveal ways in which women strategically chain trips on one day to get the most out of a transit day pass. The large group session ended with a general discussion of safety and security issues, dependent on the time of day, and was dealt with more on an individual basis during the map exercise.

The map exercise was designed to obtain information in three basic areas:

- places women visit on a *regular basis* and on an *occasional basis*;
- places women *would like to visit*, but cannot due to mobility constraints; and
- areas women feel *unsafe,* during the day, evening, or both.

Women were given maps of their community and surrounding areas to indicate their home location along with places they visit, which were color coded according to whether the place was accessed on a regular or occasional basis. To assist researchers in the analysis of data, a separate sheet of paper was used to list the names and addresses or cross streets of places of interest. Along with the labeling of place, respondents also indicated the mode used. When the mode of travel was the bus, the route number was indicated. Respondents and researchers immediately saw the relationship between distance and type of visit. Places visited on a regular basis were more often closer to the home location than those places visited on an occasional basis. The places women would have liked to visit, but could not due to mobility constraints, were indicated on maps with a blue color and were places of entertainment (movie theatres), recreation (ports, campsites, and beaches), discount shopping (warehouse stores and shopping centers), and intra-urban transit stations (train or bus). Mobility constraints included the bus not providing service to the place of interest, and bus scheduling conflicts including untimely bus transfers and limited service at certain times of day. In some cases women, through their social networks, were able to utilize a private vehicle as a passenger. Incorporating GIS mapping to verify bus routes and schedules, and to showing travel times from home locations to places of interest aided in validating accessibility challenges identified during the FG discussions. The GIS environment also supported the use of Quebec's 2001 O-D survey of 27,000 households to identify disadvantaged women, with limited transport options and from low job categories, to benchmark the results from the small detailed sample of FG participants. The O-D data provides the larger picture of identifying accessibility challenges among all women who are heads of households, alone or in a couple.

This multi-method approach allows policy makers to see in a meaningful way how transportation policies and land-use development limit the activity and action space of individuals who are primarily dependent on public transportation. Instead of focusing on the use of the roadway system, this methodology places a focus on the places to which women need access, and the challenges of reaching those activities when the scheduling and routing of the bus system are not supportive of their accessibility needs. The methodology also allows room for addressing some of the individual characteristics that impact a low-income woman's activity space, i.e. age, income, type of household, fear of travel, and attitudes towards public transit.

Drawbacks of the methodology include the time and organizational challenges of extracting and analysing meaningful information from the self mapping of activities, verifying activity locations, and placing the accessibility information in a GIS platform. Key to providing a larger context, not limited to the 34 women participating in the focus groups, is exploring creative ways of incorporating the larger dataset of 27,000 households. Developing this crucial qualitative/quantitative link has the potential of adding an invaluable dimension to the accessibility debate, helping demonstrate how perceptions shape the travel patterns and activity space of the socially excluded. A summary of results will be found in McCray and Brais, (2005).

Adding cyberspace to diaries of activities and travel

Arguably, "travel" in cyberspace is an increasingly serious omission from most data collection on spatio-temporal behaviour, but there are some recent developments to correct this. Kwan (2003), for instance, used a logging program installed on respondents' computers to directly record individuals' activities on the Internet as part of an activity-travel diary survey. This was a pilot study conducted in Columbus, Ohio (USA), in which data were collected from a sample of Internet users. In the study, an activity-Internet diary was developed and used to collect detailed data about the space-time characteristics of an individual's physical and cyberspatial activities. The data collected include the location, timing, duration and space-time fixity of an individual's Internet activities and physical activities over a 24 hour day. The active recording of these data was complemented by the automatic logging of respondents' Internet activities for the survey period.

Preliminary results suggest that both men and women expand their activity space and access to opportunities with the use of the Internet, mainly owing to the increased opportunities that the Internet provides for transactions (Kwan, 2003). However, women in the sample still face considerable space-time constraints associated with certain activities they need to perform in the physical world when compared to men. They seem to have a tendency to fit their Internet activities into their existing schedule instead of adopting fundamentally new ones. The data also indicate that women do not use the Internet primarily to help reduce their space-time constraints (e.g. through e-shopping). They instead use the Internet mainly for social and recreational activities that are not temporally fixed.

The method used in this pilot study seems to have some potential for studying the interaction between people's activity-travel behaviour and their use of ICTs. It is based on the abovementioned behavioural model of human cyberspatial cognition formulated by Kwan (2000, 2001). As argued in Kwan (2001), human behaviour in cyberspace bears certain similarities with spatial behaviour in the physical world. Behavioural theories about spatial learning, spatial cognition, as well as choice and decision-making behaviour, may be helpful for understanding the patterns of individual activities in cyberspace. For instance, many features of spatial learning and wayfinding have their equivalents in cyberspace (e.g. using bookmarks as landmarks, and building long-term memory using computer-assisted history mechanisms; Allen, 1999; Golledge, 1999). This in turn suggests that conventional models in transportation research may be extended along this line to provide the behavioural foundation for including human cyberspatial activities in the study of individual accessibility. Models that conceive the individual as an extensible agent and focus on the access interface between the individual and cyberspace will be particularly helpful. Using these models, fundamental notions in conventional accessibility measures - such as distance, impedance and attractiveness of opportunities - can be reformulated and applied to the study of individual accessibility to opportunities in both the physical world and cyberspace.

The results of the pilot study, however, reveal several difficulties specific to the use of Internet logging procedures. First, similar to the use of "active" GPS in recording people's activity locations, Internet logging methods assume that research subjects have sufficient computer skills to handle the installation of the logging program as well as to identify and save the correct log file. Different levels of computer knowledge of the subjects may render the method less reliable in certain situations. Second, there is serious concern about violation of personal privacy because the logging program keeps a complete record of one's Internet activities during the survey period. However, human subject protection procedures legally sanctioned by the government and implemented by the Institutional Review Board of research institutions (in the USA) may mitigate this concern to a certain extent. Third, in this pilot study, men were considerably under-represented (only 36% of the sample). This hints at the possibility that surveys involving Internet logging may incur selection bias in favour of female respondents. Lastly, although data are collected about people's Internet activities, how a person's activities in the physical world and in cyberspace interact can still be quite unclear. Qualitative information about the decision-making process and motivations behind people's Internet use should also be collected in future research.

Summary

The three case studies described in this section illustrate various multi-method approaches developed by the authors to address the complexity of accessibility. These approaches are more sensitive to individual differences than many traditional approaches. They are also more inclusive of groups and issues that are often ignored but are important to society (e.g. the transportation disadvantaged). It is apparent that we cannot understand everything with a single approach, be it traditional trip logging, monitoring of activity scheduling processes in space and time, or gaming-simulation, as useful as each approach may be. A holistic understanding of human activity-travel behaviour calls for a variety of innovative combinations of methods.

CONCLUSION

This paper is something of a cautionary tale. It says that to bring the broader issues of equity/accessibility into the latest travel behavioural modelling efforts, we need ways to incorporate multiple classes of microdata, some of which have hitherto been associated only with qualitative methods. The way we design microdata collection must recognise the needs of new computational opportunities (process models, spatio-temporal statistics in GIS, multi-agent microsimulation, etc.), but we need to work from the problems to the models rather than vice-versa. In a similar vein, supporting information technologies such as GPS and laptops with CASI software are transforming some survey methods, but the variable set should not default to whatever these can provide: these supports should be used in combination with other techniques, notably skilled in-depth interviewing, to keep our focus on the inclusiveness of the issues and people studied. That said, we must admit that we have proposed the collection of micro-data the full potential of which will require multidimensional analysis, modelling and decision-support methods that are also a work in progress. Some of these will come from fields that not exclusively concerned with transport, such as OLAP[9] developments in geomatics.

The modelling challenge in our field is about complexity. As Koskenoja and Pas (2002), put it, "...complexity deals with non-linear, nested structures, which lead to unexpected higher level behaviours". The recent focus on decision processes to understand such structures is

[9] OLAP: On-Line Analytical Processing

indispensable, but we need to make sure that the data collection includes respondent *perception* and *cognition*. Thus, it is very useful to efficiently track the sequencing, timing and grouping of activities, as does the instrument designed for Albatross (Arentze *et al.*, 2000). We must also ask people when and why they planned and changed activities, as does CHASE (Doherty and Miller, 2000). But we must also find ways to characterise how people perceive the limits of their own activity space/time (including activity cyberspace), and the limits of a feasible – albeit conditionally feasible – action space/time.

Finally, it remains central, as Heggie and Jones (1978) asserted long ago, to understand decisions about activities and travel in terms of the degree to which they are linked in either or both of the *spatio-temporal* and *interpersonal* dimensions. The three data collection case examples presented represent, of course, only a partial response to the data needs of contemporary models, but they are sensitive to both these dimensions. As we constantly strive to keep our methods in pace with evolving behaviour patterns, we believe that the personal and societal *meanings* of space, cyberspace and time, and of social networks, will need to be revisited often.

REFERENCES

Allen, G. L. (1999). "Spatial abilities, cognitive maps, and wayfinding: bases for individual differences in spatial cognition and behaviour". In: *Wayfinding Behaviour: Cognitive Mapping and Other Spatial Processes* (Ed R G Golledge), pp. 46-80. Johns Hopkins University Press, Baltimore.

Arentze, T., A. Borgers, D. Snellen and H. Timmermans (2000). "The activity diary". In: *Albatross, a Learning Based Transportation Oriented Simulation System* (Arentze T. and H. Timmermans). EIRASS, Technische Universiteit Eindhoven, Netherlands.

Axhausen, K. W. and T. Gärling (1992). Activity-based approaches to travel analysis: conceptual frameworks, models, and research problems. *Transport Reviews*, **12**(4), 323-341.

Axhausen, K.W, A. Zimmerman., S. Schönfelder, G. Rindfüser and T. Haupt (2002). Observing the rythms of daily life: A six-week travel diary. *Transportation*, **29**, 95-124.

Ben-Akiva, M. and S. R. Lerman (1985). Discrete Choice Analysis: Theory and Application to Travel Demand. MA: MIT Press.

Cervero, R. and K. M. Kockelman (1997). Travel demand and the three Ds: density, diversity and design. *Transportation Research Part D: Transportation and Environment*, **2**, 199-219.

Chapin, F. S. (1974). *Human Activity Patterns in the City: Things people do in time and in space.* New York: John Wiley and Sons, Inc.

Couclelis, H. (1999). From sustainable transportation to sustainable accessibility: Can we avoid a new "tragedy of the commons"? In: *Information, place, and cyberspace: Issues in accessibility* (D. Janelle and D. Hodge, eds.). Berlin: Springer-Verlag.

Crane, R. and R. Crepeau (1998). Does neighbourhood design influence travel?: A behavioural analysis of travel diary and GIS data. *Transportation Research Part D: Transportation and Environment*, **3**, 225-238.

Denmark, D. (1998). The Outsiders: Planning and Transport Disadvantage. *Journal of Planning Education and Research*, **17**(3), 231-245.

Dewey, J. and A.F. Bentley (1949). *Knowing and the Known,* Boston: Beacon.

Dijst, M. (2003). ICTs and accessibility: An action space perspective on the impact of new information and communication technologies. In: *Transport Developments and Innovations in an Evaluating World* (M. Beuthe, V. Himanen, and A. Reggiani, eds.).

Dijst, M. and V. Vidakovic (2000). Travel time ratio: The key factor of spatial reach. *Transportation,* **27**, 179-199.

Dijst, M., T. de Jong and J. R. van Eck (2002). Opportunities for transport mode change: An exploration of a disaggregated approach. *Environment and Planning B,* **29**, 413-430.

Doherty, S. T. and E. J. Miller (2000). A computerized household activity scheduling survey. *Transportation*, **27**(1), 75-97.

Forer, P. and O. Huisman (2000). Space, time and sequencing: substitution at the physical/virtual interface. In: *Information, Place, and Cyberspace: Issues in Accessibility* (D.G. Janelle and D. C. Hodge, eds.), pp. 73-90. Berlin: Springer.

Friedman, B., S. P. Gordon and J. B. Peers (1994). Effect of neotraditional neighbourhood design on travel characteristics. *Transportation Research Record*, (1466), 63-70.

Gärling, T., J. Saïsä, A. Böök and E. Lindberg (1986). The spatiotemporal sequencing of everyday activities in the large-scale environment. *Journal of Environmental Psychology*, **6**, 261-280.

Gärling, T. (1998). Behavioural Assumptions Overlooked in Travel-Choice Modelling. *Transport Modeling* (J. Ortuzar, S. Jara-Diaz and D. E. Hensher, eds.), pp. 3-18. Oxford: Pergamon.

Gärling, T., R. Gillholm and A. Gärling (1998). Reintroducing attitude theory in travel behaviour research: The validity of an interactive interview procedure to predict car use. *Transportation*, **25**, 129-146.

Golledge, R. G. (1999). "Human wayfinding and cognitive maps". In: *Wayfinding Behaviour: Cognitive Mapping and Other Spatial Processes* (R. G. Golledge, ed.), pp. 5-45. Johns Hopkins University Press, Baltimore.

Golledge, R. G., M.-P. Kwan and T. Gärling (1994). "Computational-process modeling of household travel decision using a geographical information system". *Papers in Regional Science,* **73**(2) 99-117.

Golob, Thomas F. and Amelia C. Regan (2001). Impact of information technology on personal travel and commercial vehicle operations: Research challenges and opportunities, *Transportation Research C*, **9**, 87-121.

Hägerstrand, T. (1970). What about people in regional science? The 9th European Congress of the Regional Science Association, 7-21.

Handy, S. L. (1996). A study of Austin neighborhoods. *Transportation Research Record*, 1552, 135-144.

Handy, S. L. and D. A. Niemeier (1997). Measuring Accessibility: an Exploration of Issues and Alternatives. *Environment and Planning A*, **29**(7), 1175-1194.

Hansen, W. G. (1959). How Accessibility Shapes Land Use. *Journal of the American Institute of Planners*, **25**, 73-76.

Hanson, S. (1982). The determinants of daily travel-activity patterns: relative location and sociodemographic factors. *Urban Geography*, **3**(3), 179-202.

Hanson, S. and P. Burnett (1981). Understanding complex travel behaviour: measurement issues. In: *New horizons in travel-behaviour research* (Stopher, P.R., A. Meyburg and W. Brög, eds.). D.C. Heath.

Hanson, S. and G. Pratt (1990). Geographic perspectives on the occupational segregation of women. *National Geographic Research,* **6**(4), 376-399.

Hanson, S. and G. Pratt (1995). *Gender, Work, and Space*. London: Routledge.

Harvey, A. S. and M. E. Taylor (2000). Activity settings and travel behaviour: A social contact perspective. *Transportation*, **27**, 53-73.

Holtzclaw, J. (1990). *Explaining urban density and transit impacts on auto use*. Presented by the Natural Resources Defense Council and the Sierra Club to the State of California Energy Resources Conservation and Development Commission.

Ingram, D. R. (1971). The concept of accessibility: a search for an operational form. *Regional Studies*, **5**, 101-107.

Jones, P.M. (1979). " 'HATS': A technique for investigating household decisions". *Environment and Planning A*., **11**(1), 59-70.

Jones, P. M., M. C. Dix, M. I. Clarke and I. G. Heggie (1983). *Understanding travel behaviour*. Gower: Aldershot.

Joud, F., C. Leclerc, and M. Lee-Gosselin (2003). Approche méthodologique sur les processus comportementaux de l'organisation spatio-temporelle des activités des ménages de la région de Québec, Congrès annuel, Association francophone pour le savoir, Rimouski, Québec.

Kim, H.-M. and M.-P. Kwan (2003). Space-time accessibility measures: A geocomputational algorithm with a focus on the feasible opportunity set and possible activity duration. *Journal of Geographical Systems,* **5**(1), 71-91.

Kitamura, R. and M. Kermanshah (1983). Identifying time and history dependencies of activity choice. *Transportation Research Record*, **944**, 22-30.

Kitamura, R., P. Mokhtarian and L. Laidet (1997). A micro-analysis of land use and travel in five neighbourhoods in the San Francisco Bay Area. *Transportation*, **24**, 125-158.

Koskenoja, P. and E.I. Pas (2002). Complexity and activity-based travel analysis and modelling. In: *In Perpetual Motion: Travel Behaviour Research Opportunities and Application Challenges* (H. Mahmassani, ed.). Elsevier, Oxford.

Kwan, M.-P. (1998). Space-time and integral measures of individual accessibility: a comparative analysis using a point-based framework. *Geographical Analysis*, **30**, 191-217.

Kwan, M.-P. (1999a). Gender and individual access to urban opportunities: a study using space-time measures. *The Professional Geographer,* **51**, 210-227.

Kwan, M.-P. (1999b). Gender, the home-work link, and space-time patterns of nonemployment activities. *Economic Geography,* **75**, 370-394.

Kwan, M.-P. (2001). Cyberspatial cognition and individual access to information: The behavioural foundation of cybergeography. *Environment and Planning B*, **28**(1), 21-37.

Kwan, M.-P. (2002). Time, information technologies and the geographies of everyday life. *Urban Geography*, **23**(5), 471-482.

Kwan, M.-P. (2003). Internet use, space-time constraints and human activity patterns: A time-geographic perspective. Paper presented at a workshop on "The Domestic Impact of ICT", Estes Park, Colorado, USA, June 5-8.

Lee-Gosselin, M. (1989). In-depth research on life style and household car use under future conditions in Canada. In: *Travel Behaviour Research* (International Association for Travel Behaviour, ed.), 102-118. Aldershot, UK: Gower.

Lee-Gosselin, M.E.H. (1996). Scope and potential of Interactive Stated Response data collection methods. *Conference on Household Travel Surveys: New Concepts and Research Needs.* Transportation Research Board, Conference Proceedings 10, National Academy Press, Washington DC.

Lee-Gosselin, M.E.H. and J.W. Polak (1997). Summary of workshop on data needs, *Activity-Based Travel Forecasting Conference,* New Orleans, Travel Model Improvement Program, US DOT/USDOE/EPA,173-182.

Lee-Gosselin, M.E.H and T. Turrentine (1997). Investigating past and future choice-sets of car-users, with special reference to perceptions of safety. 8th International Conference on Travel Behaviour Research, Austin, Texas, September.

Lee-Gosselin, M.E.H.(2003). You can get there from here: reflections on Stated Response survey quality and innovation. In: *Transport Survey Quality and Innovation* (Stopher. P.R and P.M. Jones, eds.). Elsevier, Oxford.

Lee-Gosselin, M.E.H. (2004). A data collection strategy for perceived and observed flexibility in the spatio-temporal organisation of household activities and associated travel. EIRASS Conference on Progress in Activity-Based Analysis, Maastricht, The Netherlands, 28-31 May.

Lenntorp, B. (1976). Paths in Space-Time Environments: A Time-Geographic Study of the Movement Possibilities of Individuals. Lund Studies in Geography B: Human Geography. Lund: Gleerup.

Lenntorp, B. (1978). A time-geographic simulation model of individual activity programmes. In: *Human Activity and Time Geography* (T. Carlstein, D. Parkes, and N. Thrift, eds.), pp. 162-180. London: Edward Arnold.

Miller, H. (1999). Measuring space-time accessibility benefits within transportation networks: basic theory and computational procedures. *Geographical Analysis*, **31**, 1-26.

McCray, T. M. (2000). Delivering Healthy Babies: Transportation and Healthcare Access. *Planning Practice and Research*, **15**(1/2), 17-29.

McCray, T. M. (2004). An issue of culture: the effects of daily activities on prenatal care utilization patterns in rural South Africa. *Social Science & Medicine*, **59**, 1843-1855.

McCray, T. M., N. Brais and P. James (2003). Analysing Travel Behaviour of Disadvantaged Women in Quebec City: A Multi-method Approach Involving Focus Groups, Map-making, and Benchmarking From OD Data. Paper presented at the *Association of American Geographers Annual Meeting*, New Orleans, LA, March 5-8.

McCray, T.M. and N. Blais (2005). "Filling the gap: developing activity-based methodologies to better understand disadvantaged women's travel patterns in Quebec City, Proceedings of Conference on Women's Issues in Transportation, Chicago, November.

McFadden, D. (1974). Conditional logit analysis of qualitative choice behaviour. In: *Frontiers in Econometrics* (P. Zarembka, ed.), pp. 105-142. New York, Academic Press.

Miller, E. J. (2005). Propositions for Modelling Household Decision-Making. *In this volume.*

Mohammadian, A. and E.J. Miller (2003). Dynamic Modeling of Household Automobile Transactions, presented at the *82nd Annual Meeting of the Transportation Research Board*, Washington, DC, January.

Mokhtarian, P. L. (1990). A typology of relationships between telecommunications and transportation. *Transportation Research A,* **24**, 231-242.

Mokhtarian, P. L. and R. Meenakshisundaram (1999). Beyond tele-substitution: disaggregate longitudinal structural equations modeling of communication impacts. *Transportation Research C,* **7**, 33-52.

Mokhtarian, P. L. (2003). Telecommunications and Travel: the Case for Complementarity. *Journal of Industrial Ecology*, **6**(2), 43-57.

Moore, M. T. (2003). Walk Can't Walk. *USA Today*, April 23, pp. 1-2.

Morris, J. M., P. L. Dumble and M. R. Wigan (1979). Accessibility indicators for transport planning. *Transportation Research A*, **13**, 91-109.

Niemeier, D. A. and J. G. Morita (1996). Duration of trip-making activities by men and women: a survival analysis. *Transportation*, **23**, 353-371.

Openshaw, S. and P.J. Taylor (1979). A Million or So Correlated Coefficients: Three Experiments on the Modifiable Areal Unit Problem. In: *Statistical Applications in the Spatial Sciences* (Wrigley, N. and R.J. Bennet, eds.). Pion, London.

O'Sullivan, D., A. Morrison and J. Shearer (2000). Using desktop GIS for the investigation of accessibility by public transport: An isochrone approach. *International Journal of Geographical Information Science,* **14**, 85-104.

Palm, R. (1981). Women in nonmetropolitan areas: A time-budget survey. *Environment and Planning A,* **13**(3), 373-78.

Pas, E. I. (1984). The Effect of Selected Sociodemographic Characteristics on Daily Travel Activity Behaviour. *Environment and Planning A*, **16**(5), 571-581.

Pendyala, R.M., R. Kitamura, A. Kikuchi, T. Yamamoto and S. Fujii (2004). FAMOS: The Florida Activity Mobility Simulator. 83rd Annual Meeting of the Transportation Research Board, Washington DC.

Pickup, L. (1984). Women's gender-role and its influence on their travel behaviour. *Built Environment*, **10**, 61-81.

Pickup, L. (1985). Women's travel need in a period of rising female employment. In: *Transportation and Mobility in an Era of Transition* (G.R.M. Jansen, P. Nijkamp and C.J. Ruijgrok, eds.), 97-113. Amsterdam, North Holland.

Pirie, G. H. (1979). Measuring accessibility: a review and a proposal. *Environment and Planning A*, **11**, 299-312.

Ramadier, T., M.E.H. Lee-Gosselin, F. Frenette and F. Joud (2005). Conceptual Perspectives for Explaining Spatio-Temporal Behaviour in Urban Areas. *In this volume*

Roorda, M., S. T. Doherty and E. Miller (2005). Operationalising household activity scheduling models: Addressing assumptions and the use of new sources of behavioural data. *In this volume*

Rosenbloom, S. and E. K. Burns (1993). Do Environmental Measures and Travel Reducation Programs Hurt Working Women? Final Report, U.S. Department of Labor, Washington, D.C.

Rosenbloom, S. (1995). *Travel by women.* Washington, D.C.: Federal Highway Administration, U.S. Department of Transportation.

Salomon, I. (1986). Telecommunications and travel relationships: A review. *Transportation Research A*, **20A**(3), 223-238.

Séguin, A.-M. and M. Thériault (2000). Residential trajectories of professional workers in the metropolitan area of Quebec City, *Congrès annuel de l'Association canadienne de sciences régionales*, Toronto, June.

Shen, Q. (1998). Spatial technologies, accessibility, and the social construction of urban space. *Computing, Environment, and Urban Systems*, **22**(5), 447-464.

Stead, D. (2001). Relationships between land use, socioeconomic factors, and travel patterns in Britain. *Environment and Planning B: Planning and Design*, **28**, 499-528.

Stopher, P. R. (1992). "Use of an activity-based diary to collect household travel data." Transportation, **19**(2), 159-176.

Stopher, P. R., O. Jiang and P. Bullock (2003). Making Sense of It All: Applications for Managing and Analysing In-Vehicle GPS Data. *Computer Aided Civil and Infrastructure Engineering*, in press.

Stopher, P. R. and A. H. Meyburg (1975). *Urban Transportation Modeling and Planning*, 125-173. Lexington, MA: Lexington Books.

Tivers, J. (1985). *Women Attached: The Daily Lives of Women with Young Children*. London: Croom Helm.

Vovsha, P., E. Peterson and R. Donneley (2004). Impact of Intra-Household Interactions on Individual Daily Activity Travel Patterns. 83rd Annual Meeting of the Transportation Research Board, Washington DC.

Weber, J. (2001). Evaluating the Effects of Context and Scale on Individual Accessibility: A Multilevel Approach. Ph.D. Dissertation, Department of Geography, Ohio State University.

Weber, J. (2003). Individual accessibility and distance from major employment centers: An examination using space-time measures. *Journal of Geographical Systems*, **5**(1), 51-70.

Weber, J. and M.-P. Kwan (2002). Bringing time back in: A study on the influence of travel time variations and facility opening hours on individual accessibility. *The Professional Geographer*, **54**, 226-240.

Weber J. and M.-P. Kwan (2003). Evaluating the effects of geographic contexts on individual accessibility: A multilevel approach. *Urban Geography*. Forthcoming.

Wen, C. H. and F. S. Koppelman (2000). A conceptual and methodological framework for the generation of activity-travel patterns. *Transportation*, **27**, 5-23.

Wilson, W. (1996). *When Work Disappears*. New York: Knopf.

Wilson, W. C. (2001). "Activity patterns of Canadian women: an application of ClustalG Sequence Alignment software". *Annual Meeting of the Transportation Research Board*, Washington DC, January, 28.

U.S. DOT (U.S. Department of Transportation) (1993). Transportation implications of telecommuting. Washington, DC: U.S. Department of Transportation.

ACKNOWLEDGEMENTS

This paper was primarily supported by the Social Sciences and Humanities Research Council of Canada, through a Major Collaborative Research Initiative, 2000-2005: *Access to activities and services in urban Canada: behavioural processes that condition equity and sustainability*, and GEOIDE, the Network of Centres of Excellence in Geomatics, with additional support from the Québec Ministry of Transport. The authors would like to thank the reviewers for their guidance, one of whom inspired a substantial and beneficial revision of the original paper, which had been presented at the 2003 IATBR Conference in Lucerne, Switzerland.

6

MODELLING LEARNING AND ADAPTATION IN ROUTE AND DEPARTURE TIME CHOICE BEHAVIOUR: ACHIEVEMENTS AND PROSPECTS

Apivat Jotisankasa and John W. Polak
Centre for Transport Studies, Imperial College London, UK

INTRODUCTION

During the past decade, there has been an increasing interest in developing integrated land-use and transport models for the evaluation of various transport policies. In order to develop such models, there is a need to understand the interaction between transport system, travel behaviour and land use.

Regarding the studies into the interaction between transport system and travel behaviour, a lot of attention has been given to the analysis of the dynamics of commuters' route and departure time choice decisions, as these decisions determine the major flow patterns in transport networks during peak periods. Of course, route and departure time choice are not the only choice dimensions available to commuters or the only ones which various transport policies are trying to influence. However, they constitute the short term behavioural responses to congestion and various transport policies, while the time frames for other choice dimensions i.e. mode shifts, and residence and job relocations tend to be of a longer term nature than these two dimensions.

During the past two decades, there have been two types of studies into the dynamics of commuter's route and departure time choice decisions. The first type of study focuses on the day-to-day dynamics of these decisions in interaction with the performance of the transport system, while the second focuses on the dynamics of these decisions under the provision of Advanced Traveller Information Systems (ATIS). This chapter presents an overview of the first type of study. As the development in this area is still in its infancy, the purpose of the chapter is two-fold. First, it is intended as a guidepost along the wayfor the subject that is still very much under active development. Second, based on the review, it identifies key issues and areas where there is a need for further research.

The rest of the chapter is organised as follows. In the next section, we review the existing approaches to modelling day-to-day dynamics of route and departure time choice decisions in interaction with the performance of the transport system. In the third section, we identify key issues and areas where there is a need for further research, and this is followed by a brief summary.

EXISTING MODELLING APPROACHES

A number of models have been proposed to represent the day-to-day dynamics of route and departure time choice behaviour. Two of the key mechanisms underlying these behavioural dynamics are the processes of learning and adaptation of travellers. In this review, we classify the existing models into two groups based on the underlying learning mechanisms in the models. The first group consists of the models which contain explicit updating mechanisms for the perception of the attributes of the network (i.e. travel time and travel time uncertainty); while the second group consists of the models which do not contain this type of updating mechanisms. This section is divided into three parts. The first and the second parts review the existing approaches used in the first and the second type of model, respectively. The third part provides a review of some additional learning mechanisms which can be added to both types of model.

Models with Updating Mechanisms for the Perception of Network Attributes

This type of model is comprised of two components, namely the learning component and the decision-making component. The learning component concerns travellers' perceptions of the attributes of the network, while the decision-making component concerns the strategies that travellers apply in their decision-making process. In this type of model, it is generally assumed that travellers will update their perception as they gain some experience in the network and that they then will make decisions based on this perception. Two of the most important attributes of the network in modelling route and departure time choice are travel time and travel time uncertainty. This section provides a review of the existing approaches to modelling the learning component including the mechanisms for updating the perceptions of travel time and of travel time uncertainty, and the existing approaches to modelling the decision-making component.

Mechanisms for updating travel time perception. The existing models can be classified into three groups, namely: the weighted average approach; the adaptive expectation approach; and the Bayesian approach.

a) Weighted average approach

In this approach, it is generally assumed that a traveller formulates the perception of travel time in the current time period based on weighted average of travel times in previous time periods. In the most basic form, the model can be expressed as

$$\hat{C}_r^{i,t} = \sum_{k=1}^{t-1} w_k(t-1)C_r^{i,k} \tag{1}$$

where $\hat{C}_r^{i,t}$ is the travel time on alternative r as perceived by individual i before making decision in time period t, $C_r^{i,k}$ is the actual travel time on alternative r in time period k, and $w_k(t-1)$ is a non-negative weight attached to time period k satisfying $\sum_{k=1}^{t-1} w_k(t-1)=1$.

A number of models have been proposed based on the above equation. However, they are different in terms of the assumptions regarding the travel times stored in the memory, the length of the memory, the function of weights, and the error component.

Horowitz (1984) proposed three learning models in which travellers are assumed to have unlimited memory. The first two models assume that travellers can acquire all the information about travel times from both the alternatives chosen and not chosen, while the third model assumes that travellers can acquire information only thought their own experiences. Cascetta (1989) relaxed the assumption regarding unlimited memory by assuming the travellers can possess only a finite number of travel times from the previous time periods. However, he still assumed that travellers can acquire the information about travel times from the alternatives not chosen as well. The uniform weights were assumed in the empirical study. Polak and Hazelton (1998) proposed the model similar to Cascetta (1989), but the weights are parameterised according to a geometrically declining function.

Recently, Nakayama *et al.* (1999) proposed a model in which the perceived travel time is updated from the weighted average of n recalled experienced travel times and the maximum and minimum expected travel times on each alternative. The assumption regarding the weights in this study is quite different from the ones previously mentioned. In this study, the range of the perceived travel time elements is divided equally into L intervals, and a weight is assigned to each interval. This is equal to assuming that the weights depend upon the values of experienced travel times. Thus, the model is capable of representing different attitudes towards risk among different groups of travellers. Risk-averse travellers may have weights that are larger in higher ranges of experienced travel times, while risk-prone travellers may have weights that are lower in higher ranges. Another form of model which can be used to capture the attitudes towards risk in travel time perception was proposed by Nakayama *et al.* (2001). It is simply assumed that travellers update their perception of travel time based on the weighted average of the mean experienced travel times in the past and the difference between the maximum and minimum experienced travel times in the past. Risk-averse travellers may attach a larger weight to the difference between the maximum and minimum experienced travel times in the past, while risk-prone travellers may attach a smaller weight. Two types of model in which travellers possess limited and unlimited length of memory were proposed.

More recently, Ettema *et al.* (2003a) proposed a model in which the perceived travel time is updated based on the weighted average of the experienced travel times in the same class. It is assumed that travellers will first classify the experienced travel times from the previous time periods into mental classes. For each class, the experienced travel times will be ordered with respect to their ages and then assigned a rank order. The perceived travel time for each mental class is then updated based on the weighted average of the experienced travel times in that class where the weight is a decay function of the rank of the travel time in the class. Ettema *et al.* (2003b) extended the above model by assuming that the weight is not only a function of the rank of the travel time in the class but also a function of the deviation between that travel time and the perceived travel time for that class. This model is also capable of representing the attitude towards risk among group of different travellers.

While this type of model is applied widely, it is surprising to note that the empirical investigation of the model especially in the context of weight set is very limited. This is probably due to the complexity of the analysis. Another weakness of the model was pointed out by Iida *et al.* (1992). They argued that this type of model may not necessarily reflect the actual traffic phenomena when the choice is made repeatedly because other factors such as a magnitude of the discrepancy between the actual and the perceived travel times and the regret over past choices seem to be important in the actual choice behaviour.

b) Adaptive expectation approach

This type of model is probably the most frequently encountered in the literature. It is assumed that a traveller updates their perception of travel time on the basis of the magnitude of the discrepancy between the actual and the perceived travel times from the previous day. In the most basic form, the model can be expressed as

$$\hat{C}_r^{i,t} = \tau C_r^{i,t-1} + (1-\tau)\hat{C}_r^{i,t-1} \qquad (2)$$

where τ is the learning parameter and $0 \leq \tau \leq 1$.

Note that in the most basic form, this model is equivalent to maintaining a memory of the exponentially weighted moving average of the past experiences or, in other words, the weighted average model where $w_k(t-1) = \tau(1-\tau)^{t-k-1}$.

A number of studies have been using this model for the travel time perception updating process; see for example, Cascetta and Cantarella (1991), Ben-Akiva *et al.* (1991), Dimitropoulos (1993), Van der Mede and Van Berkum (1993), Vaughn *et al.* (1993), Dimitrakopoulou (1994), Emmerink *et al.* (1995), Axhausen *et al.* (1995), Van Berkum and Van der Mede (1998), Oladeinde (2000) and Polak and Oladeinde (2000). Iida *et al.* (1992) extended the above model by including a constant or the learning bias into the model. The adaptive expectation model was also used to model queue length perception updating in Van Berkum and Van der Mede (1998).

The popularity of this type of model is based on its computational and conceptual simplicity; and on the plausible psychological interpretation of the model (Axhausen *et al.*, 1995; Polak and Oladeinde, 2000). Unlike other type of weighted average models, many empirical analyses of the parameters of the model were reported; see for example, Iida *et al.* (1992), Dimitropoulos (1993), Van der Mede and Van Berkum (1993), Vaughn *et al.* (1993), Dimitrakopoulou (1994), Axhausen *et al.* (1995), Van Berkum and Van der Mede (1998), Oladeinde (2000) and Polak and Oladeinde (2000). One of the disadvantages of this model is that the value of the perceived travel time is bounded by the values of the last experience and the prior perception; thus, the travellers cannot use their experiences on the previous days to help estimate the future travel conditions (Axhausen *et al.*, 1995).

c) Bayesian approach

An alternative approach to modelling travel time perception updating was proposed by Jha *et al.* (1998) and Chen and Mahmassani (2004). They developed a Bayesian updating model to describe how travellers update their perception of travel time on the basis of their previous perception and their experience. In the proposed framework, the mean travel time and the experienced travel time as perceived by each traveller are represented by random variables, whose variances are indicator of traveller's confidence of the source of information. Jha *et al.* (1998) proposed a model in which travellers are assumed to update their perception every day, while Chen and Mahmassani (2004) assumed that travellers may not update their perception every day and proposed three different mechanisms for triggering and terminating the updating process.

It is of interest to note some similarity between the adaptive expectation approach and the Bayesian approach where travellers are assumed to update their perception everyday, and both the prior distribution of the mean travel time and the distribution of the experienced travel time as perceived by each traveller are assumed to be normal, and the variance of travel time is assumed to be known as in Jha *et al.* (1998). If we define the initial variance of the prior distribution of the mean perceived travel time as α and assume that the variance of the experienced travel time as perceived by each traveller is fixed at β every day (i.e the confidence of experience is equal every day), the Bayesian updating formula for the mean of the posterior distribution of the mean perceived travel time can be written in a form of an adaptive expectation model where the weight is a function of α , β, and time as follows:

$$\hat{C}_r^{i,t} = \tau_t C_r^{i,t-1} + (1-\tau_t)\hat{C}_r^{i,t-1} \tag{3}$$

where $\hat{C}_r^{i,t}$ and $\hat{C}_r^{i,t-1}$, in this case, are the means of the posterior and the prior distributions of the mean perceived travel time respectively, and

$$\tau_t = \frac{\alpha}{(t-1)\alpha + \beta} \tag{4}$$

Note that even though this approach can treat variance (or the level of confidence) whereas the weighted average approach and adaptive expectation approach cannot, it also has a number of limitations. First, the variance of the mean perceived travel time always decreases from one day to the next. Second, the underlying travel time is assumed to have a steady mean value. As a result, the model will not be applicable if there is any drastic change in the network. Moreover, some psychological experiments (e.g. Edwards, 1982) have shown that human prediction is consistently different from the one based on the Bayesian updating model.

Mechanisms for updating the perception of travel time uncertainty. The travel time uncertainty is generally expressed as the variance of travel time. To the best of our knowledge, there have been only three studies in modelling the mechanisms for updating the perception of the variance of travel times. They are all based on the weighted variance approach. Ettema *et al.* (2003a) and Ettema *et al.* (2003b) assumed the same weights as they used for the travel time perception updating. For the reason of computational convenience, Van Ber Kum and Van Der Mede (1998) proposed an adaptive variance expectation model which is an approximation of the exponentially weighted variance.

It is of interest to note that while it is possible to apply the Bayesian updating approach to variance perception updating, we are not aware of any study using this approach to model variance perception updating.

Mechanisms for modelling route and departure time choice. Previous approaches to modelling route and departure time choice can be classified into four groups, namely: the utility maximisation approach; the expected utility maximisation approach; the satisficing approach; and the utility-based satisficing approach.

a) Utility maximisation

A number of choice models have been proposed based on the utility maximisation. It is assumed that the decision-maker's preference of an alternative is captured by a utility and the decision-maker chooses the alternative with the highest utility. For the route choice model, Nakayama *et al.* (1999) and Nakayama *et al.* (2001) simply assumed that travellers will select their route based on the minimum perceived travel time. Dimitropoulos (1993), Van der Mede and Van Berkum (1993), Dimitrakopoulou (1994), and Van Berkum and Van der Mede (1998) proposed route choice models based on the logit models, while Cascetta (1989) and Polak and Hazelton (1998) proposed route choice models based on the probit specification.

For the route and departure time choice model, Jha *et al.* (1998) proposed a nested logit model where the decision at the higher level of the hierarchy is the departure time choice and the decision at the lower level of the hierarchy is the route choice.

While this type of model is applied widely, it has been argued that choices made repeatedly over time, especially commuters' route and departure time choices, may not require the high level of cognitive process, e.g. utility maximization, all the time (Van Berkum and Ven der Mede, 1998). In order to account for the habitual effect, Van Berkum and Van der Mede (1998) proposed a logit model that includes an attribute representing the habitual choice behaviour. It is assumed that after a certain route is chosen it becomes more likely that the same route will be chosen again, and the likelihood that a route is chosen depends on the number of times it was chosen previously. The perception of the variance of travel time and the perception of queue length were also included in their utility function. An alternative model which account for the habitual effect was also proposed by Polak and Hazelton (1998). It is simply assumed that a proportion α of travellers will consider re-routing and the rest of the travellers will continue using the same routes as those in the previous time period.

b) Expected utility maximisation

In the expected utility approach, it is assumed that travellers account for uncertainty in travel time by weighting each outcome by its probability. Thus, the travellers must be assumed to have a perception of the probability of travel time (through the mean and variance of travel time) rather than only a perception of the travel time. This approach has been used by Ettema *et al.* (2003a) and Ettema *et al.* (2003b) to model departure time choice under uncertainty.

It is of interest to note that while expected utility has many attractions and the resulting model is attractively simple, it was argued that the actual choice will depend not only on the perception of the probability of travel time but also the attitude towards risk (Bonsall, 2003). If travellers are risk-seeking, they would probably exaggerate the probabilities of short travel times. On the other hand, if they are risk-averse, they would probably put more weight on the probabilities of long travel times. Prospect Theory (Kahneman and Tversky, 1979), and more recently, Cumulative Prospect Theory (Tversky and Kahneman, 1992) can be used in this context as it allows for a subjective weighting of perceived probabilities reflecting the decision makers' attitude towards risk (Bonsall, 2003).

c) Satisficing approach

In this approach, it is assumed that travellers make decisions about route and departure time choice based on the concept of bounded rationality in which they will maintain their choices from the previous day as long as the outcomes do not exceed some threshold values. If the previous outcomes are considered unacceptable, they will adjust the previous decisions through some mechanism. These threshold values or the indifference bands can also vary from day-to-day depending on the travellers' experiences. Jou and Mahmassani (1998) proposed joint models for indifference bands for both departure time switching and route switching. The joint indifference bands for departure time switching is a function of initial bands, age, gender, number of unacceptable early and late arrival, and a ratio of the difference between travel times and the amount of departure time that traveller adjusted between the last two days. The joint indifference bands for route switching is a function of initial bands, standard deviation of travel times and the number of unacceptable early and late arrivals. Mahmassani and Jou (2000) proposed slightly different models where (a) the joint indifference bands for departure time switching are a function of initial bands, a ratio of the difference between travel times and the amount of departure time that traveller adjusted between the last two days, and (b) the standard deviation of travel times, and the joint indifference bands for route switching is a function of initial bands, and standard deviation of travel times. The models for the indifference bands which incorporate trip chaining behaviour were also proposed by Mahmassani and Jou (1998).

Regarding the mechanisms for departure time adjustment, Mahmassani and Chang (1986) proposed two models for this behavioural adjustment process. The first model or the myopic adjustment model assumes that travellers will adjust their departure time based on the experience travel time and the actual schedule delay from the previous day. The second model or the learning model assumes that travellers will adjust their departure time based on the weighted average of the travel time experienced in the past. Chang and Mahmassani (1988) and Mahmassani (1990) proposed a model in which travellers update their departure times on the basis of the experienced travel times on the latest and immediately preceding day, schedule delay, and the number of unacceptable arrivals in the previous time period. Note that in all models, the travellers will adjust their departure times if the arrival times exceed the indifference bands.

Another type of departure time adjustment models have been proposed by Oladeinde (2000) and Polak and Oladeinde (2000). They assumed that travellers will adjust their departure time based on an adaptive expectation model with no assumption regarding the indifference bands.

d) Utility-based satisficing approach

A number of models which combine the theory of utility maximisation and the satisficing approach have been proposed. Emmerink *et al.* (1995) proposed a model in which travellers will choose the current route if the difference between the experienced travel time and the perceived travel time before the trip was made does not exceed some threshold value. If that difference exceeds the threshold value, the travellers will switch to the best route. Chen and Mahmassani (2004) proposed a model in which travellers will choose the current route if the difference between the travel time in the current route and the travel time in the best route does not exceed some threshold value. If that difference exceeds the threshold value, the travellers will switch to the best route.

When the departure time dimension is included in the analysis, Cascetta and Cantarella (1991) proposed a mixed bounded rationality random utility model which assumes that travellers will choose the current route and departure time if the actual observed utility is within a prefixed percentual range of the perceived observed utility before the trip was made. If it is not, the travellers will select their route and departure time choice based on a nested approach where the decision at the higher level of the hierarchy is the route choice, and the decision at the lower level of the hierarchy is the departure time choice, a process that fits a nested logit model.

Rossetti *et al.* (2002) proposed a model in which travellers are assumed to use the current route if the difference between the travel time in the current route and the travel time in the best route does not exceed some threshold value. If that difference exceeds the threshold value, the travellers will switch to the best route. It is also assumed that travellers will use the current departure times as long as their arrival times do not exceed some threshold value for lateness. If their arrival times exceed the threshold value, they will simply adjust their departure times by the difference between the actual arrival time and the acceptable arrival time from the previous day.

Hu and Mamassani (1995; 1997) assumed that travellers will maintain their choices from the previous day as long as the outcomes do not exceed the indifference bands. If the outcomes exceed, they will adjust their route and departure time choices in a manner that can be represented by a joint logit model.

Models without Updating Mechanisms for the Perception of Network Attributes

The previous section presented the models which contain the updating mechanisms for the perception of the network attributes. In this section, the models which do not contain such mechanisms are presented. These models can be classified into two groups; namely the models that have been developed based on the theory of Inductive Learning, and the models that have been developed based on the theory of Machine Learning.

Inductive Learning. The approaches mentioned in the previous section are based on the deductive learning. Recently, Nakayama and Kitamura (2000) proposed a model in which drivers are assumed to reason and learn inductively based on the theory of cognitive psychology. The model system is basically a production system, which is a compilation of if-then rule in which the rules are revised by using a genetic algorithm. It is assumed that travellers remember the routes and their experienced travel times for *m* days. Before making trips, they will match their memory with the if-then rules they have, and select a route based on the rule that is activated. If there is more than one rule activated, the rule with the lowest inferiority value will be chosen. Then, the rules and their inferiority value will be updated based on the weighted average of the previous inferiority, and the experienced travel time. The learning process is represented by assuming that the travellers will update their set of the if-then rules using principles like those in genetic algorithms.

While this approach seems to be solidly supported by research in cognitive psychology, it was argued that it is difficult to observe how routes are chosen in this model because there are numerous combinations of the if-then rules (Nakayama *et al.*, 2001).

Machine Learning. This section presents models that were developed based on machine learning theory. In this approach, it is generally assumed that travellers will learn about the outcomes of their actions through rewards, and thus there is no learning about the attributes of the network.

Ozbay *et al.* (2001) developed a route choice model using stochastic learning automata theory. The route choice probabilities are updated on the basis of the experience of travellers. It is assumed that once the travellers selected their route choice, and a favourable output results, the probability of choosing that particular route for the next time will be increased and the probabilities of all other routes will be decreased. On the other hand, if an unfavourable output results, the probability of choosing that particular route for the next time will be decreased and the probabilities of all other routs will be increased. This mechanism is also called a Linear Reward-Penalty reinforcement scheme. In their subsequent work (Ozbay *et al.*, 2002), the departure time choice dimension was also included in the analysis.

Klügl and Bazzan (2003) proposed a model in which travellers are assumed to update the probability of choosing a certain route periodically according to the cumulative rewards obtained from the route they have chosen.

Arentze and Timmermans (2003) developed a framework for modelling dynamic choice based on the theory of reinforcement learning. The proposed framework consists of three components namely: the reward function, the incremental action value function, and the action selection method. The reward function represents feedback on action; the action value function models how travellers integrate rewards received in the past to assess the value of the action; and the action selection method defines a choice of action given the action values. This framework was applied to modelling repetitive choices of destination, mode, and route for conducting a given activity under varying conditions of available time, time of day, and road congestion. If the rewards are travel times, it is of interest to note the similarity between this model and the models with explicit updating mechanisms for the perception of the attributes of the network; the incremental action value function and the action selection method follow any of the travel time perception updating models, and the choice models, respectively.

Additional Components in Learning

This section presents a review of other existing approaches which have been used as an additional learning component in both the models with and without explicit updating mechanisms for the perception of the attributes of the network.

Evolutionary methods. Genetic algorithms (GAs) are a part of evolutionary computing. The algorithms are inspired by Darwin's theory of evolution. In transport applications, GAs have been used mainly in the area of optimisation. Recently, a number of studies have applied these algorithms to the area of learning and adaptation. As noted above, Nakayama *et al.* (1999) applied GAs to the weighted average approach by assuming that travellers will update their weight set based on these algorithms. Nakayama and Kitamura (2000) applied GAs to the inductive learning approach by assuming that travellers will update their if-then rules on the basis of these algorithms. Nakayama *et al.* (2001) proposed a two-stage route choice model in which travellers are assumed to choose a route choice rule in the first stage, and then the rule will instruct the travellers which route to take in the second stage. There are four rules available to the travellers, namely no switching, random switching, experience-based, and rational rules. In experience-based and rational rules, it is assumed that the travellers update their perception of travel time based on the weighted average of the mean experienced travel times in the past and the difference between the maximum and minimum experienced travel times in the past. Both rules are different in terms of the length of traveller memory. The experience-based rule assumes that travellers possess a limited number of recalled travel times, whereas the rational rule assumes that the travellers possess all the information. In this study, the rules, the length of the memory (for the experienced-based rule), and the weight (for both the experienced-based rule and the rational rule) are updated using GAs.

The advantage of this approach is that there is no need to specify the values of the parameters or rules in the models beforehand, as the GAs will find the optimum values for those parameters or rules automatically. However, while this approach seems to be very convenient from an operational point of view, it is doubtful whether the real learning behaviour follows the above mechanism.

Decision Tree Induction Methods. Another approach which has been used as an additional component in learning is the decision tree induction methods. In this approach, it is assumed that the learning process also depends on a set of conditions. However, given the limitation in the capability of travellers to process and store information, travellers will not process and store the information for all possible conditions, but will only distinguish between condition states which are significantly different in terms of the outcome of the event. This process can be modelled using CHAID algorithms. The algorithms use a Chi-square based measure of significance of difference in distributions between groups as a criterion. Arentze and Timmermans (2003) applied these algorithms to the reinforcement learning approach to discriminate between contexts (i.e. route, destination, mode, time-of-day, and available time for conducting an activity) which tend to produce different rewards. Ettema *et al.* (2003a) and Ettema *et al.* (2003b) applied these algorithms to the weighted average approach to discriminate between departure times that tend to produce different travel times.

Route choice learning. Another approach which has been used as an additional component in learning is the route choice learning. An attempt to include route choice learning mechanisms into a large scale multi-agent simulation has been conducted by Nagel and his colleagues (Nagel and Raney, 2003; Raney and Nagel, 2003; Raney *et al.*, 2003; Raney and Nagel, 2002). In those studies, the system begins with an initial set of routes, one per agent (i.e. traveller), based on the free-flow travel time. For each agent, this initial route will be stored in the agent database which gives an agent a memory of his/her past strategies (i.e. routes chosen) and the outcome of those strategies (i.e. travel time). Once the traffic simulation is executed, the outcome of the initial strategy (i.e. travel time) will be stored in the database. Then the new strategy (new route) based on the updated link travel time from the last simulation will be added to the database of 10% of the population. Those who received the new strategy will be forced to choose the new route, while those who did not receive the new strategy will make a choice from the previously tried routes from the database. The latter is represented by using a multinomial logit model in which the utility for any path is a function of memorized travel time from the database. Once the simulation is executed, the previous outcome is replaced by the new one. This cycle is then run for 50 times.

The above mechanism is interesting in the way that it allows the travellers to update the route choice set as the system evolves, rather than to specify the set of alternatives from the beginning. As a result, each traveller will have a different set of alternatives. However, while this type of model seems to be very convenient from an operational point of view, it is not very promising from a behavioural point of view. Further discussion regarding this issue is presented in the next section.

KEY MODELLING ISSUES AND GAPS FOR FURTHER RESEARCH

In this section, we identify key issues in modelling day-to-day dynamics of route and departure time choice decisions and make some suggestions regarding the areas for further research. The key issues and gaps presented here are mainly based on the review of the models with explicit updating mechanisms for the perception of the attributes of the network, since this type of model can be calibrated and validated using available advanced statistical and econometric techniques. Six important issues which have to be taken into account when developing the models for these behavioural dynamics are listed as follows:

- Learning component
- Choice modelling
- Habitual choice
- Choice set formation
- Data collection methods
- Evolution and convergence of the systems

Learning Component

One of the key mechanisms underlying the day-to-day dynamics of route and departure time choice behaviour is the process by which travellers gradually learn about network conditions as a result of repeated day-to-day experiences. According to the previous section, a number of different models have been proposed; however, there is still a large opportunity to further elaborate these models. First, while travel time and travel time uncertainty are two of the key attributes of the network which travellers normally learn, most of the studies in this area have been focusing only on the travel time perception updating process. The updating process for the perception of travel time uncertainty should also be included in the modelling framework.

Second, in the real world, travellers may not only learn about the travel time, but also the conditions which lead to the outcomes as well. As a result, they may have different sets of perceived travel times for each condition; for example, the travel time in case of bad weather, the travel time during the school break, or the travel time when there is a major accident on the network. However, as mentioned earlier, given the limitation in the capability of travellers to process and store information, the travellers may not process and store the information for all possible conditions, but will only distinguish between condition states which are significantly different in terms of the outcome of the event. Hence, the mechanism which distinguishes between the outcomes from different conditions should be included in the framework as well. Moreover, once they have different sets of travel time perception in their memory, it is likely that before they make decisions about route and departure time choice a particular day, they would probably first make some predictions about the condition which might occur on that day, and then choose the travel time on the basis of those predictions. This mechanism should be included in the framework as well.

Third, the traveller's attitude towards risk also needs to be taken into account when developing the learning model. In the case of the travel time perception updating models, this could be accommodated through the heterogeneity in the weight sets.

Fourth, after the travellers have used the network for a while, it is likely that the new experiences may not have as much effect on the learning process as those from when they first used the network. This is supported by the results from the previous studies (e.g. Van Berkum and Van der Mede, 1998; Oladeinde, 2000) which reported that the learning parameter in the adaptive expectation model estimated using the data from the travel diaries (where the participants have been travelling on the network for a long time) is much smaller than the one estimated using the data from the experiments (where the participants are new to the network). This again could be accommodated in the existing models through the dynamic weight sets. In addition, the mechanism which triggers or terminates the learning process could also be included in the modelling framework, as the travellers may not learn or update their perception everyday.

Fifth, while most of the existing travel time perception updating models simply assume that travellers will update only the perception of the alternative previously chosen, it is likely that they may update their perception of related alternatives as well; for example, they may update their perception of an overlapping route, or on the adjacent departure time intervals.

Last but not least, more empirical studies are needed, as most of the studies seem to be focusing on the choice behaviour, but not the learning behaviour. Designing the travel diary or the experiment to collect the data necessary for estimating the models is also another challenging area for further research. This is discussed in more detail in the section regarding data collection methods.

Choice Modelling

Another key mechanism underlying these day-to-day behavioural dynamics is the adaptation process. In this section, the general issues regarding the route and departure time choice modelling are presented. The issues regarding the habitual choice are presented in the next section.

Discrete choice models have been widely used in the analysis of travel behaviour, including the analysis of route and departure time choice. According to the previous section, most of the studies in the route choice analysis are typically based on the multinomial logit (MNL) models. This type of model has been very popular due to its simplicity and tractability; however, it imposes restrictions on the covariance structure. In the route choice analysis where there are highly overlapping routes, this could be a problem. Recently, the modifications of MNL models were proposed to overcome this problem; see for example, C-Logit Model (Cascetta *et al.*, 1996), Path-Size Logit Model (Ben-Akiva and Bierlaire, 1999), Link-Nested Logit Model (Vovsha and Bekhor, 1998), and Mixed Logit Model (Bekhor *et al.*, 2002). For a comprehensive review on discrete choice models, see for example, Train (2003) and Ben-Akiva and Bierlaire (1999).

In the departure time choice analysis, an important issue regarding the use of discrete choice models choice is that the inherently continuous departure time choice needs to be transformed into a set of mutually exclusive departure time intervals. The size of these intervals will affect the magnitude of the residual correlations among the choice alternatives. These correlations cannot be ignored especially when the time intervals are short. Small (1987) proposed an Ordered Generalised Extreme Value (OGEV) models which captures the correlation among adjacent time intervals.

When route and departure time choices are modelled together, the correlation among alternatives is even more complicated. Moreover, in a dynamic route and departure time choice model, there will be some correlation between days as well. In this case, a Mixed Logit model or Mixed Generalised Extreme Value model may be more appropriate.

Habitual Choice

As mentioned earlier, it has been argued that choices made repeatedly over time, especially commuters' route and departure time choices, may not require the high level of cognitive process, e.g. utility maximization, all the time (Van Berkum and Ven der Mede, 1998). According to the previous section, several approaches have been proposed to account for this habitual behaviour. Van Berkum and Van der Mede (1998) proposed a logit model which includes an attribute representing the habitual choice behaviour. It is assumed that after a certain route is chosen it becomes more likely that the same route will be chosen again and the likelihood that a route is chosen depends on the number of times it was chosen previously. This approach is quite simple and straightforward; however, the intrinsic correlation mentioned earlier is not captured by such a model.

Another approach is the utility-based satisficing approach which is presented in the previous section. This approach consists of two stages; the first stage being the choice whether to switch, and the second stage being the choice of route and departure time. While this type of model seems to be more appealing from the behavioural point of view, splitting the models into two separate stages can cause selectivity biases in model estimation.

An alternative approach is to combine those two stages in a nesting structure. An example of the nesting structure can be found in Srinivasan and Mahmassani (2001) which investigated alternative structural mechanisms in commuters' day-to-day departure time adjustment behaviour. The mechanisms considered include utility maximisation from unordered alternatives; ordinal responses mechanism; sequential greedy search mechanism; and a two-stage nested adjustment process where the choice alternatives are zero adjustment, 1-5, 6-10, 11-15, and >15 minute adjustment. The advantage of using a nesting structure is that it avoids the selectivity issue by allowing for a simultaneous estimation of the first and the second stage decisions.

In order to include the habitual choice behaviour in the combined route and departure time choice model, the nesting structure is even more complicated, as there will be another dimension of the nesting structure where the alternatives for not switching, switching only route, switching only departure time, and switching both route and departure time, belong to different nests. This nesting structure is also dynamic as it depends upon the alternative previously chosen. More empirical studies need to be conducted in order to quantify the appropriate modelling structure.

Choice Set Formation

Another important issue which has to be taken into account when developing the models using discrete choice models is choice set generation. In the context of route choice set generation, different approaches have been proposed in the literature (see for example a review by Cascetta *et al.*, 2002). These approaches can be classified into two categories, namely the exhaustive approach and the selective approach. In the exhaustive approach, all the loop-less paths are included in the route choice set, while in the selective approach, only a subset of the loop-less paths is included.

The models in the selective approach can be grouped into two types of models, namely the selective implicit enumeration models, and the selective explicit enumeration models. In the selective implicit enumeration models, the route choice set can be obtained through some heuristic rules associated with an algorithm that performs the choice set generation and the choice fraction computation processes at the same time (e.g. Dial, 1971), while in the selective explicit enumeration models, the route choice set is generated explicitly. Note that the exhaustive approach and the implicit selective approach can produce an unrealistic choice set, and cannot be calibrated.

The selective explicit enumeration models can be further grouped into two types of models, namely the rule-based choice set generation models, and the probabilistic choice set generation models. In the rule-based choice set generation models, the paths in the choice set can be obtained as those satisfying some rules calibrated by maximising measures of goodness of fit with respect to the sample of observations related to actual choices. At the individual level, the measures of goodness of fit could be the coverage of routes actually chosen (e.g. Ben-Akiva *et al.*, 1984; Russo and Vitetta, 1996; Bekhor *et al.*, 2002), while at the network level, these could be the sum square errors of flows counted on the network (e.g. Cascetta *et al.*, 1997). In the probabilistic choice set generation models, the probability of including a given route in the user's choice set is modelled explicitly. Cascetta *et al.* (2002) proposed a two-stage model in which, in the first stage, a set of feasible routes containing all perceived routes, is generated using a coverage factor maximising processes; in the second stage, the probability that a path belongs to the choice set is modelled as a binomial logit. Note that the estimates of this type of model also require information regarding the routes the users consider to be available to them.

While the probabilistic choice set generation models seems to be theoretically more appealing, the application of this type of model to the route choice analysis (for example in the form of latent choice class models, Ben-Akiva and Boccara, 1995) can cause computational complexity as the number of feasible choice sets increases exponentially with the number of alternatives. Recently, Castetta and Papola (2001) have proposed implicit availability/perception random utility (IAPRU) models in which the intermediate perception degrees of alternatives are included in the utility function. This type of model is computationally more convenient than explicit probabilistic choice set generation models. While this type of model has been developed for the static case, it can be extended to the dynamic case by defining the intermediate perception degrees of the alternatives as a function of past experiences on the networks. This provides an opportunity to model another aspect of learning process in which people can update their choice sets as they gain some more experiences on the networks.

In the context of departure time choice, the choice set specification is more complicated as it consists of two tasks; defining an acceptable range of departure time intervals considered by each individual, and defining the size of the departure time intervals. Both of these depend very much upon the trip purpose. For example, for a commuting trip where people are required to be at their workplace at a particular time, they do not have as many choices as they do for a shopping trip. Thus, the range and the size of departure time intervals for the commuting trip are likely to be smaller than those for the shopping trip. With regard to the choice set for a particular trip propose, most studies into the analysis of departure time choice assume that all the decision-makers have the same choice set. This assumption is quite unrealistic and can cause significant misspecification problems. Recently, Ben-Akiva and Bierlaire (1999) proposed a common procedure which is based on the preferred arrival time for defining an acceptable range of departure time intervals considered by each individual. The interval of acceptable departure times is calculated by subtracting the maximum and minimum travel times from the feasible arrival time interval. While the procedure is quite simple, it results in a "crisp" choice set. In addition, it does not allow the decision-makers to update their choice sets as they gain some more experiences on the network. In this regard, the dynamic version of

IAPRU models as mentioned earlier could be of better use. However, more empirical studies still need to be conducted.

Regarding the size of the departure time intervals, the choice of the size of the intervals can have significant impacts on both model performance and accuracy. While the use of intervals that are too large results in excessive aggregation of data, the use of intervals that are too small can lead to model complexity. In addition, the size of the intervals will also affect the magnitude of the residual correlations among the choice alternatives, and different discretisation strategies can potentially lead to different optimal choices of model structure. Given the importance of these effects, it is surprising that little effort has been devoted to quantifying the effect of different levels of temporal aggregation. Thus, this is another area that needs further investigation.

Data Collection Methods

The methodologies which have been used to collect the data in the previous studies can be grouped into two categories; namely field-survey approach and experimental approach.

Field survey approach. In the field-survey approach, the data of the behaviour of each individual is obtained through an elaborate multiday-travel diary. Some examples of the studies using this approach can be found in Jou and Mahmassani (1996), Oladeinde (2000), Van Berkum and Van der Mede (1998), and Fujii and Kitamura (2000). This approach has an advantage in that it provides the observations of the actual behaviour in a real network.

However, there are some disadvantages as follows:

- the data regarding the prevailing conditions on the network cannot be obtained directly from the survey as the participants can only report the traffic conditions on their chosen alternative
- the researchers have less control over the situation which the participants are facing, thus the effects of each attribute on the behaviour of each individual may not be clear; for examples, the participants may change route and departure time because they have to drop off or pick up someone on the way, because they may have an important meeting in the early morning, or because they may acquire some other information on the network which is unknown to the researchers (however, this can be overcome by a well-designed travel diary)
- in the case where the participants have been travelling on the network for a long time and there is no change in the network, it is likely that they may not exhibit the switching or learning behaviour in the survey
- the length of the survey is normally limited to around 10-15 days in order to minimise the burden of the participants

Nowadays, with the development of new technologies such as GPS and related vehicle tracking systems, some of the above problems can be resolved. Recently, Li *et al.* (2004) use GPS-based morning commute data collected from ongoing instrumented vehicle projects in Atlanta to examine the day-to-day variability of the journey-to-work, with regard to route choice and departure time choice behaviour, and trip chaining.

Experimental approach. In the experimental approach, travellers' behaviour is observed under controlled situations. Generally, the experiments can be described as follows. First, participants are asked to state their choices (and in some cases, their expected travel time as well) on a hypothetical network (or in some studies, on a real network). After making the decision, they are then provided with the outcome of their travel choice which is the travel time on the network. This process is repeated for a certain period of time or number of repetitions.

The numbers of the repetitions of the experiment in the previous studies range between 6 up to 200 simulated days depending on the objective of the experiment and its experimental settings. Those involving a lot of tasks are normally repeated for a shorter period, while those involving a simple task; for example, those involving participants stating only the choice of route on a two-link network, can be repeated for a longer period.

The travel time can be determined using three different approaches. It can be predetermined by the researchers (either simulated from a probability distribution during the experimental design stage or taken from other studies), simulated from a probability distribution at the time the survey is being conducted or determined by a traffic simulation model in which the inputs of the model are the actual decisions of every participant in the experiment. The examples of the first approach can be found in Van Berkum and Van der Mede (1993), Dimitropoulos (1993), Dimitrakopoulos (1994), Uno *et al.* (1996), and Iida *et al.* (1999); while the examples of the second approach can be found in Oladeinde (2000), Ozbay *et al.* (2001), Avineri and Prashker (2002), and Avineri and Prashker (2003); and the examples of the third approach can be found in Iida *et al.* (1992), Mahmassani and Chang (1985), Mahmassani and Tong (1986), Mahmassani and Stephan (1988), Mahmassani and Liu (1999), Srinivasan and Mahmassani (2002).

The advantages of the first two approaches are that the experiment is easy to develop and conduct, and it can be self-administered. In addition, the researchers have better control over the information available to the participants and thus the stimuli that might trigger their dynamic behaviour. However, these two approaches do not allow the researchers to investigate the evolution and convergence of the systems.

The third approach, on the other hand, addresses day-to-day evolution and convergence. However, it comes with a price of higher computational costs, and significantly greater complexity in the experimental set-up. Moreover, the researchers have less control over the feedback and do not know how long it will take for the experiment to converge to a steady state (if at all). In the early days when computer networking was still in its infancy, this type of experiment had to be administered on a daily basis (see for examples, Mahmassani and Chang (1985), Mahmassani and Tong (1986), and Mahmassani and Stephan (1988)). Nowadays, this approach can be conducted in a computer laboratory in which the feedback can be supplied to the participants immediately after all of them have made a decision (see for example, Mahmassani and Liu (1999), Srinivasan and Mahmassani (2002)). However, the number of the participants in each session will normally be limited by the number of the computers available in the laboratory. Regarding this issue, Internet-based experiments could be of better use as the experiments can be conducted at any location with computers and Internet access, and this provides an opportunity to conduct large coordinated experiments.

As one can see, most of the disadvantages of the field-survey approach mentioned above can be overcome by the experimental approach. However, an issue regarding the loss of realism associated with an experimental approach, as compared to a field-survey approach, remains to be studied. In addition, the comparisons between different experimental approaches also need to be investigated, as it may be the case that different experimental approaches may result in different learning behaviours. For example, in the second approach where feedbacks are simulated from a probability distribution, the participants may learn about the distribution of travel times, whereas in the third approach where feedbacks are the results of every participant's decision, the participants may learn about the pattern of the evolution of transport systems.

Evolution and Convergence of the Systems

As already raised in connection with experiments, another important issue that has to be taken into account when developing the framework and the models for modelling these day-to-day dynamics is the mechanisms of evolution and convergence of transport systems. As this type of model does not assume the equilibrium states *a priori*, whether the systems actually reach those states is always a question.

According to the literature, two unconventional concepts of equilibrium have been proposed. Mahmassani and Chang (1987) proposed the concept of Boundedly Rational User Equilibrium which is obtained when all users are satisfied with their current travel choices and thus do not want to improve their outcomes by changing decisions. The satisfaction of choices is defined in relation to some indifference band acceptable to the users. The results from a series of controlled laboratory-like experiments and actual surveys were found consistent with this concept. However, the results also indicate that the BRUE is not likely to be unique depending on the evolutionary path followed (Mahmassani, 1997).

Alternatively, Nakayama *et al.* (1999) proposed the concept of Deluded Equilibrium in which the drivers are locked in delusions and do not believe that they can improve their perceived travel times by changing decisions; which is a result of bad experiences in the past. Likewise, the final state in Deluded Equilibrium is not unique and it depends on its initial state and its history.

The difference between those two concepts is the mechanism for convergence. In the former, the systems converge because of the indifference band derived form the notion of satisficing, while in the latter, the deluded beliefs and heterogeneity in drivers' perceptions of travel times seem to be the cause of convergence (Nakayama *et al.*, 1999).

It is of interest to note that while there are quite a number of different approaches being proposed in the literature, the studies into the comparison of the evolutions and the convergences of transport systems with regard to these approaches are very limited.

The initial state is also another issue that has to be taken into account in the modelling framework. Van der Mede and Van Berkum (1993) suggested that in the case of familiar trips, the mean travel time could serve as an initial value; however, if a new network is travelled for the first time, the free-flow travel time could represent the initial state before the trip is made. Hu and Mahmassani (1995) proposed the use of initial set of paths which is system optimal, in terms of minimizing total trip time, to overcome the problem of arbitrary starting point. It is of interest to note that most studies did not state the assumption regarding their initial values explicitly. Thus, this is another area that needs further investigation.

SUMMARY

During the past decade, the task of understanding and modelling the dynamics of individual travel choice has emerged as one of the most the most challenging topics in travel behaviour research. Of particular interest are the day-to-day dynamics of route and departure time choice decisions, as these decisions constitute the short term behavioural responses to congestion and various transport policies, while the time frames for other choice dimensions tend to be of a longer term nature. In this chapter, we have provided a brief overview of the existing approaches to modelling the day-to-day dynamics of route and departure time choice decisions. The existing modelling approaches are classified into two groups based on the underlying learning mechanisms in the models. The first group consists of the models which contain explicit updating mechanisms for the perception of the attributes of the network, while the second group consists of the models which do not contain this type of updating mechanisms. Based on the review, we have discussed key modelling issues including the issues regarding the learning component, choice modelling, habitual choice, choice set formation, data collection methods, and the evolution and convergence of the systems, and have made some suggestions regarding the areas for further research. A better understanding of these behavioural dynamics could contribute to the development of integrated land-use and transport models.

REFERENCES

Arentze, T. and H. Timmermans (2003). Modeling learning and adaptation processes in activity-travel choice: A framework and numerical experiment. *Transportation*, **30**, 37-62.

Avineri, E. and J. N. Prashker (2002). Sensitivity to uncertainty: Travellers' learning perspective, *Proceedings of the 13th Mini-EURO Conference: Handling Uncertainty in th Analysis of Traffic and Transportation Systems*, Bari, Italy.

Avineri, E. and J. N. Prashker (2003). "The impact of travel time information on travellers' learning under uncertainty", paper presented at EC-Workshop "Behavioural Responses to ITS", Eindhoven.

Axhausen, K. W., E. Dimitrakopoulou and I. Dimitropoulos (1995). Adapting to change: Some evidence from a simple learning model. *Proceedings of the 23rd European Transport Forum*, PTRC, London.

Bekhor, S., M. Ben-Akiva and R. S. Ramming (2002). "Adaptation of logit kernel to route choice situation". Paper presented at the 81st Annual Meeting of the Transportation Research Board, Washington D.C.

Ben-Akiva, M., M. J. Bergman, A. J. Daly and R. Ramaswamy (1984). Modelling inter urban route choice behaviour. In: *Proceedings of the 9th International Symposium on Transportation and Traffic Theory* (J. Volmuller and R. Hamerslag, eds.), pp. 299-330. VNU Science Press, Utrecht, The Netherlands.

Ben-Akiva, M. and B. Boccara (1995). Discrete choice models with latent choice sets. *International Journal of Research in Marketing*, **12**, 9-24.

Ben-Akiva, M. and M. Bierlaire (1999). Discrete choice methods and their applications to short term travel decisions. In: *Handbook of Transportation Science* (R. W. Hall, ed.). Kluwer Academic Publishers, Boston.

Ben-Akiva, M., A. de Palma and I. Kaysi (1991). Dynamic network models and driver information systems. *Transportation Research* , **25A**(5), 251-266.

Bonsall, P. (2003). "Traveller behaviour: Decision-making in an unpredictable world". Paper presented at EC-Workshop "Behavioural Responses to ITS", Eindhoven.

Cascetta, E. (1989). A stochastic process approach to the analysis of temporal dynamics in transportation networks. *Transportation Research*, **23B**(1), 1-17.

Cascetta, E. and G. E. Cantarella (1991). A day-to-day and within-day dynamic stochastic assignment model. *Transportation Research*, **25A**(5), 277-291.

Cascetta, E., A. Nuzzolo, F. Russo and A. Vitetta (1996). A modified logit route choice model overcoming path overlapping problems. *Proceeding of the second International Symposium on the Theory of Road Traffic Flow*, Lyon, France.

Cascetta, E. and A. Papola (2001). Random utility models with implicit availability/perception of choice alternatives for the simulation of travel demand. *Transportation Research*, **9C**, 249-263.

Cascetta, E., F. Russo and A. Vitetta (1997). Stochastic user equilibrium assignment with explicit path enumeration: Comparison of models and algorithms. *Preprint of the 8th IFAC Symposium on Transportation Systems*, Chania, Greece.

Cascetta, E., F. Russo, F. A. Viola and A. Vitetta (2002). A model of route perception in urban road networks. *Transportation Research*, **36B**, 577-592.

Chang, G.-L. and H. S. Mahmassani (1988). Travel time prediction and departure time adjustment behavior dynamics in a congested traffic system. *Transportation Research*, **22B**(3), 217-232.

Chen, R. B. and H. S. Mahmassani (2004). "Travel time preception and learning mechanisms in traffic networks". Paper presented at the 83rd Annual Meeting of the Transportation Research Board, Washington D.C.

Dial, R. B. (1971). A probabilistic multipath traffic assignment algorithm which obviates path enumeration. *Transportation Research*, **5**(2), 83-111.

Dimitrakopoulou, E. (1994). "The analysis of travel time perception: Updating process and effects". MSc dissertation, Intercollegiate MSc-course in Transport, Imperial College London and University College London, London.

Dimitropoulos, I. (1993). "Updating process of travel time perception in a simulated commuting environment". MSc dissertation, Intercollegiate MSc-course in Transport, Imperial College London and University College London, London.

Edwards, W. (1982). Conservatism in human information processing. In: *Judgement Under Uncertainty: Heuristics and Biases* (D. Kahneman, P. Slovic and A. Tversky, eds.), pp. 359-369. Cambridge University Press, Cambridge.

Emmerink, R., K. W. Axhausen, P. Nijkamp and P. Rietveld (1995). Effects of information in road transport networks with recurrent congestion. *Transportation*, **22**, 21-53.

Ettema, D., H. Timmermans and T. Arentze (2003a). "Modelling perception updating of travel times in the context of departure time choice under ITS". Paper presented at EC-Workshop "Behavioural Responses to ITS", Eindhoven.

Ettema, D., G. Tamminga and H. Timmermans (2003b). "A micro-simulation model system of departure time and route choice under travel time uncertainty". Paper presented at the 10th International Conference on Travel Behaviour Research, Lucerne.

Fujii, S. and R. Kitamura (2000). Anticipated travel time, information acquisition, and actual experience: Hanshin expressway route closure, Osaka-Sakai, Japan. *Transportation Research Record*, **1725**, 79-85.

Horowitz, J. L. (1984). The stability of stochastic equilibrium on a two-link transportation network. *Transportation Research*, **18B**(1), 13-28.

Hu, T.-Y. and H. S. Mahmassani (1995). Evolution of network flows under real-time information: Day-to-day dynamic simulation assignment framework. *Transportation Research Record*, **1493**, 46-56.

Hu, T.-Y. and H. S. Mahmassani (1997). Day-to-day evolution of network flows under real-time information and reactive signal control. *Transportation Research*, **5C**(1), 51-69.

Iida, Y., T. Akiyama and T. Uchida (1992). Experimental analysis of dynamic route choice behavior. *Transportation Research*, **26B**(1), 17-32.

Jha, M., S. Madanat and S. Peeta (1998). Perception updation and day-to-day travel choice dynamics in traffic networks with information provision. *Transportation Research*, **6C**, 189-212.

Jou, R.-C. and H. S. Mahmassani (1996). Comparability and transferability of commuter behavior characteristics between cities: Departure time and route-switching decisions. *Transportation Research Record*, **1556**, 119-130.

Jou, R.-C. and H. S. Mahmassani (1998). Day-to-day dynamics of urban commuter departure time and route switching decisions: Joint model estimation. In: *Travel Behaviour Research: Updating the State of Play* (J. D. Ortuzar, D. A. Hensher and S. R. Jara-Diaz, eds.), pp. 365-384. Pergamon, Oxford.

Kahneman, D. and A. Tversky (1979). Prospect theory: An analysis of decision under risk. *Econometrica*, **47**(2), 263-292.

Klügl, F. and A. L. C. Bazzan (2003). "Route decision behaviour in a commuting scenario". Paper presented at EC-Workshop "Behavioural Responses to ITS", Eindhoven.

Li, H., R. Guensler, J. Ogle and J. Wang (2004). "Using GPS data to understand the day-to-day dynamics of the morning commute behavior". Paper presented at the 83rd Annual Meeting of the Transportation Research Board, Washington D.C.

Mahmassani, H. S. (1990). Dynamic models of commuter behavior: Experimental investigation and application to the analysis of planned traffic disruptions. *Transportation Research,* **24A**(6), 465-484.

Mahmassani, H. S. (1997). Dynamics of commuter behaviour: Recent research and continuing challenges. In: *Understanding Travel Behaviour in an Era of Change* (P. R. Stopher and M. Lee-Gosselin, eds.), pp. 279-313. Pergamon, Oxford.

Mahmassani, H. S. and G.-L.Chang (1985). Dynamic aspect of departure-time choice behavior in a commuting system: Theorectical framework and experimental analysis. *Transportation Research Record,* **1037**, 88-101.

Mahmassani, H. S. and G.-L. Chang (1986). Experiments with departure time choice dynamics of urban commuters. *Transportation Research,* **20B**(4), 297-320.

Mahmassani, H. S. and G.-L. Chang (1987). On boundedly rational user equilibrium in transportation systems. *Transportation Science,* **21**(2), 89-99.

Mahmassani, H. S. and R. Herman (1990). Interactive experiments for the study of tripmaker behaviour dynamics in congested commuting systems. In: *Developments in Dynamic and Activity-Based Approaches to Travel Analysis* (P. Jones, ed.), pp. 272-298. Avebury, Aldershot.

Mahmassani, H. S. and R.-C. Jou (1998). "Commuter behavior dynamics: Transferring insights from laboratory experiments to actual systems". Paper presented at the 77th Annual Meeting of the Transportation Research Board, Washington D.C.

Mahmassani, H. S. and R.-C. Jou (2000). Transferring insights into commuter behavior dynamics from laboratory experiments to field surveys. *Transportation Research,* **34A**, 243-260.

Mahmassani, H. S. and Y.-H Liu (1999). Dynamics of commuting decision behaviour under advanced traveller information systems. *Transportation Research,* **7C**, 91-107.

Mahmassani, H. S. and D. G. Stephan (1988). Experimental investigation of route and departure time choice dynamics of urban commuters. *Transportation Research Record,* **1203**, 69-84.

Mahmassani, H. S. and C.-C. Tong (1986). Availability of information and dynamics of departure time choice: Experimental investigation. *Transportation Research Record,* **1085**, 3-47.

Nakayama, S., R. Kitamura and S. Fujii (1999). Drivers' learning and network behavior: Dynamic analysis of the driver-network system as a complex system. *Transportation Research Record,* **1676**, 30-36.

Nakayama, S. and R. Kitamura (2000). Route choice model with inductive learning. *Transportation Research Record,* **1725**, 63-70.

Nakayama, S., R. Kitamura and S. Fujii (2001). Drivers' route choice rules and network behavior: Do drivers become rational and homogeneous through learning? *Transportation Research Record,* **1752**, 62-68.

Nagel, K. and B. Raney (2003). "Large scale multi-agent simulations for transportation applications". Paper presented at EC-Workshop "Behavioural Responses to ITS", Eindhoven.

Oladeinde, F. A. (2000). "A model of traveller learning with applications to traffic information systems". PhD Dissertation, Centre for Transport Studies, Imperial College London.

Ozbay, K., A. Datta and P. Kachroo (2001). Modeling route choice behavior with stochastic learning automata. *Transportation Research Record,* **1752**, 38-46.

Ozbay, K., A. Datta and P. Kachroo (2002). "Application of stochastic learning automata for modelling departure time and route choice behaviour". Paper presented at the 81st Annual Meeting of the Transportation Research Board, Washington D.C.

Polak, J. W. and M. L. Hazelton (1998). The influence of alternative traveller learning mechanisms on the dynamics of transport systems. *Proceedings of the 26th European Transport Forum*, PTRC, London.

Polak, J. W. and F. A. Oladeinde (2000). An empirical model of travellers' day-today learning in the presence of uncertain travel times. In: *Reliability of Transport Networks* (M. G. H. Bell and C. Cassir, eds.), pp. 11-30. Research Studies Press, Baldock.

Raney, B. and Nagel, K. (2002) "Iterative route planning for modular transportation simulation". Paper presented at the Swiss Transportation Research Conference, Monte Verita, Switzerland.

Raney, B., N. Cetin, A. Vollmy, M. Vrtic, K. W. Axhausen and K. Nagel (2003). An agent-based microsimulation model of Swiss travel: First results. *Networks and Spatial Economics*, **3**, 23-41.

Raney, B. and K. Nagel (2003). "Truly agent-based strategy selection fro transportation simulations". Paper presented at the 82nd Annual Meeting of the Transportation Research Board, Washington D.C.

Rossetti, R. J. F., R. H. Bordini, A. L. C. Bazzan, S. Bampi, R. Liu and D. van Vliet (2002). Using DBI agents to improve driver modelling in a commuter scenario. *Transportation Research*, **10C**, 373-398.

Russo, F. and A. Vitetta (1996). Network and assignment models for the Italian national transportation systems. *Proceeding of the Seventh World Conference on Transport Research*, Sydney, Australia.

Small, K. A. (1987). A discrete choice model for ordered alternatives. *Econometrica*, **55**(2), 409-424.

Srinivasan, K. and H. S. Mahmassani (2001). "Dynamics in departure time choices of commuters: A comparison of alternative adjustment mechanisms". Paper presented at the 80th Annual Meeting of the Transportation Research Board, Washington D.C.

Srinivasan, K. and H. S. Mahmassani (2002). "Dynamic Decision and Adjustment Processes In Commuter Behavior Under Real-Time Information". Centre for Transportation Research, University of Texas at Austin, SWUTC/02/167204-1.

Train, K. (2003). *Discrete Choice Methods with Simulation.* Cambridge University Press, Cambridge.

Tversky, A. and D. Kahneman (1992). Advances in prospect theory: Cumulative representation of uncertainty. *Journal of Risk and Uncertainty*, **5**, 297-323.

Uno, N., Y. Iida and A. Kubo (1996). An analysis of driver's pre-trip and en-route route choice behaviour under travel time information: Experimental results. *Proceedings of the First Conference of Hong Kong Society for Transportation Studies*, pp. 269-283.

van Berkum, E. C. and P. H. J. van Der Mede (1998). The impact of dynamic traffic information: Modelling approach and empirical results. In: *Travel Behaviour Research: Updating the State of Play* (J. D. Ortuzar, D. A. Hensher and S. R. Jara-Diaz, eds.), pp. 401-421. Pergamon, Oxford.

van Der Mede, P. H. J. and E. C. van Berkum (1993). "The impact of traffic information: Dynamics in route and departure time choice". Doctoral thesis, Proefschrift Technische Universiteit Delft.

Vaughn, K. M., M. A. Abdel-Aty, R. Kitamura, P. P. Jovanis and H. Yang (1993). "Experimental analysis and modeling of sequential route choice under ATIS in a simple traffic network". Institute of Transportation Studies, University of California, Davis, UCB-ITS-PRR-93-12.

Vovsha, P. and S. Bekhor (1998). "The link-nested logit model of route choice: Overcoming the route overlapping problem". Paper presented at the 77th Annual Meeting of the Transportation Research Board, Washington D.C.

Integrated Land-Use and Transportation Models: Behavioural Foundations
M. Lee-Gosselin and S. Doherty (Editors)

7

THE INFLUENCE OF MOBILE AND IMMOBILE COMMUNICATIONS ON ACTIVITY AND TRAVEL SCHEDULING

Anthony J. Richardson and Rita K. Seethaler
The Urban Transport Institute, Taggerty, Australia

INTRODUCTION

In much of the research on activity scheduling, there is a distinction drawn between "planned" and "spontaneous" activities. Furthermore a sequence of planned activities spread over time may often be thought of as part of an overall "project". In an era of rapid deployment of mobile communications (mobile phones, SMA devices, smart PDAs etc), the distinction between planned and spontaneous activities may not be so clear-cut. The use of mobile communications can give rise to many situations in which spontaneous activities are created as a result of a communication episode, even though that activity might be part of a larger overall planned project.

This exploratory paper will consider some of the issues surrounding the possession and use of mobile phones (we will use the Australian terminology of "mobile phones", although the same devices are called cell-phones in America and Handy-phones in parts of Europe), the way in which communication activities are related to travel activities and the issues that are likely to arise in the design of data collection programs and in the analysis of the resulting data, when proper attention is paid to the role of mobile communications in the activity scheduling behaviour of people, especially those who are heavy users of mobile communications.

SOME SCENARIOS

Consider the following possible scenarios of interactions between mobile phones and travel:

"Malcolm and Helen have been planning a dinner party for some weeks. On the night of the dinner party, as Malcolm prepares the meal he realises that they are short of some wine to go with the dinner. He makes a mobile phone call to Helen who is already driving home from work. This causes her to make a detour to pick up a few extra bottles of wine. Is this extra trip a "spontaneous" trip or is it a "planned" trip as part of a larger "planned" project?"

"Sam and Janet have gone to separate meetings in the city. They plan to meet up at a street corner after their meetings and drive home together. Janet's meeting runs over-time. She calls Sam on his mobile phone, and tells him to drive home alone and she will get a taxi later. In this case, the mobile phone has increased the number of vehicle trips, reduced the waiting time (for Sam), and changed one planned trip into two spontaneous trips."

"Sue is a teenage girl who likes to go night-clubbing. Jack is her worried father. He is not worried about the night-clubbing, but rather the trip home later at night (early next morning). Sue is not licenced to drive, but there is a good tram service from the night-club area to about half a kilometre from her home. Jack is worried about the walk home from the tram stop. Rather than stop Sue from going to the clubs, Jack buys her a mobile phone on the condition that she phones him on the way home on the tram so that he can pick her up and drive her home. Sue gets to go clubbing, Jack's mind is at ease. The mobile phone has enabled the planning of the trip home in a way that would not have been possible without the mobile phone."

"Ellen and Andrew both work and have a 4-year-old daughter Rosalie who is in pre-school care during the day. Pre-school ends at 4.30pm and kids must be picked up by 5pm sharp, otherwise financial penalties are imposed by the pre-school (to pay for teachers waiting until all kids have been picked up). Because of the nature of their jobs, Ellen and Andrew never know whether they will be forced to work a bit late. Therefore, while there is a long-standing plan for one of them to pick up Rosalie before 5pm, they never know who it will be until their regular 4.30pm mobile phone call to work things out. If neither of them can make it, they reluctantly phone Ellen's mum, who lives near the pre-school and who can walk across to collect Rosalie. The mobile phone enables a spontaneous decision to be made within the context of a long-term activity schedule."

"Mick and Bruce both love Australian Rules football. In the days before mobile phones, they lived near each other and used to car-pool to the footy together. Since then they have moved houses, but Mick used to still go round to Bruce's house to pick him up (a detour of over 10 kilometres) to go to the footy together. Since both getting mobile phones, they now go separately to the footy (Mick still drives but Bruce takes the bus), and when Bruce gets to the football ground (where the crowd is often over 70,000), he phones Mick to find out where he is seated, so they can watch the game together. Depending on what Mick is doing after the game, he sometimes gives Bruce a ride home in his car. The mobile phones have changed one planned car-pool trip into a planned car trip and a planned bus trip, and an occasional spontaneous car-pool trip".

The above scenarios have illustrated just some of the ways in which the rise of mobile phones has changed the way in which we travel and interact with other people. If we are to fully understand the impact of mobile phones on travel patterns, and the ways in which mobile phones affect the scheduling and re-scheduling of activities and travel, we need to explore these interactions more closely.

THE POSSESSION AND USE OF MOBILE PHONES

While mobile communications, especially mobile phones, are becoming an increasingly ubiquitous item of apparel in Western society, relatively little factual information is available about the demographics of the users or about the way in which mobile phones are used. There are clearly mountains of information collected about mobile phone users and the use of mobile phones, since telecommunications companies need such information for billing purposes. Undoubtedly, such information is also used for market research purposes within the telecommunications company. However, because of the extremely competitive nature of the telecommunications industry, little of this information is publicly available.

Mobile Phone Ownership Rates

There are, however, some relatively aggregate statistics about mobile phone ownership and use that give an indication of the scale andpenetration of mobile phones. The following table of mobile phone ownership rates in different regions of the world has been gleaned from various web sources, especially the International telecommunications Union website (www.itu.int) (the figures are indicative only and may refer to slightly different time periods using slightly different definitions).

Table 7.1 Mobile Phone Ownership by Region

Region	1995 Mobile Phones (millions)	2001 Mobile Phones (millions)	Annual Growth Rate (1995-2001)	2001 Mobiles per Capita
Western Europe	22.7	276.3	52%	0.72
Eastern Europe	1.4	74.1	93%	0.18
Africa	0.6	23.5	82%	0.03
Middle East	0.9	15.8	60%	0.09
Central Asia	3.8	154.2	85%	0.06
South East Asia	18.4	166.6	44%	0.23
Australasia	2.6	13.6	32%	0.58
Oceania	0.0	0.2	66%	0.03
North Americas	36.4	136.9	25%	0.43
Central Americas	1.2	29.3	71%	0.17
South Amercias	2.7	55.7	66%	0.16
TOTAL	90.7	946.3	48%	0.16

It would appear that mobile phone ownership is highest in Western Europe, followed by Australasia and North America. However, it should also be noted that the individual countries with the highest ownership rates are in South-East Asia (Taiwan and Hong Kong). The annual growth rate is generally very high, especially in those regions with low current ownership rates. Thus while mobile phones were about equal in number to fixed phones in 2001, their high growth rate means that they will soon outstrip fixed phones on a global basis (indeed they probably already have by the time this paper was written).

Demographics of Mobile Phone Owners

The demographics of mobile phone users are even more sketchy than the ownership rates, but some trends are becoming apparent in several countries. As shown in Figure 7.1 for the USA (www.upoc.com), mobile phone ownership tends to rise then fall with increasing age. The highest levels of ownership appear to be for people around 30 years of age, but range between 50% and 68% for all age groups.

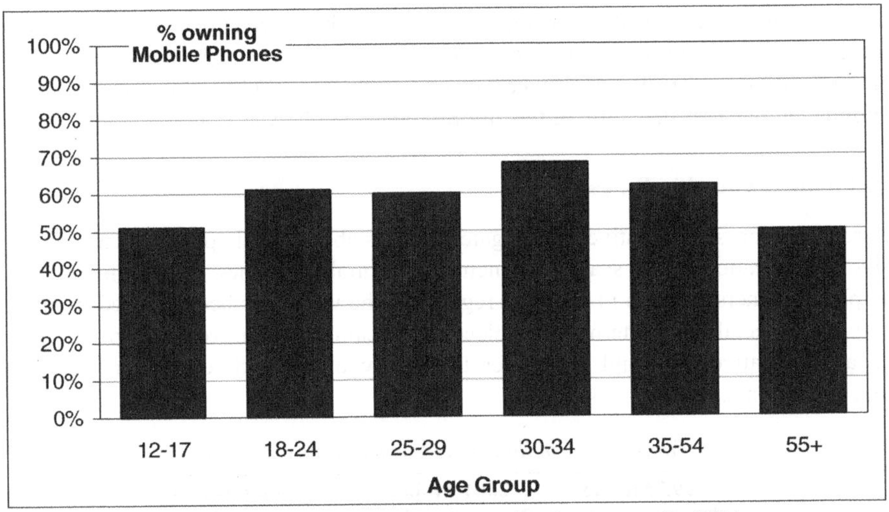

Figure 7.1 Mobile Phone Ownership by Age in the USA

This same pattern is repeated using data for Japan (Video Research Ltd., www.ddd.com.jp), as shown in Figure 7.2. Despite the overall level of mobile phone ownership for people over 12 being higher in Japan (70%) than in the USA (60%), a similar trend with age exists. However, since the Japan data also contains ownership rates by gender, another trend emerges. Young women are more likely to have mobile phones than young men, while older women are less likely to have mobile phones than older men.

The differences in mobile phone availability can be observed from a slightly different perspective using data from a recent survey in Melbourne, Australia. This survey was a roadside intercept survey of motorists' driving exposure, and as part of the interview they were asked whether they had a mobile phone with them at that time. The results of this survey, by age and gender, are shown in Figure 7.3. The fall in mobile phone possession with increasing age is apparent. In addition, women under 25 were more likely to have a mobile phone with them than men of the same age. However, unlike the Japanese results, women over 60 in Melbourne were also more likely to have a mobile phone with them than men over 60.

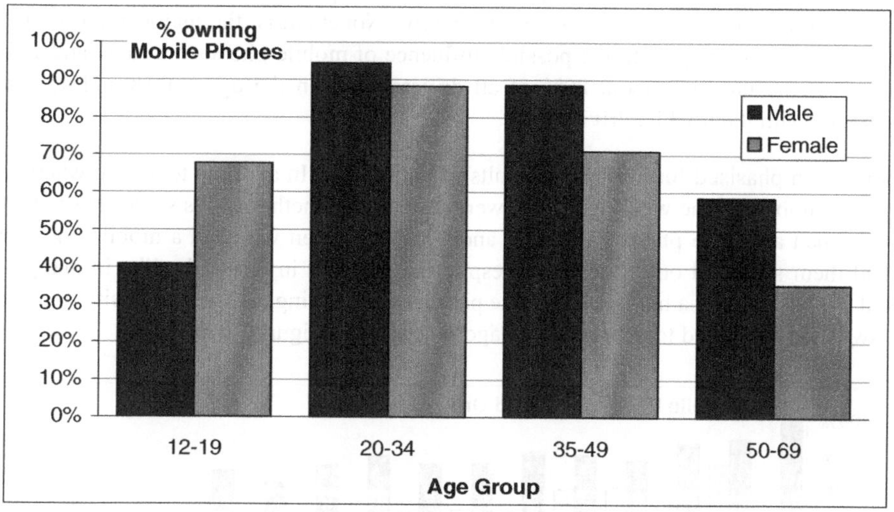

Figure 7.2 Mobile Phone Ownership by Age and Gender in Japan

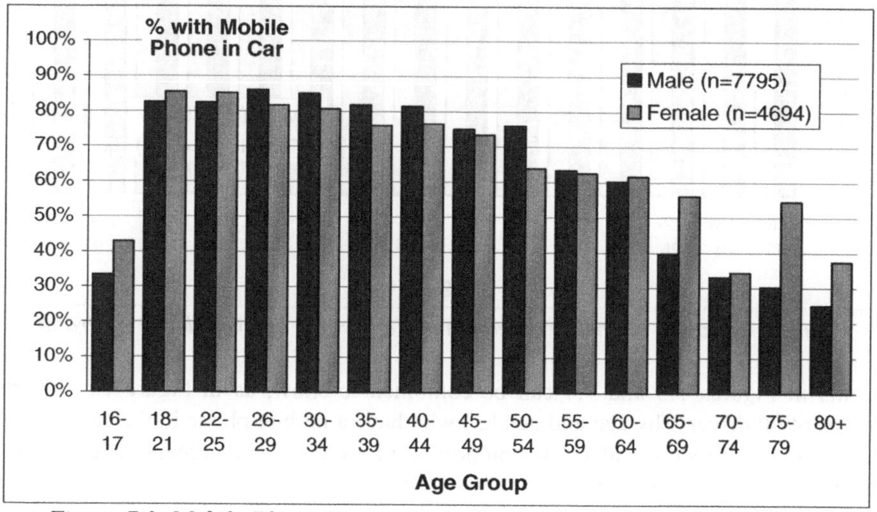

Figure 7.3 Mobile Phone Possession in Car by Age and Gender in Australia

It should be realised, however, that the results in Figures 7.2 and 7.3 are not totally comparable. Figure 7.2 refers to ownership of mobile phones while Figure 7.3 refers to possession of a mobile phone while driving. It could be argued that the real phone ownership rate is higher than shown in Figure 7.3, since some people may have left their mobile phone at home. On the other hand, it could be argued that Figure 7.3 is an over-estimate of the mobile phone ownership of the population, since car drivers are probably an atypical sample of the population with respect to mobile phone ownership. Nonetheless, the major point to emerge from Figure 7.3, with respect to the possible influence of mobile phones on travel and activity scheduling, is the fact that about 80% of all drivers between the ages of 18 and 45 have a mobile phone with them while driving.

This point is emphasised further by the results in Figure 7.4. In addition to asking whether the driver had a mobile phone with them, they were also asked whether it was switched on. 89% of males who had a mobile phone with them and 77% of women who had a mobile phone with them had them switched on at the time (despite the fact that in Australia, like in many other places, it is illegal to use a hand-held mobile phone while driving!). The propensity to have the phone switched on tended to decrease with age, as shown in Figure 7.4.

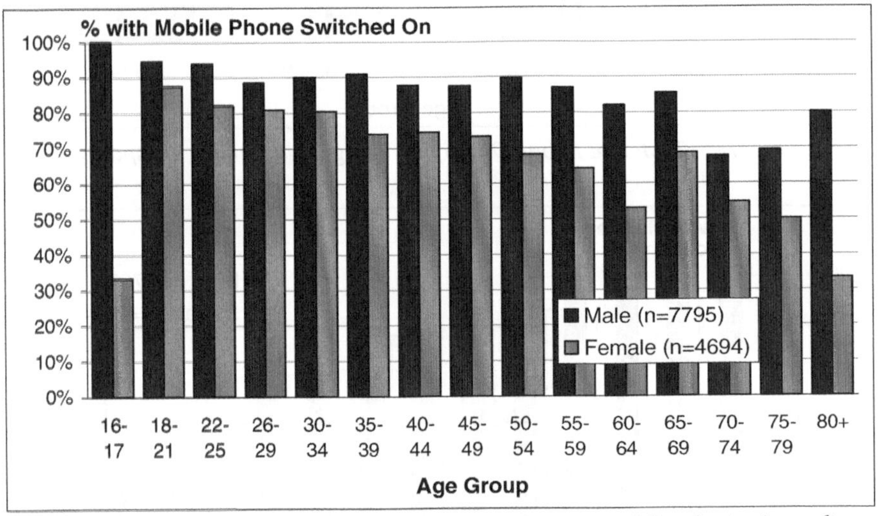

Figure 7.4 Mobile Phone Switched-On in Car by Age and Gender in Australia

The results in Figures 7.3 and 7.4 can be combined to show, as in Figure 7.5, the overall proportions of all drivers (by age and gender) who have a mobile phone in the car switched on and ready to make or receive calls. This proportion of drivers are all susceptible to having their travel and activity schedules disrupted by the use of the mobile phone. 70% of male drivers and 58% of female drivers are in this situation.

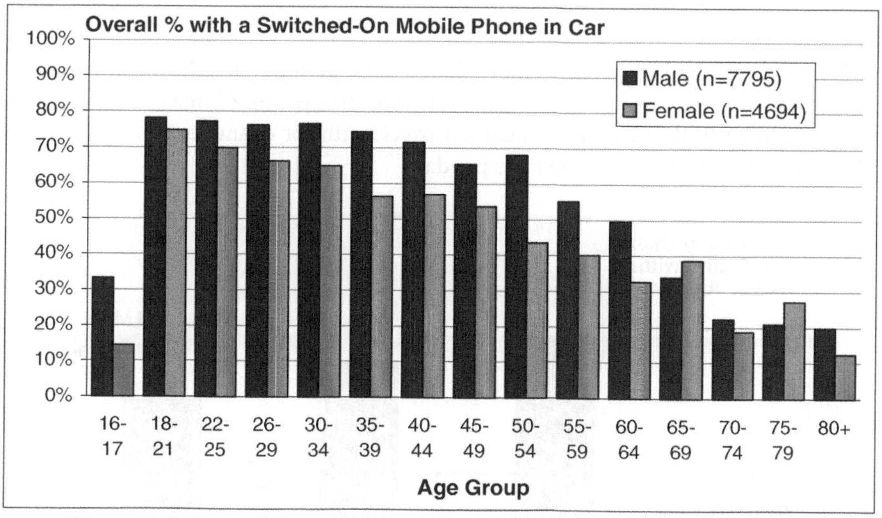

Figure 7.5 Overall Proportions with Mobile Phone Switched-On in Car

Usage of Mobile phones

If information about the demographics of mobile phone users is sketchy, then solid data about mobile phone usage is even more scarce. An exception to this is the Japanese data quoted above (Video Research Ltd.) which also contains some interesting facts about mobile phone usage. The overall levels of usage are shown in Figure 7.6, where it is seen that the most frequent usage level is one or two calls (in or out) per day. Some users hardly use their mobile phone at all (less than once or twice per week), while others use it extensively (more than 10 calls per day).

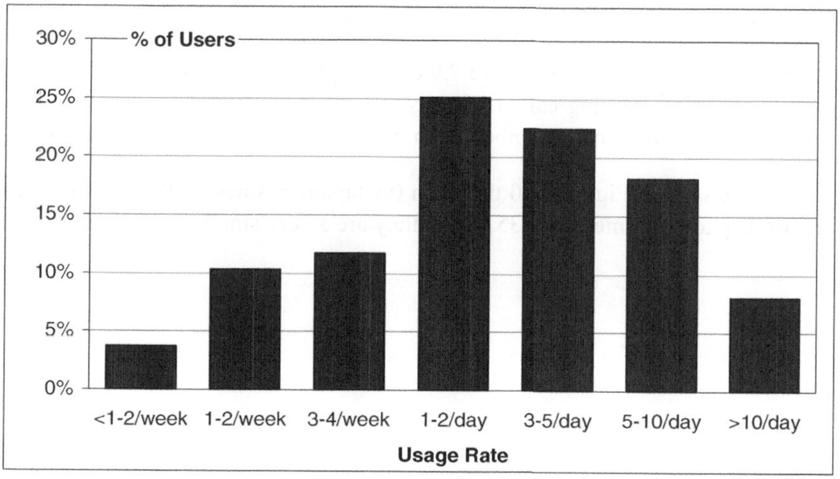

Figure 7.6 Mobile Phone Usage in Japan

The characteristics of the heavy users (more than 5 calls per day) are shown in Figure 7.7. It can be seen that men (in Japan) are heavier users than women at all ages, and that people in their twenties and early thirties are the heaviest users, with for example 39% of men between 20 and 34 making more than 5 mobile calls per day.

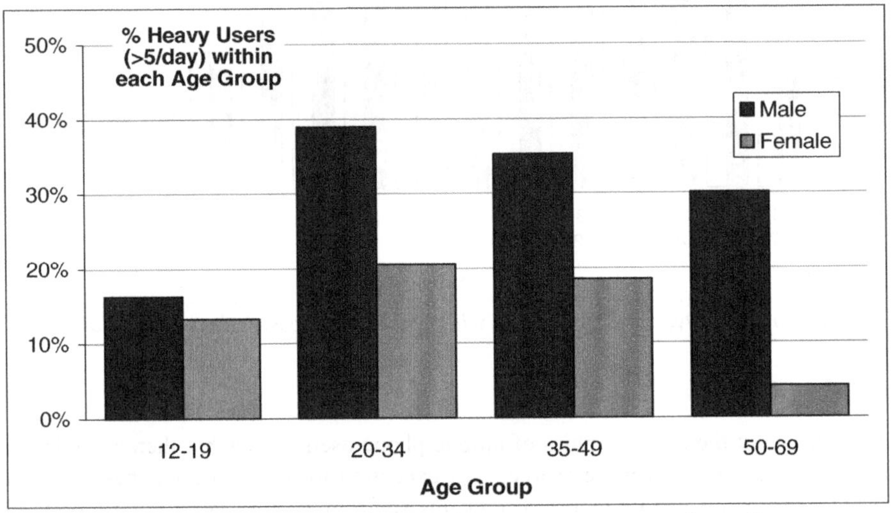

Figure 7.7 Mobile Phone Heavy Users by Age and Gender in Japan

The reasons why people make mobile phone calls are shown in Figures 7.8 through 7.10. It can be seen in Figure 7.8, for example, that people under 35 (especially teenage boys) tend to use the mobile phone for making calls to friends, with for example 88% of all mobile calls made by boys between 12 and 19 being made to friends.

On the other hand, it can be seen in Figure 7.9 that people over 35 (especially women) tend to use the mobile phone for making calls to family members. Indeed women at all ages tend to make more of their calls to family members than men make calls to family members.

However, it can be seen in Figure 7.10 that men (in Japan) business calls tend to dominate the use of the mobile phone for men over 35, while they are a very small proportion for women.

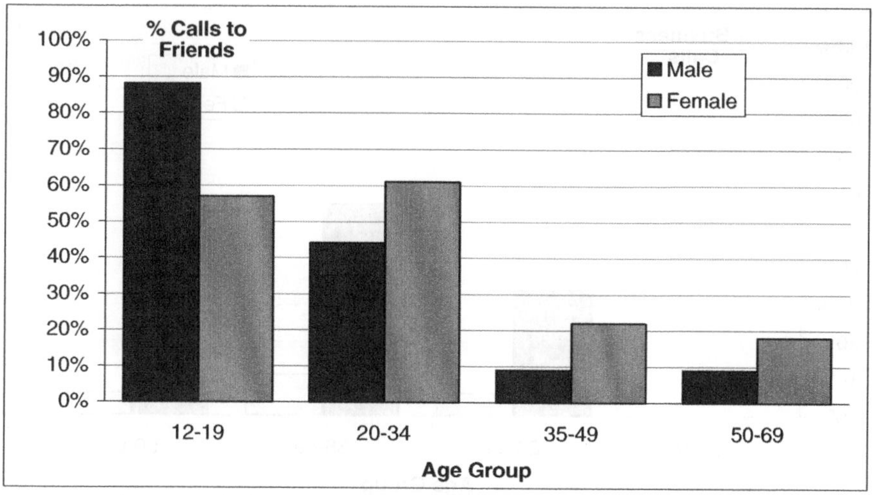

Figure 7.8 % Calls to Friends by Age and Gender in Japan

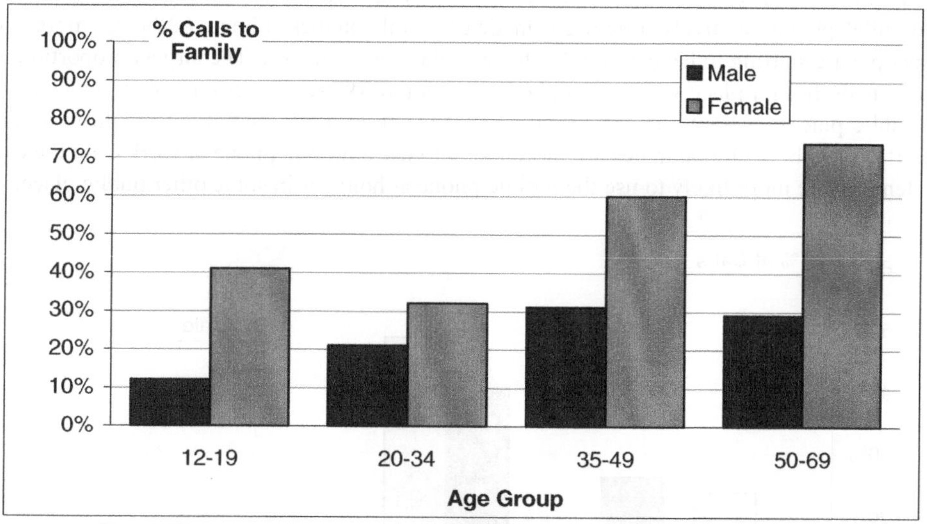

Figure 7.9 % Calls to Family Members by Age and Gender in Japan

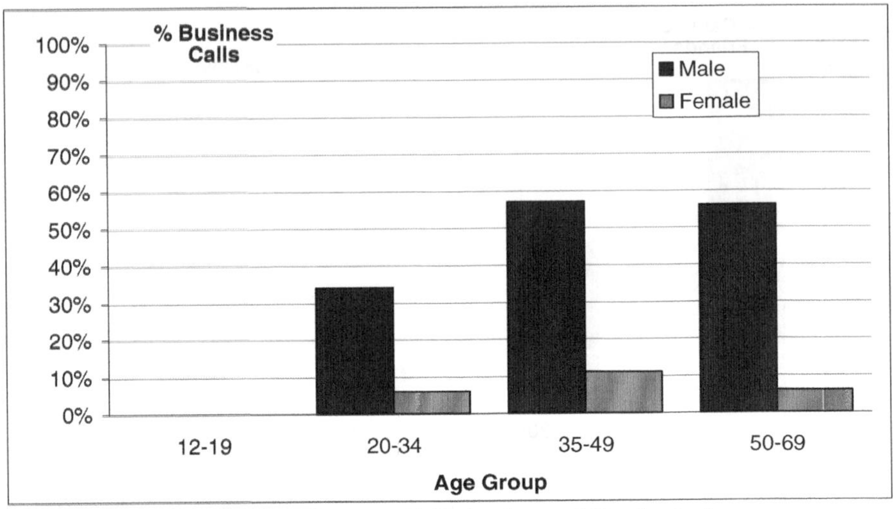

Figure 7.10 % Business Calls by Age and Gender in Japan

The place where calls are made (or received) is shown in Figure 7.11. In line with the nature of a "mobile" phone, relatively little use is made of mobile phones at home (although more and more people are starting to have a mobile phone as their only phone). The largest proportion of calls are made from a place which is neither home nor work (such as from a shop, or in some other public place). About 10% of all calls are made from a car, and about 3% from public transport. It can be seen that males are more likely to use a mobile phone at work or in the car, while females are more likely to use the mobile phone at home or in some other public place.

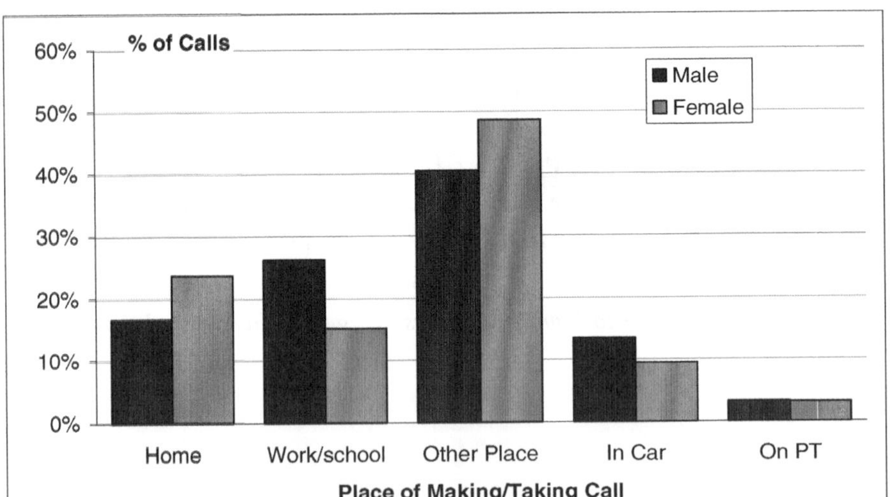

Figure 7.11 Place of Making or Taking a Call by Gender in Japan

The place where mobile phones are used, however, varies with the age of the user, as shown in Figure 7.12. Younger people are more likely to use a mobile phone at home (teenagers are probably using the mobile phone to avoid sitting on the main land-line phone when others are wanting to use it), while those in their twenties may only have a mobile home (even at home). Older people are more likely to use the mobile phone in public places, and also while in a car. Indeed, 15-18% of all mobile phone calls from people over 35 are made to or from a car. This has significant implications for the interaction between mobile phone use and travel and activity scheduling.

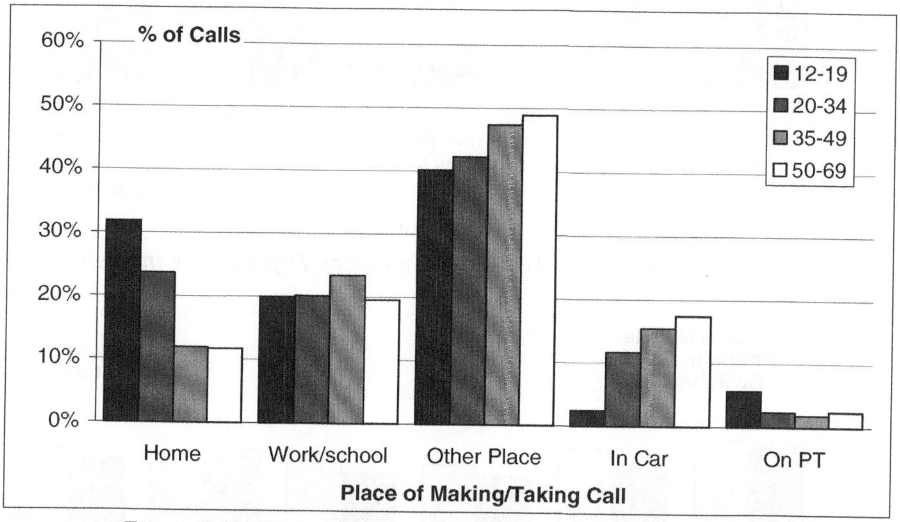

Figure 7.12 Place of Making or Taking a Call by Age in Japan

The interaction between mobile phones and travel can also be considered using the Australian data, by examining the trip situations in which mobile phones are more available. Figure 7.13 shows the proportion of trips, of various purposes, on which a mobile phone was available (in the car and switched on) for males and females. It can be seen that mobile phone availability is highest for trips to and from work and while driving for work purposes, and is lowest on shopping trips. For all trip purposes, except shopping, males are more likely to have a mobile phone available than females.

Mobile phone availability is shown as a function of the timing of the trip in Figure 7.14. It can be seen that, on weekdays, mobile phones are more likely to be available on trips in the morning and in the evenings, and less available for trips during the day. On weekends, however, there is little variation in mobile phone availability across the entire day.

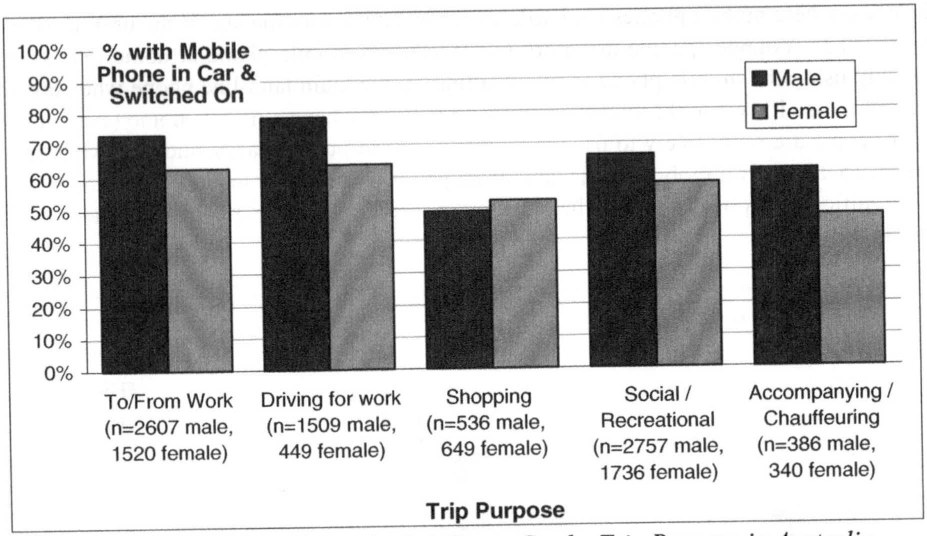

Figure 7.13 Mobile Phone Availability in Car by Trip Purpose in Australia

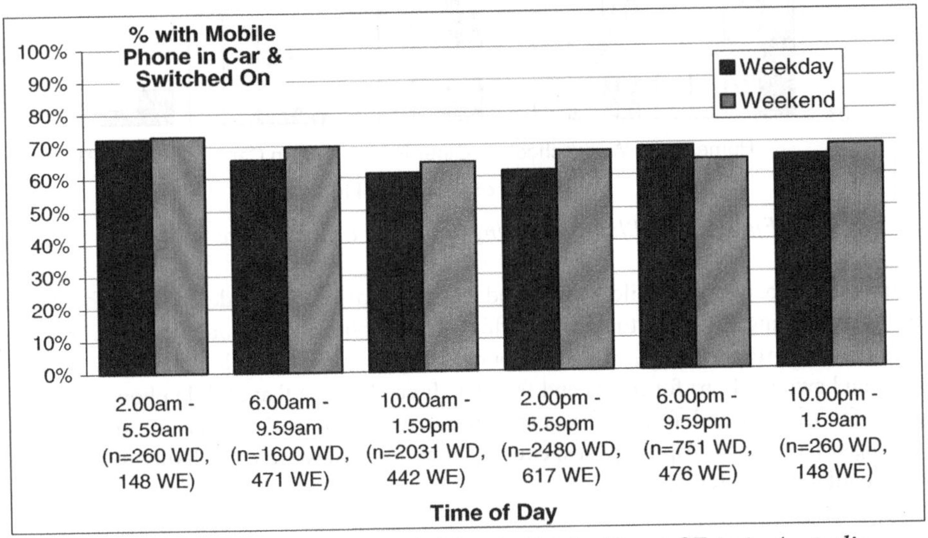

Figure 7.14 Mobile Phone Availability in Car by Time of Trip in Australia

COMMUNICATIONS AND TRAVEL

The information of mobile phone use patterns and their possible relationship to travel and activity scheduling is mostly speculative. While there are plausible reasons for how mobile phones might influence travel patterns, there is no hard evidence. What is needed is a single data set that combines travel and mobile communications in the same framework. Unfortunately, there are very few examples. Mokhtarian and Salomon (2002) provide some examples of studies that have investigated the relationships between travel and telecommunications, but with one exception (Yim, 1994) the studies have concentrated on the use of conventional fixed phones. While Yim (1994) concluded that "cellular communication generated additional trips rather than substituting for them", Mokhtarian and Salomon (2002) cast doubt on this conclusion, citing problems with "attribution of causality" to the cell-phone. They raise the possibility that "it was not the cell phone per se but rather any phone that caused the trip".

Zumkeller (2001, 2002) has been one of the pioneers in examining, quantitatively, the relationships between telecommunication and transport, and in designing surveys that address both aspects of interaction. However, once again, mobile communications do not receive special attention, with all types of phone calls being classified as a "phone call".

It was therefore with some excitement that, while working on data from the 1997 Australian Time Use Survey (ABS, 1999) for another purpose, we realised that the 1997TUS recorded phone calls as part of the time use survey, and also recorded the type of phone being used and the purpose of the phone call. It was therefore with some disappointment that we found after some analysis of the data that of the 6556 phone calls recorded in the data as primary or secondary activities, only 32 were made with a mobile phone and only 4% had a specific purpose recorded (the other 96% of calls were recorded as simply "talking on the phone").

Since time use surveys would appear to be a potentially fruitful data sources for looking at the relationships between mobile communications and travel, we therefore inquired in the Time Use Research community as to what data sources might be best suited for such analysis. Unfortunately, the situation was summed up most succinctly by Dr. Andy Harvey (Director, Time Use Research Program, St. Mary's University, Halifax, Canada) when he stated that "Unfortunately I do not know of any. Technology has been essentially a wash in time use studies. Hopefully if we get our grant, we will focus on that in our survey".

It therefore appears that with current data sets, there is little possibility of quantitatively exploring the relationships between mobile communications and travel. However, rather than abandon all hope at this point, we decided to follow up on the suggestion of Mokhtarian and Salomon (2002) that "it was not the cell phone per se but rather any phone that caused the trip". That is, if we can't explore the relationships between mobile phone use and travel in the 1997TUS data, then let's see what can be gleaned from an examination of the relationships between phone communications (in general) and travel.

In exploring these relationships, there are a number of possible hypotheses that could be tested:

- Phone calls are a substitute for travel
- Phone calls stimulate travel
- Phone calls are used to plan travel
- Phone calls may disrupt planned travel

The following analyses attempt to cast some light on the validity, or otherwise, of these various hypotheses.

From a sample of 14,315 people in the 1997 Australian TUS, the average number of trips per person per day was 4.1, and the distribution is as shown in Figure 7.15. About 11% of people were immobile on their survey day. The distribution of trips from this time-use survey was very similar to that found in more conventional travel surveys.

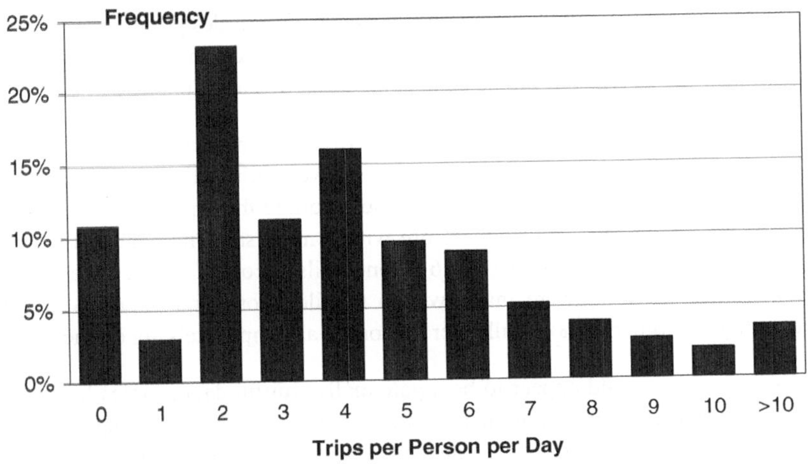

Figure 7.15 Frequency Distribution of Trips per Person per Day

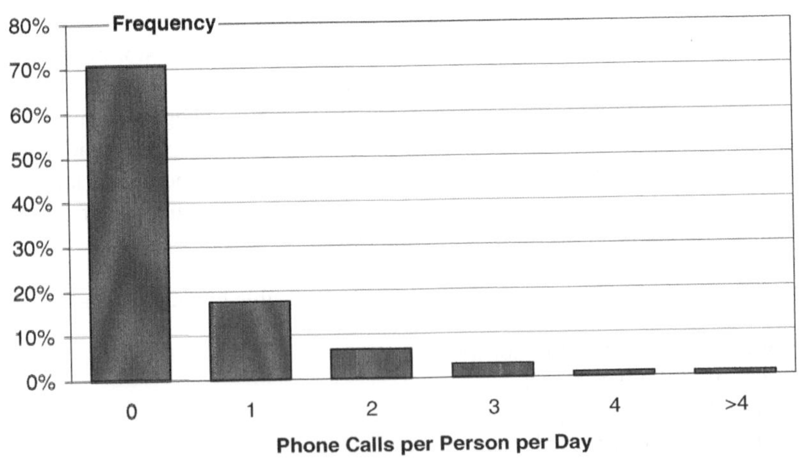

Figure 7.16 Frequency Distribution of Phone Calls per Day

The average number of phone calls per person per day was 0.5, and the distribution is as shown in Figure 7.16. About 70% of people made no phone calls on their survey day.

One way of testing whether trips and phone calls are substitutes or complements is to examine whether those making more trips make more or fewer phone calls. As shown in Figure 7.17, it appears that trips and phone calls are complementary, since those making more trips also make more phone calls. This does not imply that the phone calls cause more trips, but merely that those making more trips also make more phone calls.

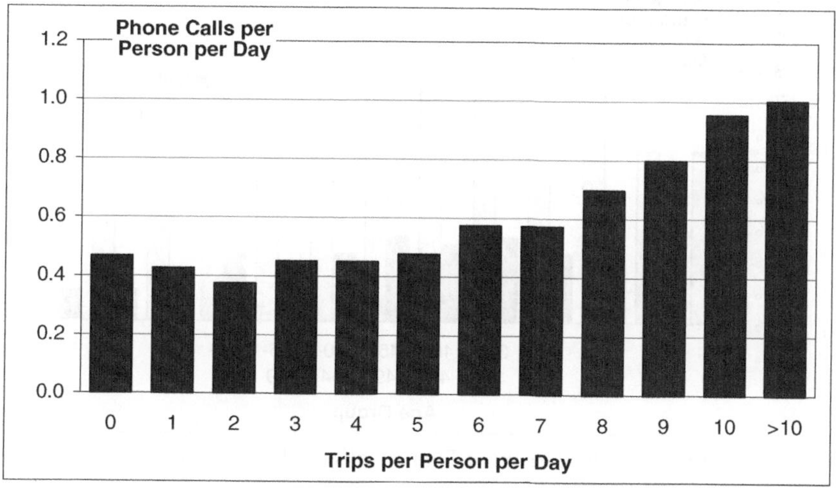

Figure 7.17 The Relationship between Trips per Day and Phone Calls per Day

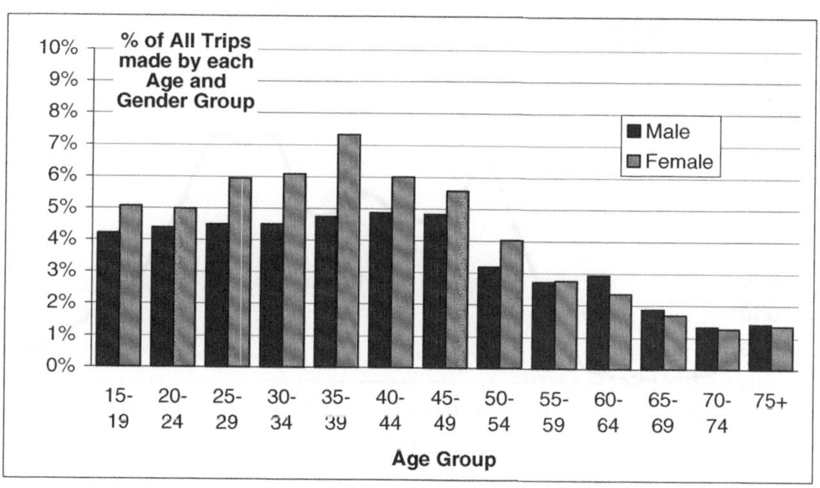

Figure 7.18 Percent of Trips by Age and Gender

The demographics of tripmakers and phone callers, in terms of age and gender, are shown in Figures 7.18 and 7.19. Each figure shows the proportion of all trips (phone calls) made by each age and gender group.

The incidence of both trips and phone calls peaks for people in their late thirties. Interestingly, females make more trips and phone calls than males in nearly all age groups. This is especially the case for phone calls where females make about 75% of all phone calls.

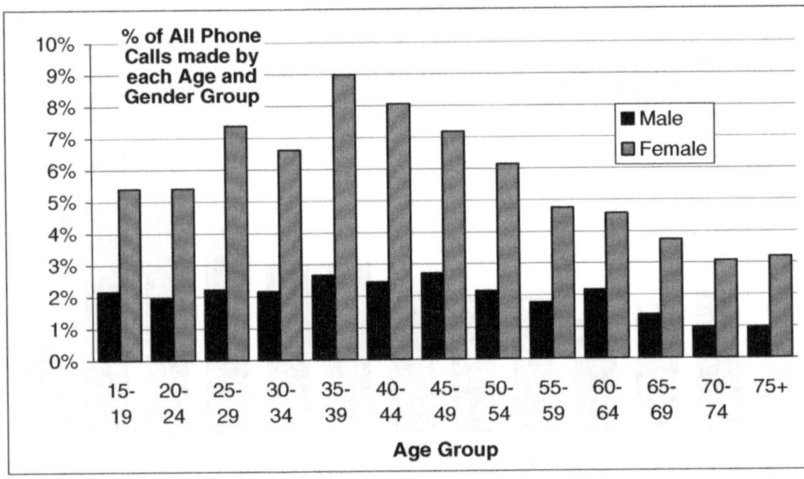

Figure 7.19 Percent of Phone Calls by Age and Gender

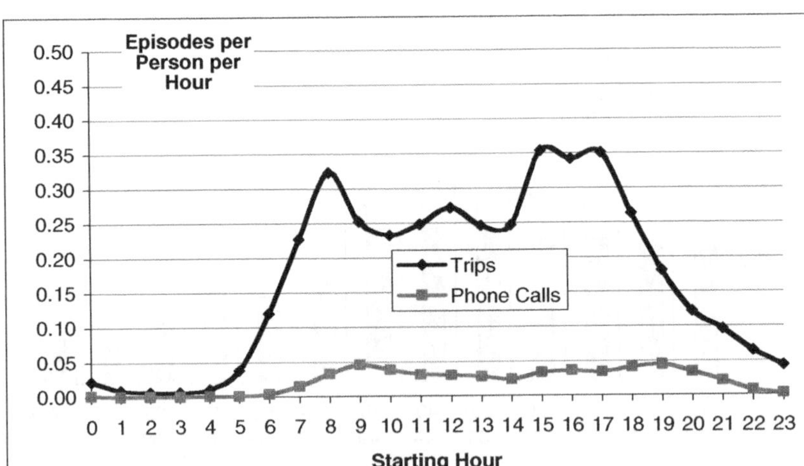

Figure 7.20 Trips and Phone Calls per Person per Hour

The distributions of trips and phone calls (generically called episodes in time use research terminology) across the day are shown in Figure 7.20. The main feature of this graph is that trips are almost an order of magnitude greater in number than phone calls.

The shapes of the distributions are better appreciated by calculating the proportion (rather than the absolute number) of trips and phone calls per hour, as shown in Figure 7.21. It can be seen that both trips and phone calls go through two major peaks; one in the morning and one in the evening. However, it is clear that the phone calls lag the trips by an hour or two in both the morning and the evening.

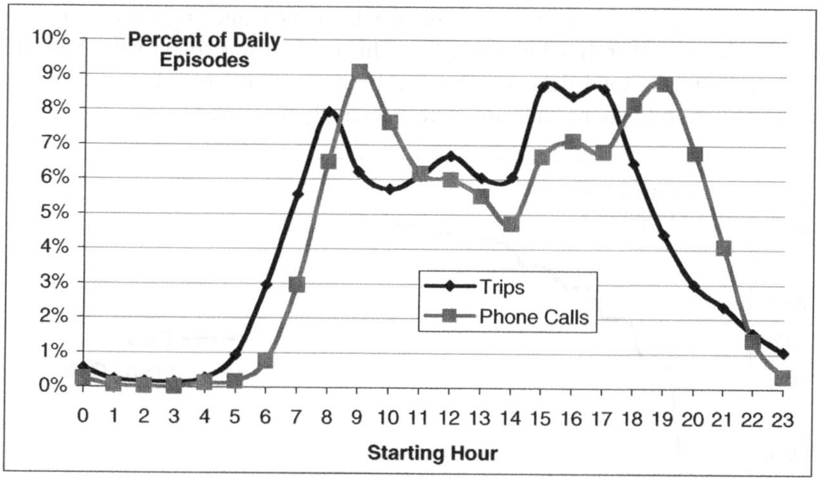

Figure 7.21 Percent of Trips and Phone Calls by Hour of Day

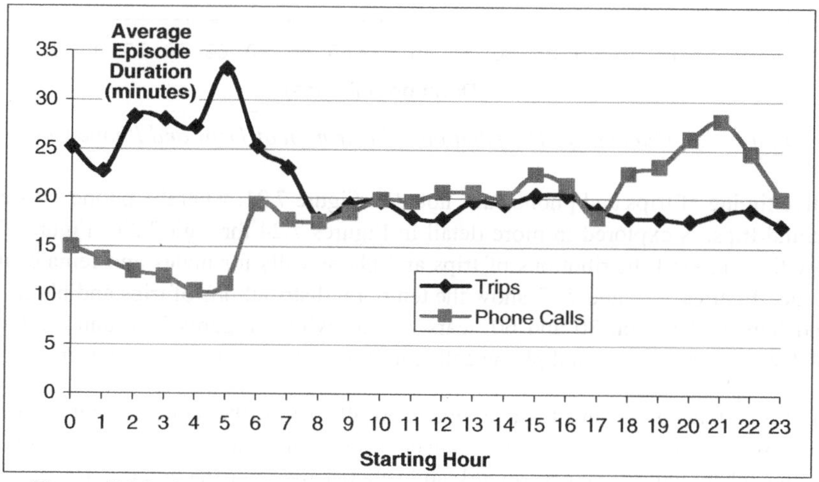

Figure 7.22 Average Duration of Trips and Phone Calls by Hour of Day

The average duration of trips and phone calls across the day are shown in Figure 7.22. It can be seen that during the day, trips and phone calls have about the same average duration of about 20 minutes. However, early in the day, trips are longer, and during the evening, phone calls are longer.

The cumulative distributions of trip and phone call durations are shown in Figure 7.23. It can be seen that phone calls are slightly longer overall. One important aspect of the time use data becomes evident from a consideration of the trip and phone call durations. There are nearly no trips or phone calls with durations less than 5 minutes are recorded. For trips, this probably means that short trips such as walking trips are under-estimated. For phone calls, however, the impact is probably greatest for mobile phone calls, which one could argue are more likely to be much shorter in duration than fixed phone calls. This could also explain why there are so few mobile phone calls in the total data set. Such a limitation must be addressed in future data collection efforts if mobile phone calls are to be effectively measured.

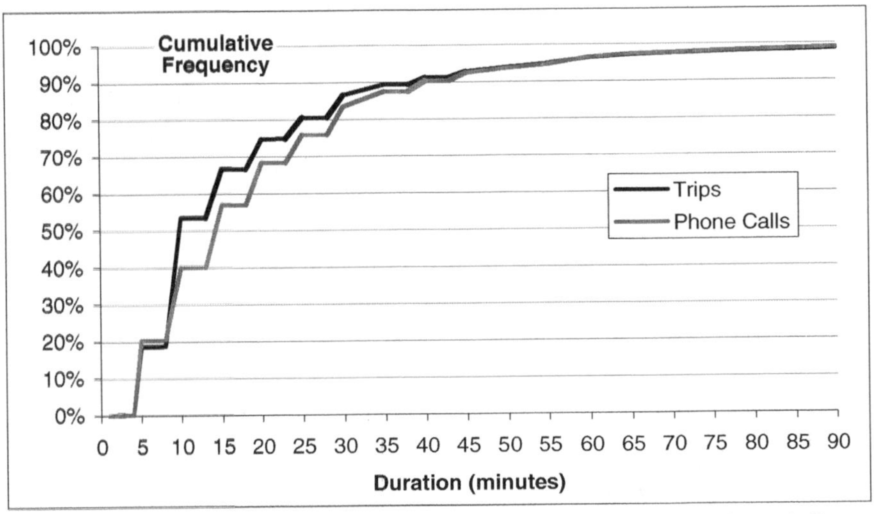

Figure 7.23 Cumulative Distributions of Duration of Trips and Phone Calls

The relative timing of trips and phone calls noted in Figure 7.21, whereby phone calls appeared to lag behind trips, is explored in more detail in Figures 7.24 through 7.29. Figures 7.24 and 7.25 show the temporal distributions of trips and phone calls for males and females under 30 years of age. Figures 7.26 and 7.27 show the temporal distributions of trips and phone calls for males and females between 30 and 60 years of age, while Figures 7.28 and 7.29 show the temporal distributions of trips and phone calls for males and females over 60 years of age.

It would appear that males and females under 60 all exhibit the same lag between trips and phone calls, with phone calls tending to be made after the morning travel peak (after arriving at work or after others have gone off to work and school) and after the evening travel peak (after arriving home at night).

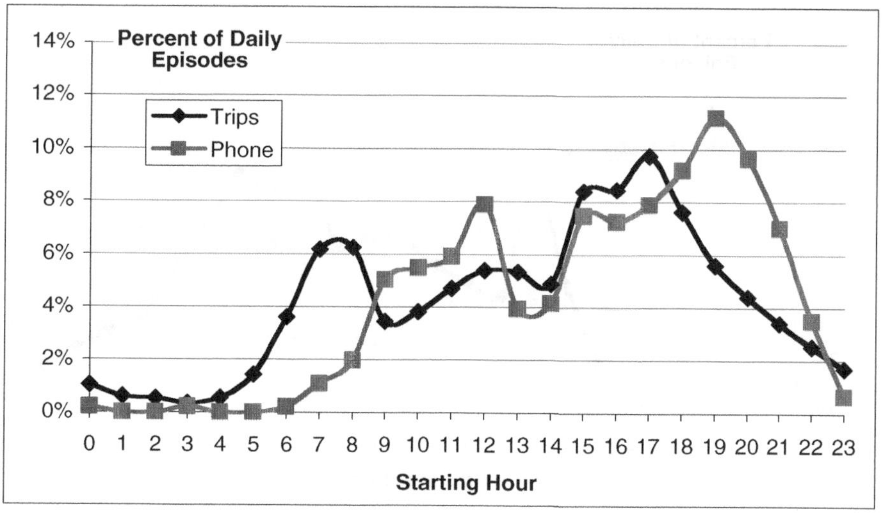

Figure 7.24 Trips and Phone Calls by Hour of Day by Males under 30

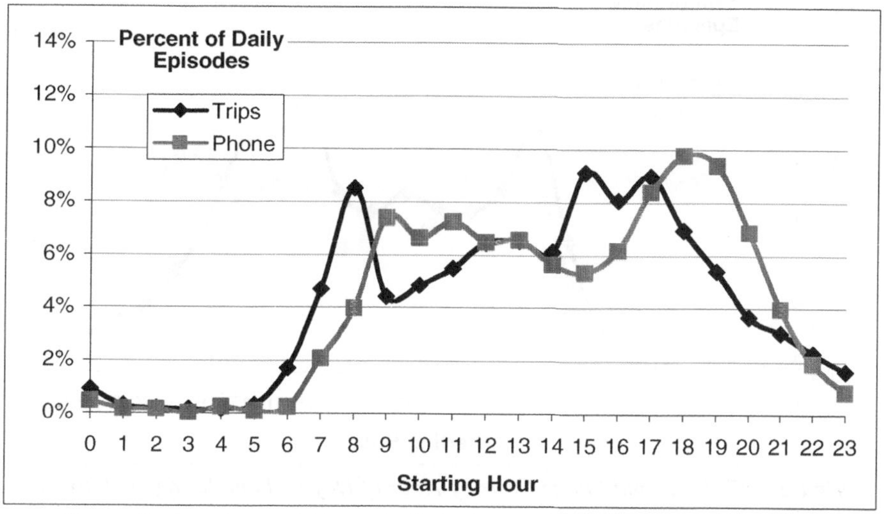

Figure 7.25 Trips and Phone Calls by Hour of Day by Females under 30

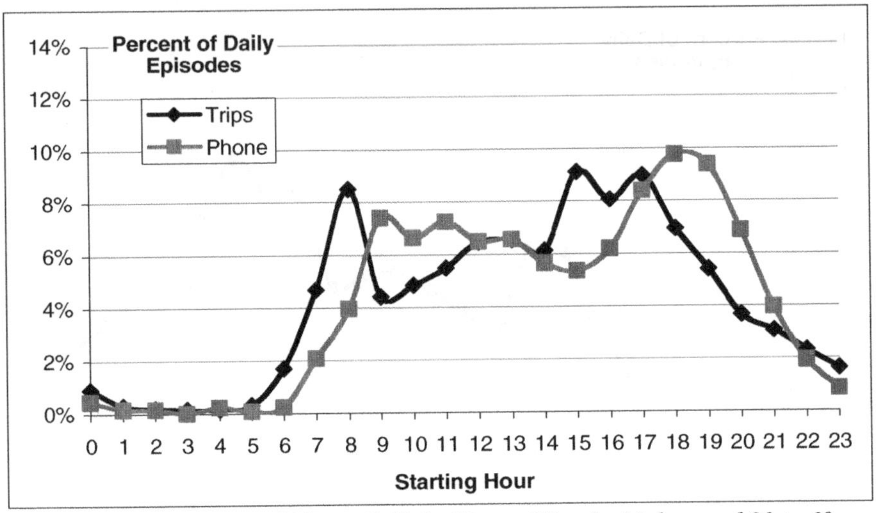

Figure 7.26 Trips and Phone Calls by Hour of Day by Males aged 30 to 60

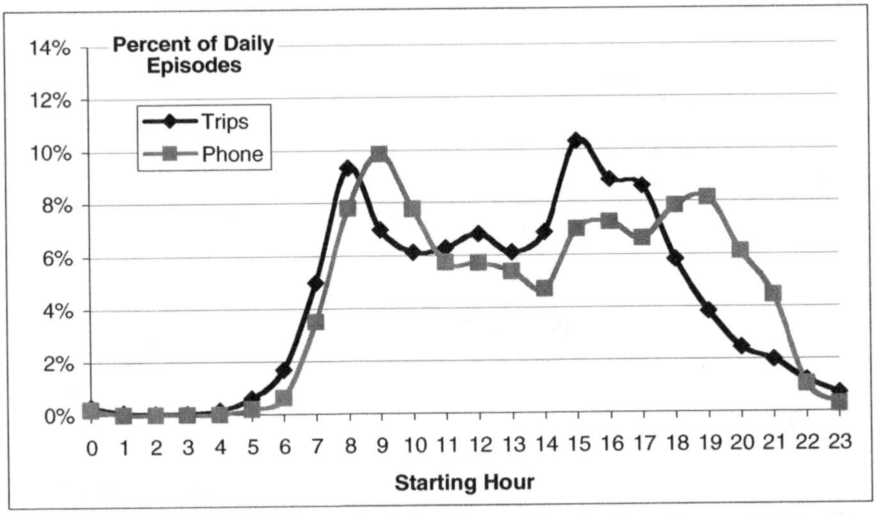

Figure 7.27 Trips and Phone Calls by Hour of Day by Females aged 30 to 60

On the other hand, males and females over 60 tend not to show a lag between trips and phone calls. People over 60 tend to have trip and phone call peaks in the morning, with both activities declining during the day. In the morning, the trip and phone call activity tends to rise together, suggesting a different relationship between trips and phone calls than demonstrated for those under 60.

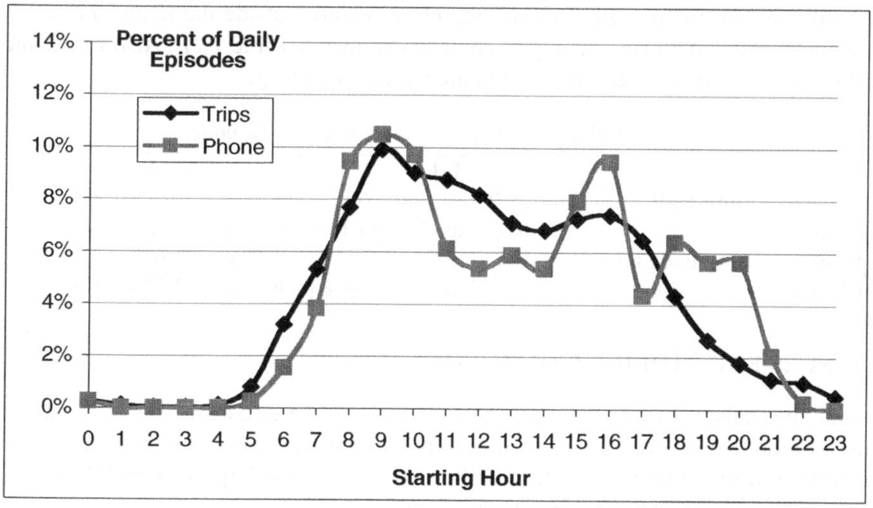

Figure 7.28 Trips and Phone Calls by Hour of Day by Males aged over 60

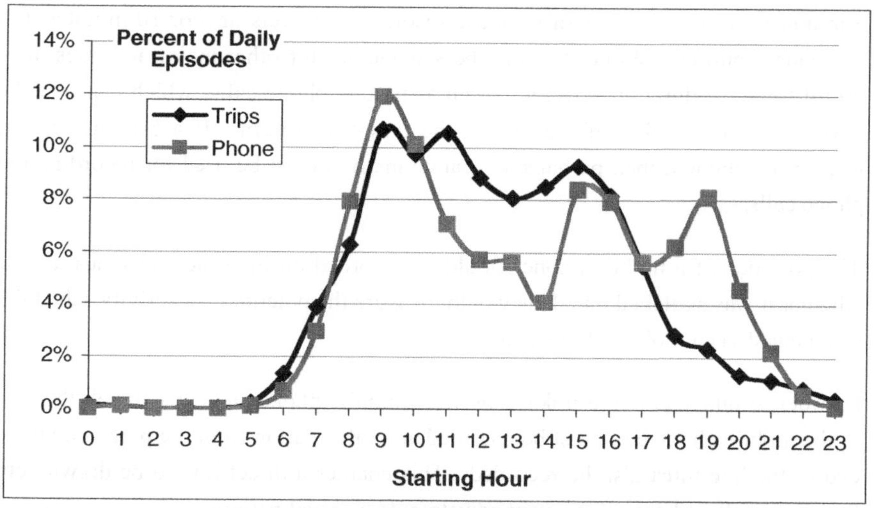

Figure 7.29 Trips and Phone Calls by Hour of Day by Females aged over 60

Note, however, that these results are based on two marginal distributions of the timing of trips and phone calls. Greater understanding of the relationship between trips and phone calls requires the collection of more detailed data about the joint distribution between the two activities, whereby the timing and reasons for trips and phone calls can be investigated within any one respondent, rather than relying on aggregate distributions for the entire sample.

Finally, unlike mobile phone calls, which take place mainly outside the home (Figure 7.11), fixed line phone calls take place mainly at home (as shown in Table 7.2). Also unlike mobile phone calls, the proportion of calls made at home increases with age.

Table 7.2 Place of Phone Call by Age and Gender

	Male			**Female**		
Place of Phone Call	Under 30	30->60	60+	Under 30	30->60	60+
Home	91.9%	89.2%	94.1%	92.2%	95.7%	99.0%
Workplace if outside home	2.4%	5.7%	0.5%	1.8%	1.6%	0.0%
Elsewhere	5.7%	5.1%	5.4%	6.0%	2.7%	1.0%

DATA ISSUES FOR MOBILE COMMUNICATIONS

Even though the analysis of the 1997 Australian Time-Use Survey data did not shed much light on the use of mobile phones and their relationship with travel, it did highlight some issues that must be addressed in future surveys that wish to investigate the impact of mobile phones on travel and activity scheduling. The major issues are:

- If a time-use survey is used for data collection, the time interval used must be finer than 5 minutes. Many time-use surveys use recording intervals of 5 or 10 minutes, to ease respondent burden. While this may be satisfactory for other major activities, it is not useful for short duration activities (such as mobile phone calls) which can nonetheless have a profound effect of the rest of the time-use pattern. If a 5-minute recording interval is retained, then perhaps a separate method must be used for recording mobile phone calls.

- The presence of a mobile phone should be recorded during other activities, as well as whether it was switched on. This would measure the potential for activity scheduling to be changed via a mobile phone call.

- In order to understand the linkage between mobile phone calls and travel, the purpose of the mobile phone call must be recorded and information from the person on the other end of the line must also be recorded. This enables a direct link to be drawn between the phone call and any subsequent adjustments to travel patterns.

- A distinction must be made between calls made and call received (the so-called active and passive modes of communication (Zumkeller, 2002)). This will enable identification of the triggering mechanism for any changes in travel patterns.

- Other mobile communications, such as SMS text messaging, should be recorded as well as audio phone calls.

Designing a data collection instrument with the above points in mind will greatly facilitate the investigation of the impact of mobile phones on travel and activity scheduling, but could also greatly increase the complexity of the survey instrument and the burden placed on the respondents. Because of this increase in complexity, it is feasible to collect information about the interactions between travel and communications only in specially designed surveys, rather than adding the communications questions to activity surveys which are already pushing the bounds of acceptable respondent burden.

SUMMARY

This paper started with the intention of examining the impact of mobile phones on travel and activity scheduling. The study was hampered firstly by a lack of accessible data on mobile phone ownership and general usage patterns, and secondly by a scarcity of any data sets which jointly record travel and the use of mobile phones. As a means of obtaining some insight into the design of future data sets that cover travel and the use of mobile phones, an analysis was performed on the 1997 Australian Time-Use Survey data, concentrating on the use of fixed phones and their relationships to travel patterns. It was found that trip patterns and phone use patterns could be analysed in a similar fashion and that there were some consistent patterns to their usage in terms of timing and the demographics of users. Finally some recommendations were made for the design of future data sets which would truly enable a more detailed study of the impact of mobile phones on activity scheduling.

REFERENCES

Australian Bureau of Statistics (1999). Information Paper: Time Use Survey, Australia: Confidentialised Unit Record File, 1997. ABS Catalogue No. 4151.0, ISBN 0 642 27501 7.

International Telecommunications Union website: www.itu.int

Mokhtarian, P.L. and I. Salomon (2002). Emerging Travel Patterns: Do Telcommunications make a Difference? In: *In Perpetual Motion: Travel Behaviour Research Opportunities and Application Challenges* (Hani Mahmassani, ed.), Chapter 7, pp. 143-182. Pergamon: Oxford UK.

Upoc Inc. (USA) website: www.upoc.com

Video Research Ltd (Japan) website: www.ddd.com.jp

Yim, Y. (1994). The Effects of Mobile Telephones on Transportation and Urban Form. *33rd Annual Meeting of the Western Regional Science Association*, Tuscon, Arizona.

Zumkeller, D. (2001). The Impact of Telecommunication and Transport on Spatial Behaviour. In: *Travel Behaviour Research: The Leading Edge* (David Hensher, ed.), Chapter 54, pp. 917-930. Pergamon: Oxford UK.

Zumkeller, D. (2002). Transportation and Telecommunication: First Comprehensive Surveys and Simulation Approaches. In: *In Perpetual Motion: Travel Behaviour Research Opportunities and Application Challenges* (Hani Mahmassani, ed.), Chapter 8, pp. 183-207. Pergamon: Oxford UK.

Integrated Land-Use and Transportation Models: Behavioural Foundations
M. Lee-Gosselin and S. Doherty (Editors)

8

THE CREATION OF SIMULATED HOUSEHOLD TRAVEL SURVEY DATA BASED ON AVAILABLE DEMOGRAPHIC DATA FROM HOUSEHOLDS

Peter Stopher, John Rose and Phillip Bullock
Institute of Transport Studies, University of Sydney, Australia

INTRODUCTION

Household travel surveys (HTSs) are expensive to undertake, costing up to AU$300 per respondent in Australia, depending on the technique employed for data collection. Elsewhere around the world, similar high costs are being experienced. CATI surveys in the U.S. routinely cost around US$150 and up per completed household (Cambridge Systematics, 1996). Given the large sample sizes (2,000 to 3,000 households are considered a minimum prerequisite) required in order to capture representativeness of existing household types within both large and small metropolitan areas, the costs of HTSs soon become prohibitive to all but large, well funded government organisations. Yet, despite the absolute and relative costs, the data provided by HTSs are critical to travel-demand forecasts used to estimate the impact of proposed transport policies, as well as to determine regional travel characteristics. This is accomplished by the collection of household demographic and socio-economic data along with associated travel or activity data for each of the surveyed households that are in turn used to derive travel patterns for particular areas over a period of time.

Prohibitive costs are not the only barrier to carrying out HTSs. As with any survey method, public resistance to undertaking the survey task poses many problems and can result in high non-response rates and associated non-response errors (U.S. Census Bureau, 2001). In the transport field, there appears to be an acceptance of response rates of below 40 percent for household travel surveys, and some agencies report response rates as low as 15 or 20 percent. The National Household Travel Survey (NHTS) in the United States has battled with the Office of Management and Budget (OMB) over response rates that are below the standards that OMB considers should be maintained in any Federally-funded survey.

A large part of the reluctance of people to engage in household travel surveys must be laid at the door of unscrupulous marketing firms that use surveys as a pretext to get into households with items they are trying to sell. Even with up-front assurances that the agency is not selling anything, many contacted individuals and households still refuse to participate in household travel surveys (Kalfs and van Evert, 2003). In addition, increasing levels of personal mobility mean that it is more difficult to find individuals at their place of residence to undertake the survey at a time convenient to the interviewer. For face-to-face surveys, the increasing occurrence of gated communities and security buildings is making it more and more difficult for interviewers to gain access to potential respondents, while increasing levels of unlisted telephone numbers are also making telephone contact more difficult. New rules introduced in the U.S. relating to unsolicited marketing calls may also further impact telephone contacts. A further issue relating to telephone contacts is the increasing incidence of households that have only mobile telephones, where, because of charging rates, most surveys try to avoid contacting mobile phones.

In sum, it is becoming increasingly difficult to collect household travel survey data, while it is also becoming more expensive to collect. Rising non-response levels not only contribute to the problems of cost, but also raise serious concerns about the quality of the data. Increasingly, specific segments of the population are able to make themselves much more difficult to contact, thereby increasing non-response biases.

In addition to these problems that affect the collection of data in general, the demands to be more accurate in travel forecasting also leads to a desire to increase sample sizes for the surveys that are undertaken. Rather than being concerned solely with the collection of sufficient data to estimate regional models, there are increasing needs to have sufficient data to inform small area and corridor planning activities. For an average medium to large urban area, with a household travel survey of 3,000 households, the likely sample size falling within a sub-region or corridor of interest may fall as low as 100 or fewer samples. This is completely inadequate to inform the planning process. There is, therefore, a need to consider how to obtain larger samples for local planning issues.

Furthermore, using a nationwide survey, such as NPTS, to estimate forecasting models has not proved workable. The data are drawn from too large a variety of areas, and the sample size available in one metropolitan area is much too small for modelling. On the other hand, efforts to develop models and "borrow" the models in other metropolitan areas, while proposed and researched quite extensively, have not proved to be effective or acceptable to most metropolitan regions in North America and elsewhere.

SIMULATING HOUSEHOLD TRAVEL SURVEY DATA

For some years, micro-simulating households for various types of travel demand and activity modelling has been undertaken (Chung and Goulias, 1997; McNally, 1997). In that approach, a base of current households is usually used. The processes of household change are then simulated, in order to forecast the characteristics of a future cohort of households. These processes include such occurrences as births, deaths, dissolution of a marriage, children leaving home to set up their own households, etc.

In the approach that is the subject of this paper, the household demographics are assumed to be available from some exogenous source. Given these demographic characteristics, the idea is then to simulate the daily travel of the household, in terms that are useful for estimating models of travel demand. In effect, this approach provides a simulation of what it might be expected that a household with certain demographic characteristics would have provided in response to a household travel survey. To accomplish this, two basic elements are needed: a source of real household data with sufficient demographic detail, and a data set of travel characteristics of sufficient size and detail to permit the development of distributions of the characteristics for different subgroups of the population.

Demographics

The best source of demographic data on households is provided from census information, at least in Australia and the U.S. In the census products available in both of these countries, there is a subsample, called the Public Use Microdata Set (PUMS) in the U.S. and the Household Sample File (HSF) in Australia (Australian Bureau of Statistics, Inc., 1996). The U.S. PUMS comprises 1 percent and 5 percent samples from the decennial census, while the HSF is a 1 percent sample of the 5-yearly census in Australia. Both of these sources share certain other attributes that make them eminently suitable for this task. Both provide fully disaggregate data about the sampled households, meaning that complete socio-demographic data are provided for each household. In order to protect confidentiality of the data, both PUMS and HSF provide only very coarse identification of the geographic location of the household, with specificity being only to an area containing around 100,000 households or so in both cases. In the case of PUMS, the data are grouped into Public Use Microdata Areas (PUMAs), which are areas consisting usually of about 100,000 households, and are defined by local jurisdictions and provided to the U.S. Bureau of the Census. In the case of the HFS, the areas consist generally of one or two Statistical Local Areas (SLAs), which are quite variable in size, comprising usually single local government areas. In 2001, there were 14 SLAs in the Sydney Metropolitan Area, containing a total of about 1.42 million households, or an average of 100,000 households each. In the U.S. case, the households included in the sample are drawn entirely from those households that completed the "long form", thereby providing the richest possible data on the characteristics of the household. The Australian data are likewise rich in demographic details.

In the case of the PUMS data, the households are not sampled through a strict simple random sampling process, and each record is therefore provided with an expansion factor. Applying this expansion factor allows one to determine the contribution of the household to the total population, and would permit creation of an expanded population by replicating each household by the number of its expansion factor. In other words, if a particular household has an expansion factor of 27, then replicating this household 27 times would allow a reconstruction of the actual population, within the bounds of sampling error. In the HSF data, the sample is a simple random sample, so that each household represents 100 households in the full population. Included in the available demographics in both the PUMS and the HSF are:

- The number of household members;
- The number of vehicles available;
- The composition of the household by age and gender;
- The number of workers in the household by age and gender;
- Home ownership status;
- Household income;
- Etc.

The purpose of these files is to provide the source for the households that will have their travel characteristics simulated. One can draw subsamples from the HSF or PUMS that would be identical to the samples of real households that might be drawn for a household travel survey, or one can draw households at random specifically from more limited geographic areas, down to the level at which PUMS or HSF data are provided. By first expanding the HSF or PUMS data, it is also possible to draw samples up to the size of the population in the area of interest. Of course, the sample so created will probably have lower variances on most characteristics than the real population. This is a potential area to explore for future applications, in which it might be feasible to add random components to each expanded household, so as to increase variability.

Travel Characteristic Distributions

The current procedure used in simulating household travel survey data is a Monte Carlo procedure. In this procedure, it is necessary to start with distributions of the desired travel characteristics, from which a series of samples is drawn at random for a specific household. Several issues need to be addressed here. First, the appropriate characteristics describing a household's travel must be defined. Second, a source of the distributions of these characteristics must be found that also has demographic characteristics of the households, similar to those in the HSF or PUMS data, associated with the travel characteristics. Third, it is desirable to group households by demographic characteristics, so that distributions can be used that distinguish between markedly different types of households. This also requires that the source travel data must be a large sample, so that categorisation into a number of subgroups of households is feasible.

It was decided that four primary characteristics of travel by households would be the minimum necessary to achieve the goals of the simulation. These characteristics are:

- Number of trips by purpose made by the household;
- Main mode of travel used for each trip;
- Time of departure of each trip; and
- Trip duration (in minutes) for each trip.

The number of trips by purpose would permit trip production models to be estimated. Trip attraction models could also potentially be estimated from the same data, if there is sufficient specificity on the nature of the attraction end of the trip. Trip duration would permit estimation of impedance functions for a trip distribution model, although it would be desirable to have some geography of travel in order to complete estimation of such a model. At the moment, this is a future development for this approach. Main mode of travel, together with trip duration, would permit at least a fairly simple mode choice model to be estimated. Finally, the time of departure would allow trips to be classified into peak and off-peak trips for the purposes of model estimation and network assignment. Of course, network assignment yet requires geographic information about the trips, which is still a future development. Assuming that highway and public transport networks were available for the subject region, it would be relatively easy, given these simulated data, to create appropriate data on the various time and cost elements of travel by alternative modes, thereby permitting more detailed mode-choice model estimation. Without geographic information, this would require creation of ranges of times and costs that correspond to a total trip duration by mode of a specific amount.

The source selected for the travel characteristics was the 1995 Nationwide Personal Transportation Survey (NPTS) conducted by the U.S. Federal Highway Administration. This is a large household travel survey that was conducted in a manner that is similar to but not identical to that used by most household travel surveys. Nevertheless, the similarities are sufficient to permit relatively easy conversion to consistent data. The sample size is large, with the full data set, including add-ons, amounting to approximately 42,033 households. Data were collected throughout an entire year, and included travel on weekdays and weekends. For comparability to standard household travel surveys, the data on weekends were removed. In addition, problems were found (Greaves, 2000; Stopher and Greaves, 2001) with some of the NPTS data, relating to such issues as proxy reports and incomplete households. Imputation was used to complete households with low levels of person nonresponse. After all cleaning and imputation, the sample available for the source distributions consisted of 30,400 households, containing 74,500 persons, and making 305,856 trips, or slightly more than an average of 10 trips per household per day.

Having selected this data set, which also contains similar demographic data on households to that contained in the PUMS and HSF, it was desired to group households into those with similar travel characteristics. This was done for each trip purpose, each mode, each departure time category, and each trip-length group, using Correlation and Regression Tree Analysis (CART) (SPSS, 1998). The results of this are reported elsewhere (Stopher and Greaves, 2001;

Greaves and Stopher, 2000). The results were, however, to define groupings of households for each travel characteristic, with anything from 5 to 16 groupings being used for different elements.

For each defined grouping, distributions were created of the relevant characteristics. These were created as cumulative distributions, in order to assist the Monte Carlo process. For example, it might be found that households with one person and one worker had 10 percent of cases in which no work trip was made, 28 percent with one work trip per day, 55 percent with two work trips per day, 2 percent with 3 trips per day, 3.5 percent with 4 trips, and the remaining 1.5 percent making more than 4 trips. This would produce a cumulative distribution with up to zero trips at 10 percent, up to one trip at 38 percent, up to two trips at 93 percent, up to 3 trips at 95 percent, up to 4 trips at 98.5 percent, etc. In the Monte Carlo procedure, each random number drawn was converted to a percentage. Assuming that 5-digit numbers were drawn from the range 00000 to 99999, these numbers were treated as a percentage, with three decimal places. If the number drawn was 08736, this was assumed to be 8.736 percent, and would produce 0 work trips. If a number of 79264 was drawn, this would be 79.264 percent and would produce 2 work trips.

The Simulation Process

The simulation process would follow the procedure described in this section. First, the limits of the region are defined for which a sample is desired, with the constraint that the limits must correspond to the boundaries of the available socio-demographic data (PUMAs or HSF Areas). Next, a sampling strategy is decided upon, which could be a simple random sample, or some type of stratified sample, using characteristics of the households in the source data sets. For example, a sample stratified by household size and number of vehicles in the household could be drawn. Having drawn the sample of households, each household is taken in turn. Based on its demographic characteristics, random drawings are made from the appropriate distributions to determine how many trips the household will make for each trip purpose.

The remaining simulation is done based on each trip that has been simulated in the first step. For each trip, the next characteristic to be simulated is the mode of travel. The appropriate distribution for the characteristics of the household and the trip purpose are used at this step. Similarly, the third step uses the purpose, the mode, and characteristics of the household to determine the appropriate distributions of departure time, which are also selected at random and assigned to the trip. Finally, the simulation produces a trip length in time, based on the purpose, mode, time of day, and the characteristics of the household. It is important to note that this is a cascading simulation, in that the later characteristics of each trip are dependent on what has already been simulated.

The process described here is repeated for each trip for each household in the sample. We have experimented with making multiple drawings for each household and averaging the results, as is usually done in Monte Carlo simulation, but have found that this makes almost no discernible difference in the results. This is probably because we have generally applied this to samples of around 3,000 households, so that we are effectively producing the same results by simulating a large number of households and looking at aggregate statistics as we would obtain from repeated simulations for each individual household and averaging the results. For example, if a sample of 3,000 households is being used, each grouping for trip purpose is likely to contain around 200 households, so that the trips by purpose are simulated approximately 200 times from the same distribution. This will produce about 2,000 trips for these households, and the subsequent simulations are therefore repeated probably more than 200 times for some of the subsequent distributions.

It is important to note that the simulation procedure described here is intended to produce aggregate statistics and data for use in standard aggregate models. The results are not consistent at a disaggregate level, where, for example, a household may end up with independent simulations of trips that represent impossible combinations of trips by purpose, by mode, by time of day, and by trip length. Because the simulations cannot be readily made totally interdependent, it is possible, for example, for a household to end up with two work trips for one worker, with one trip being made by car, and the other by bus. Likewise, both trips might be made in the morning peak, and one might be 50 minutes long while the other is 15 minutes long. Thus, at a disaggregate level, implausible combinations of trip characteristics will be likely to arise. However, when looked at as groups of households, the results overall make sense and approximate what would be obtained in a real household travel survey, as is discussed in subsequent sections of this paper.

It would be appealing to find ways to circumvent this implausible disaggregate combination, and our intentions in this regard are to move to attempting to simulate trip tours rather than trips. In such a process, it might be expected that the first step would be to simulate the number of trip tours by type that would be made by the household, or even by individual persons in the household. The next step would be to simulate the number of stops made on the tour, and the next might be to simulate the modes used for the entire tour. Certain mode combinations would be specified as being possible, while others would not be possible. In a similar way, one might next simulate the total duration of the tour, and then the duration of each segment of the tour and each stop along the tour, with the constraint that these must add up to the total tour length. This is likely to require some sophistication in generating truncated distributions along the way, so that values can be drawn only from the plausible range at each step. Finally, a time of departure may be simulated for the beginning of the tour, contingent on the total length of the tour, its purpose and number of stops. The time of departure for each segment of the tour would then be calculated, not simulated. Other options are also possible, but these yet remain as future research.

The research reported in this paper represents a continuation of work undertaken by Greaves, Stopher and Bullock (Greaves, 1998; Greaves, 2000; Greaves and Stopher, 2000; Stopher *et al.*, 2003).

Non-Mobile Households

One additional issue that needs to be discussed at this point is that of households that report no travel on the survey day. In any household travel survey, such households are found. It is known that reporting no travel is sometimes used as a non-response mechanism, while in other cases, either a person or a household may genuinely not have travelled on the survey day. The proportion of non-mobile households in a household travel survey therefore varies significantly from place to place and from survey to survey. While there are no clear statistics to support this, it is often considered that personal non-mobility is probably on the order of 5 to 10 percent on a weekday, while household non-mobility is on the order of 2-6 percent. The simulation process, as described here, will produce non-mobile households (although trips are not designated specifically to persons, so personal non-mobility figures are not able to be determined). However, the creation of a non-mobile household is necessarily a rare event in the simulation procedure, because it requires seven independent Monte Carlo simulations to all produce zero trips as the result, there being seven trip purposes for which a number of trips is to be simulated. For example, if each purpose has only a 10 percent probability of there being zero trips made for that purpose, then the probability of a simulation producing a non-mobile household is 0.0000001, or 1 in 10 million. This is a much rarer occurrence than the reporting of non-mobility.

Surprisingly, the simulations have produced around 2 percent of non-mobile households. However, most of the real household travel surveys show non-mobility rates that range from 3 percent to 8 percent. This will immediately create a discrepancy in average trips rates for all households. Necessarily, the simulation will underestimate non-mobile households, while concomitantly overestimating trip rates.

This presents another challenge for future research. However, a move to changing the base of the simulation to trip tours may prove beneficial in this context also. If the number of trip tours by type can include zero, then this should permit the creation of non-mobile households at an appropriate rate, where there would no longer be the requirement for a series of independent drawings to all produce zero trips, in order to define a non-mobile household.

Assessing the Results of Simulation

In the exploratory stages of this research, the main focus has been to try to see how well simulation, as described in this paper, replicates the results that would be obtained from a real household travel survey. To do this, we have chosen specific locations in the U.S.A. that had household travel surveys conducted near to both the 1995 NPTS and the 1990 Census, in order that each of these two data sources could be used to provide the sample of households and simulate the travel characteristics. For each such location, the first step was to select a household sample from the local PUMS data that was sampled by the same procedure as was used in the real household travel survey, and with identical numbers of households. Given a simulation of travel characteristics for households that were selected to look as much as possible like those from the household travel survey, comparisons were made between the simulated and actual characteristics at an aggregate level.

In the very first research, conducted by Greaves (2000), Baton Rouge was selected as the locality for testing, because the Baton Rouge Personal Transportation Survey (BRPTS) was conducted in 1996 using the same exact survey instruments and protocols as the NPTS, and was conducted by the same survey firm. (Baton Rouge was originally intended to have been one of the NPTS add-on sites, but, due to contracting delays, was eventually undertaken too late to be included in the add-on sample.) Greaves concluded that simulation had performed quite well, and that trip rates, in particular, showed remarkably close results between simulation and the real survey. Time of departure was also good, but mode was a little less close, as was also the case with trip length. Nevertheless, the results overall looked good enough to pursue the topic further.

Next, the approach was extended to two other urban areas in the U.S. – Dallas-Fort Worth and Salt Lake City (Stopher and Greaves, 2001; Stopher *et al.*, 2003). Both of these were larger regions than Baton Rouge, with approximately 4 million and 1 million population, respectively, compared to 500,000 in Baton Rouge. In addition, other characteristics of the two new urban areas were significantly different from Baton Rouge and from each other. The household travel surveys of these two regions were also substantially different from each other and from the Baton Rouge and NPTS surveys. While the results were not quite as good as those obtained for Baton Rouge, they still provided quite reasonable approximations to the household travel surveys, and also showed that the simulation procedure responded to differences in the populations between the three locations. Results of each step of the simulation showed a number of differences between each of the locations.

While statements have been made here about the closeness of results, there are some problems in being able to assess this. In his initial work, Greaves used statistical tests of differences between trip rates, and percentages of trips in different categories, such as modes, departure times, and trip lengths. The result of this was to find a large number of statistically significant differences. However, this is misleading. Because of the rather large numbers of households involved in the simulation efforts, even very small numeric differences between rates and shares will produce statistically significant differences. In many cases, sampling error alone may account for most of the difference detected between a simulation rate and a real household travel survey rate. Subsequent work has continued to report the results of the statistical tests of significance, but has also concentrated more on looking at numerical differences and also in expanding the data to give regional totals of trips by characteristic (Stopher *et al.*, 2003). While the comparison of expanded data gives a more accurate picture of the extent to which the simulation and real household travel surveys would produce similar results, there is no statistical procedure to determine if these differences are statistically significant. It becomes a matter of subjective judgement.

In the long run, it is probably more appropriate to compare the results of estimating models on both the simulated and real data, to determine if parameters of the models are significantly different. However, estimating a full set of models from both data sets is a substantial undertaking and has not yet been attempted, although Greaves did show the results of doing this with Baton Rouge data for a trip production model (Greaves, 2000).

Most recently, this research has been extended outside the United States. In this effort, the metropolitan area of Adelaide was chosen. This choice was made for several reasons: the data from Adelaide were made available; Adelaide conducted its most recent household travel survey in 1999, and the Australian Bureau of Statistics provided 1996 HSF data for the Adelaide region; and Adelaide represents a medium-sized city in the context of the previous research, having a population of a little over 1 million people in 1996. The balance of this paper describes the results of the simulation applied to Adelaide, using the HSF data for the Adelaide region, but also using the NPTS distributions of travel characteristics and the classifications developed in the U.S. applications. Comparisons are made to the results obtained in the U.S. research with Dallas, Salt Lake, and Baton Rouge.

ADELAIDE SIMULATION CASE STUDY

A sample of households, with socio-demographic descriptors, was generated from the 1996 Census Household Sample File (HSF), derived from the 1996 Australian Bureau of Statistics (ABS) Census Data, and travel distributions derived from the earlier work in the United States. Travel distributions, (or key variables influencing travel behaviour) were derived from US data mainly because no national household travel survey exists in Australia. The synthesized data were then compared to the 1999 Metropolitan Adelaide Household Travel Survey (AHTS) to determine the ability of simulated data to synthesize the actual travel patterns observed in the AHTS survey.

Data Preparation

The Metropolitan Adelaide Household Travel Survey (AHTS) conducted in 1999 by Transport South Australia of the Department for Transport, Urban Planning and the Arts sampled 5,886 households with 14,004 associated individuals out of a then total population of 1,045,854 individuals. The survey collected demographic and socio-economic data for each household as well as information on travel activities undertaken over a two day period. These data were collected using several surveys of each household which required cleaning and combining for use. Several categorical variables were created for the analysis including a lifestyle variable consisting of the number of workers and children within a household. After cleaning the data and removing households that had not completed all of the surveys, 5,615 households remained.

The 1996 Census Household Sample File (HSF) represents a one percent sample from the 1996 census household data collected by the ABS. The file consisted of three files covering private dwellings, their associated household and individual records as well as a one percent sample of persons from non-private dwellings. These were combined so that the format of the final data file was similar to that of the AHTS data file. Again significant cleaning of the data was required before analysis could be undertaken.

Table 8.1 shows comparative statistics for the AHTS and the HSF for the city of Adelaide. For completeness, the table also includes the comparative statistics for the data collected for an earlier study using the 1995 U.S. Nationwide Personal Transportation Survey (NPTS) as well as HTSs undertaken in Baton Rouge (Greaves, 2000) Dallas and Salt Lake (Stopher *et al.*, 2003).

In comparing the Australian data sources, it becomes evident that possible sampling problems exist between the two. In comparison to the HSF, the AHTS appears to have under sampled the number of non-car-owning households whilst over sampling the percent of home owners. For all other variables, the statistics shown appear to be similar across data sets. It can be seen that the number of household workers in both Australian data sets is below the average represented across the U.S. data. Further, it is evident that the proportion of non-car-owning households in the Australian data is greater than that represented in the U.S. data, whilst the average number of household vehicles is below that in the U.S. data. The above suggests deviations away from the trip distributions derived in the U.S. towards a greater use of public transport and a lower number of home-work trips.

Table 8.1 Summary Statistics from the Household Travel Surveys

Statistic	AHTS	HSF	NPTS	Baton Rouge	Salt Lake	Dallas
Average Age	38	35	35	33	34	35
Average Household Size	2.46	2.44	2.63	2.71	3.14	2.47
Percent in Single Family Dwellings	N/A	N/A	74%	75%	73%	78%
Percent from Non-Car-Owning Households	9.4%	13.2%	8%	8%	4%	5%
Average Vehicles per Household	1.56	1.43	1.73	1.78	1.97	1.84
Percent Females in Sample	51.9%	50.9%	51%	52%	53%	52%
Percent Home Owners	76.5%	59.5%	64%	66%	76%	68%
Average Workers per Household	1.07	1.03	1.33	1.34	1.31	1.40

SIMULATING TRAVEL SURVEY DATA

A simulation sample was drawn from the HSF data replicating as closely as possible the AHTS sample using a procedure similar to that described in Stopher *et al.* (2003). This simulation sample was then compared to the actual trip rates in the AHTS sample. Table 8.2 compares the simulation data to the AHTS data. Table 8.2 shows that in terms of general household characteristics, the simulated data compared favourably to the AHTS data.

Table 8.2 Comparison of Statistics of the HTS and Simulation Samples

Statistic	AHTS	Simulation
Sample Size (Households)	5615	**5615**
Average Vehicles per Household	1.56	**1.42**
Average Household Size	2.46	**2.44**
Average Workers per Household	1.07	1.04

Trip Rates

Table 8.3 compares the trip rates for several trip purposes across the AHTS and simulation data using the Z test for equal population means. As can be seen from the table, home-based work, home-based school, home-based college, home-based other and other-work trips were significantly different at the 99% level between the two data sets. Total number of trips was also significantly different at the 99% level. With the exception of home-based other trips, the simulation overestimated the trip rates.

Table 8.3 Comparisons of HTS and Simulated Person Trip Rates per Household

Purpose	**AHTS**		**Simulation**	
	Mean	**Std. Dev.**	**Mean**	**Std. Dev.**
Home-Based Work	1.07	1.57	1.38**	1.71
Home-Based School	0.34	1.07	0.60**	1.38
Home-Based College	0.08	0.42	0.16**	0.63
Home-Based Shop	1.30	1.75	1.28	1.72
Home-Based Other	3.83	4.27	3.28**	3.53
Other-Work	0.80	2.34	1.00**	1.83
Other-Other	1.86	3.18	1.95	2.90
TOTAL TRIPS	9.28	8.02	9.66**	7.28

** Statistically significant difference in trip rates at the 95 percent confidence level*
***Statistically significant difference in trip rates at the 99 percent confidence level*

Table 8.4 Comparisons of HTS and Simulated Person Trip Rates per Mobile Household

Purpose	**AHTS**		**Simulation**	
	Mean	**Std. Dev.**	**Mean**	**Std. Dev.**
Home-Based Work	1.17	1.61	1.43**	1.72
Home-Based School	0.37	1.12	0.62**	1.40
Home-Based College	0.09	0.44	0.17**	0.64
Home-Based Shop	1.42	1.79	1.32**	1.73
Home-Based Other	4.20	4.31	3.39**	3.54
Other-Work	0.88	2.44	1.03**	1.85
Other-Other	2.04	3.28	2.01	2.93
TOTAL TRIPS	10.17	7.89	9.98	7.19

** Statistically significant difference in trip rates at the 95 percent confidence level*
***Statistically significant difference in trip rates at the 99 percent confidence level*

One possible explanation for the over-estimation of trip rates by the simulation is the under-estimation of the number of households not making any trips. For the AHTS survey, 492 households or 8.8% of the total sample, made no trips on the day of the survey whilst the simulation included a relatively small 180 households or 3.2% of the total sample. Correcting this by computing trip rates only for mobile households produces the results shown in Table 8.4.

This creates an interesting result. Most of the trip rates that were over-estimated by the simulation are now closer, with smaller z-scores for the difference, while those that are under-estimated by the simulation are further apart, and the z-scores have increased. However, the overall trip rate is now not significantly different and has closed in on around 10 trips per household per day. This suggests that another part of the problem may be differences in classification of trip purposes between the AHTS and the traditional procedures used in the United States. For example, the simulation produces trip rates for home-based work, home-based school, home-based college, and other-work that are higher than those observed in Adelaide. In contrast, the home-based other, and to a lesser extent the home-based shop are lower than in the Adelaide data. The other-other rate is not different between the two sets of data. This might suggest that the Adelaide survey classified more trips as being home-based other, possibly because of problems in determining the real purpose. This result bears further investigation. Somewhat similar results were also found in Dallas and Salt Lake, and may point to discrepancies between the NPTS method of classifying trip purposes and the methods used more commonly in household travel surveys.

Table 8.5 compares the above findings to that of Stopher *et al.* (2003). Given that the distributions of trips used in each simulation study were derived from the NPTS, differences between the simulation data generated for different localities demonstrates that the simulation is responsive to differences in the local population characteristic. It can be seen, from Table 8.5, that significant differences exist between Adelaide and Salt Lake City for all trip purposes at the 99% level. Significant differences between Adelaide and both Baton Rouge and Dallas do not exist for home-based college and other-other trips.

Table 8.5 also shows the significant differences between simulation and the actual HTS data for each of Baton Rouge, Dallas, and Salt Lake. It is interesting to note here that Baton Rouge had only one significant difference, which was at 95% not 99%. Dallas and Salt Lake each had most purposes showing significant differences. One of the primary differences between these surveys is that the Baton Rouge survey was conducted using identical instruments and protocols to the NPTS survey from which the distributions have been derived, while Salt Lake, Dallas, and Adelaide used markedly different survey procedures. This may suggest that there is also a methodological issue here in the methods used for the surveys that are being used to benchmark the simulations. It is also worth noting here that the number and magnitude of significant differences for Dallas, Salt Lake, and Adelaide between the simulated and actual HTS data for each region are about the same. Regarding these results, which are uncorrected for differences in mobility rates, Adelaide and Dallas each have two trip rates that are not significantly different, while Salt Lake has one, but Salt Lake also has two trip rates that are significantly different at 95% but not at 99%. All three regions show significantly different overall trip rates than their respective HTSs.

Table 8.5 Comparisons of Simulated Person Trip Rates per Household across Data Sets

Purpose	Adelaide	Baton Rouge	Dallas	Salt Lake	Adelaide		
	Simulated Mean	Simulated Mean	Simulated Mean	Simulated Mean	Diff. from Baton Rouge	Diff. from Dallas	Diff. from Salt Lake
Home-Based Work	1.38**	1.83	1.86**	1.83**	**	**	**
Home-Based School	0.6**	0.74	0.60	1.07*	**		**
Home-Based College	0.16**	0.17	0.16	0.23*			**
Home-Based Shop	1.28	1.32*	1.14**	1.38**		**	**
Home-Based Other	3.28**	3.69	3.19**	4.17**	**		**
Other-Work	1.00**	1.34	1.35**	1.33	**	**	**
Other-Other	1.95	2.02	1.86**	2.26**			**
TOTAL TRIPS	9.66**	11.11	10.17**	12.28**	**	**	**

** Statistically significant difference in trip rates at the 95 percent confidence level*
***Statistically significant difference in trip rates at the 99 percent confidence level*

Table 8.6 shows a comparison of total trips by purpose, applying the trip rates from each of the simulation and the AHTS to the expanded population of Adelaide. Overall, simulation underestimates trips in Adelaide by 2.8 percent. There are variable over- and under-estimations by purpose, some of which may indicate different trip-chaining behaviour in Adelaide, compared to the U.S. base, and some of which may indicate inconsistencies in trip-purpose definitions. They may also be consistent with the biases suspected in the Adelaide AHTS data. As an example of the differences in chaining or classification of purposes, for example, if workers in Adelaide are more likely to drop children at school on their way to or from work, this would account for the low figures for home-based work trips and part of the high figures for other-work trips in the AHTS, compared to the simulation. Similarly, if there is more trip chaining for people travelling to and from school and college, with a number of those trips being for shopping and for other purposes, this could account for most of the differences in home-based school, college, shop, and other trips.

Table 8.6 Comparison of Total Trips By Purpose

Trip Purpose	AHTS	Simulation	Difference between AHTS and Simulation
Home-work	375,808	399,122	+23,314
Home-school	131,814	190,175	+58,361
Home-college	19,174	52,079	+32,905
Home-shop	363,922	406,205	+42,283
Home-other	1,193,587	1,052,156	-141,431
Other-work	415,720	281,307	-134,413
Other-other	601,813	632,570	+30,757
Total	3,101,839	3,013,613	-88,226

Mode Share Comparisons

Table 8.7 shows a comparison of the mode shares for the Adelaide HTS and simulation data as well as comparing the Adelaide simulation mode shares with those of the simulation data from Dallas and Salt Lake City. Significant differences are also shown. Baton Rouge was not compared due to the non-availability of data. As can be seen for the Adelaide study, significant differences exist for nearly every mode share, the exceptions being for home-school drivers, home-college bike/walk, home other transit, other-work driver and other-other transit.

Examination of the differences between both US simulation data and the Adelaide simulation data show that home-school, other-other and home-college trip rates (Dallas only) and mode shares are statistically similar. Various other mode shares are also statistically similar across the simulation data.

Table 8.8 shows the expanded data for each of the modes and purposes. Also, in the table, we show the results that would have been obtained if the numbers of trips had been exactly the same in the simulation as in the actual AHTS data. Because the simulation is a cascading one, any differences in trip rates and, therefore, in aggregate trips, will continue to appear in each succeeding simulation step. Therefore, it is useful to remove these differences from earlier simulation steps, in order to see how well the mode simulation works. The simulation overestimates privately-owned vehicle driver (POV Driver), underestimates car passenger, overestimates transit, and underestimates bike and walk modes overall. After adjusting the trip totals, car driver shifts further from the AHTS, car passenger is almost exactly correct, transit drops from being too high to too low, and bike and walk changes very little. Within the purposes, there are more substantial differences, with the simulation predicting a much higher transit use for home-school trips than the AHTS found, and much lower transit use for home-other and other-work trips.

As with the trip numbers by purpose, examination of the expanded data shows smaller numeric differences between the simulation and the actual HTS than might be expected from looking at the percentages of mode use by purpose. This tends to bear out the problems of making comparisons just on rate data.

Table 8.7 Comparisons of Simulated Data by Mode and Purpose

Trip Purpose	Mode	Adelaide		Dallas	Salt Lake	Adelaide Simulation	
		HTS	Simulation	Simulation	Simulation	Diff From Dallas	Diff From Salt Lake
Home-Work	Driver	82.56%	86.53%**	88.76%	89.92%	**	**
	Passenger	7.51%	9.33%**	7.44%	7.42%	**	**
	Transit	6.43%	2.21%**	2.16%	1.22%		**
	Bike/Walk	3.49%	1.92%**	1.63%	1.43%		*
Home-School	Driver	1.61%	2.28%	3.75%	5.43%	**	**
	Passenger	58.86%	37.3%**	38.99%	36.40%		
	Transit	13.67%	51.11%**	49.16%	49.68%		
	Bike/Walk	25.84%	9.3%**	8.09%	8.48%		
Home-College	Driver	47.97%	73.34%**	71.51%	68.90%		
	Passenger	11.48%	10.66%	11.02%	8.27%		
	Transit	31.98%	6.85%**	7.81%	7.85%		
	Bike/Walk	8.55%	9.14%	9.64%	14.98%		**
Home-Shop	Driver	64.79%	67.75%**	73.04%	72.33%	**	**
	Passenger	17.20%	24.41%**	21.66%	22.65%	**	*
	Transit	3.51%	2.16%**	1.38%	1.22%	**	**
	Bike/Walk	14.47%	5.66%**	3.90%	3.80%	**	**
Home-Other	Driver	49.52%	60.47%**	62.38%	59.40%	**	
	Passenger	36.05%	31.74%**	30.66%	33.62%	*	**
	Transit	1.90%	2.11%	1.87%	1.45%		**
	Bike/Walk	12.51%	5.66%**	5.07%	5.54%	*	
Other-Work	Driver	82.54%	82.79%	85.09%	85.12%	**	**
	Passenger	4.94%	10.4%**	8.91%	9.72%	**	
	Transit	1.22%	1.86%**	1.50%	1.13%		**
	Bike/Walk	11.28%	4.93%**	4.48%	4.04%		*
Other-Other	Driver	48.23%	61.88%**	63.22%	60.26%		*
	Passenger	35.50%	30.57%**	29.76%	32.99%		**
	Transit	1.96%	1.97%	1.74%	1.46%		**
	Bike/Walk	14.28%	5.56%**	5.26%	5.28%		
All Purposes	Driver	56.35%	64.36%**	68.24%	63.80%	**	
	Passenger	27.92%	25.1%**	22.53%	25.55%	**	
	Transit	3.27%	5.21%**	4.69%	5.70%	**	**
	Bike/Walk	12.44%	5.31%**	4.52%	4.95%	**	*

* *Statistically significant difference in trip rates at the 95 percent confidence level*
***Statistically significant difference in trip rates at the 99 percent confidence level*

Table 8.8 Mode by Purpose for the Simulation and Adjusted Simulation versus the AHTS

Mode	Source	Home-work	Home-school	Home-college	Home-shopping	Home-other	Other-work	Other-other	Total
POV Driver	AHTS	313,797	1,642	11,176	226,135	576,476	291,481	278,483	1,699,191
	Simulation	340,055	3,159	33,553	270,717	617,173	229,566	374,109	1,868,330
	Adjusted	319,825	2,190	12,365	241,738	698,586	338,813	355,455	1,968,972
POV Passenger	AHTS	29,164	86,515	3,501	60,820	393,206	27,171	199,142	799,520
	Simulation	38,424	69,965	5,768	103,852	340,001	31,905	198,717	788,632
	Adjusted	36,138	48,494	2,127	92,735	384,852	47,088	188,809	800,243
Transit	AHTS	16,533	2,004	1,302	8,032	52,439	57,838	22,668	160,816
	Simulation	11,334	96,340	5,594	11,809	26,922	4,802	18,987	175,787
	Adjusted	10,660	66,775	2,063	10,545	30,473	7,087	18,040	145,593
Bike/ walk	AHTS	15,883	41,653	3,195	67,736	168,346	38,601	100,522	435,937
	Simulation	9,309	20,712	7,083	19,828	67,635	14,977	40,532	180,075
	Adjusted	8,755	14,356	2,612	17,705	76,557	22,104	38,511	180,600
Total	AHTS	375,378	131,814	19,174	362,723	1,190,468	415,091	600,815	3,095,463
	Simulation	399,122	190,175	51,998	406,205	1,051,731	281,249	632,345	3,012,824
	Adjusted	375,378	131,814	19,174	362,723	1,190,468	415,091	600,815	3,095,408

Departure Time Comparisons

Table 8.9 shows the departure time comparisons for all trip purposes. The Kolmogorov-Smirnov D-value provides a test to determine if the distribution of trips over the day is significantly different between the HTS and simulated data. This shows that for all but home-based school trips and other-work trips significant differences are observed between the two over the day. However the Z test between the individual fractions in each time period demonstrate slightly better results with 14 out of the 32 fractions showing statistically different results. Interestingly, the differences largely occur in the home-based shopping, home-based other and other-other trips suggesting that for all other categories Australians and Americans share similar travel time patterns. Clearly examination of the percentages suggest that residents of Adelaide prefer to, or are forced by necessity to shop between business hours whereas in America, a greater percentage of shopping activity probably occurs outside this period.

Table 8.9 Comparisons of HTS and Simulated Departure Times by Trip Purpose

Purpose	Time Period	AHTS	Simulation	D Stat
Home-Based Work	6.01 a.m. - 9.00 a.m.	35.3%	35.7%	**
	9.01 a.m. - 4.00 p.m.	24.5%	24.2%	**
	4.01 p.m. - 7.00 p.m.	28.3%	25.3%*	**
	7.01 p.m. - 6.00 a.m.	11.9%	14.8%	**
Home-Based School	6.01 a.m. - 9.00 a.m.	52.7%	49.4%	
	9.01 a.m. - 4.00 p.m.	42.4%	45.2%	
	4.01 p.m. - 7.00 p.m.	4.6%	4.9%	
	7.01 p.m. - 6.00 a.m.	0.3%	0.5%	
Home-Based College	6.01 a.m. - 9.00 a.m.	25.4%	36.0%*	**
	9.01 a.m. - 4.00 p.m.	44.8%	40.4%	**
	4.01 p.m. - 7.00 p.m.	22.3%	13.7%	**
	7.01 p.m. - 6.00 a.m.	7.6%	9.9%	**
Home-Based Shop	6.01 a.m. - 9.00 a.m.	5.5%	5.9%	**
	9.01 a.m. - 4.00 p.m.	67.4%	53.9%**	**
	4.01 p.m. - 7.00 p.m.	21.2%	24.9%**	**
	7.01 p.m. - 6.00 a.m.	5.9%	15.2%**	**
Home-Based Other	6.01 a.m. - 9.00 a.m.	11.3%	12.0%	**
	9.01 a.m. - 4.00 p.m.	47.6%	40.8%**	**
	4.01 p.m. - 7.00 p.m.	24.6%	26.6%*	**
	7.01 p.m. - 6.00 a.m.	16.4%	20.4%**	**
Other-Work	6.01 a.m. - 9.00 a.m.	11.7%	13.0%	
	9.01 a.m. - 4.00 p.m.	65.9%	64.2%	
	4.01 p.m. - 7.00 p.m.	17.8%	18.2%	
	7.01 p.m. - 6.00 a.m.	4.6%	4.5%	
Other-Other	6.01 a.m. - 9.00 a.m.	6.1%	6.8%	**
	9.01 a.m. - 4.00 p.m.	68.1%	61.2%**	**
	4.01 p.m. - 7.00 p.m.	17.3%	18.8%	**
	7.01 p.m. - 6.00 a.m.	8.6%	13.1%**	**
TOTAL TRIPS	6.01 a.m. - 9.00 a.m.	6.1%	16.3%**	**
	9.01 a.m. - 4.00 p.m.	68.1%	47.0%**	**
	4.01 p.m. - 7.00 p.m.	17.3%	22.2%**	**
	7.01 p.m. - 6.00 a.m.	8.6%	14.3%**	**

** Statistically significant difference in trip rates at the 95 percent confidence level*
***Statistically significant difference in trip rates at the 99 percent confidence level*

Table 8.10 provides the comparisons on expanded data for departure time. In the unadjusted simulation, home-work trips are reasonably well distributed by time of day, except for the 7 pm to 6 am time slot, where the simulation overestimates rather significantly. The adjusted figures for this purpose show an underestimate for 6-9am and an overestimate for 7pm–6am. After adjustment, home-school trips fall quite close to the AHTS distribution by time. The same is true for home-college trips. Home-shopping trips show an under estimate in the 9am-4pm period and an overestimate for 7pm-6 am, which, as noted before, may be due to the shorter shopping hours in Adelaide than are common in the U.S. After adjustment, home-other trips are too high in the night period, and too low in both the 6-9 am and 9am-4pm periods. Other-work trips are overestimated in the early morning and underestimated in the early evening. Other-other trips are also overestimated in the overnight period, while the early morning and midday periods are underestimated. Again, however, the overall results show a reasonably close approximation to the AHTS, especially after adjustment for the mis-prediction of the number of trips.

Table 8.10 Time of Day by Purpose for Simulation and the AHTS

Departure Time	Source	Home-work	Home-school	Home-college	Home-shopping	Home-other	Other-work	Other-other	Total
6am–9am	AHTS	144,871	69,708	4,817	26,015	173,910	40,036	43,713	503,070
	Simulation	145,129	93,406	15,198	23,356	124,029	39,664	40,690	481,472
	Adjusted	136,385	64,742	5,595	20,922	140,586	58,606	38,703	465,539
9am–4pm	AHTS	86,511	55,782	7,644	228,326	547,247	264,740	416,639	1,606,890
	Simulation	97,392	84,216	22,822	225,352	449,553	180,612	402,361	1,462,308
	Adjusted	91,524	58,372	8,402	201,865	509,566	266,866	382,717	1,519,312
4pm–7pm	AHTS	99,246	5,976	4,184	85,660	287,655	90,502	98,140	671,362
	Simulation	97,498	11,000	8,151	94,241	273,356	49,451	111,153	644,851
	Adjusted	91,624	7,624	3,001	84,419	309,848	73,067	105,726	675,309
7pm–6am	AHTS	44,445	349	2,529	23,867	183,802	20,372	43,195	318,559
	Simulation	59.102	1.553	5.907	63.256	205.217	11.580	78.367	424.983
	Adjusted	55,541	1,076	2,175	56,663	232,613	17,110	74,541	439,719
Total	AHTS	375,074	131,814	19,174	363,868	1,192,614	415,650	601,687	3,099,881
	Simulation	399,122	190,175	52,079	406,205	1,052,156	281,307	632,570	3,013,613
	Adjusted	375,073	131,814	19,174	363,868	1,192,613	415,650	601,688	3,099,880

Adjustment should also be made, in this case, for mode. However, to do this becomes rather complex and has not been attempted at this time. It is possible that this would lead to closer approximations to the actual departure time distributions.

Trip-Length Comparisons

Table 8.11 shows the comparison for the Adelaide HTS and simulated trip lengths. Statistical differences are observed for all trips purposes with the exception of home-based shopping trips. With the exception of home-based school trips, where a difference exists, the simulation underestimates the trip length. One of the critical shortcomings of the simulation is that it does not take into account physical geography.

Table 8.1 Comparison of HTS and Simulated Vehicle Trip Lengths (minutes)

Purpose	Adelaide		Simulation	
	Mean	Std. Dev.	Mean	Std. Dev.
Home-Based Work	23.63	16.64	19.46**	16.44
Home-Based School	14.08	12.04	16.34**	13.76
Home-Based College	29.35	20.44	19.02**	15.32
Home-Based Shop	12.10	11.64	12.32	12.00
Home-Based Other	16.64	18.87	14.01**	15.75
Other-Work	17.26	18.93	15.11**	17.42
Other-Other	15.50	20.87	14.29**	16.49
TOTAL TRIPS	16.66	18.34	14.97**	15.77

* *Statistically significant difference in trip rates at the 95 percent confidence level*
***Statistically significant difference in trip rates at the 99 percent confidence level*

Table 8.12 shows the aggregate comparisons with adjustment again for the overall numbers of trips by purpose. Figure 8.1 shows the results for home-based work trips, and Figure 8.2 for home-based other trips. The figures are a little easier to interpret. For home-based work trips, Figure 8.1 shows that the distribution is quite similar, except for the unexpected peak at 26-30 minutes. For the rest, the simulation and the original AHTS data show a similar decay pattern for trip making. For home-based other trips, Figure 8.2 shows a very similar decay pattern between the AHTS and the simulation results. Again, this shows that the statistical tests for significant differences really fail, because the similarity of the trip pattern is very much closer than the statistics would lead one to expect.

CONCLUSIONS

While the simulation produced a considerable number of statistical differences to the HTS, that warrant further investigation as to how the simulation procedure can be improved, the overall performance of the simulation was not much different from that found in application to Dallas and Salt Lake in the U.S. As has been stated, there do appear to be significant differences between the US and Australia in terms of reliance upon privately owned automobiles. As a result, the simulated results are less close than would be desired when applied to the Australian sphere. This suggests that in the future, perhaps a small local survey be conducted and the results used to update the distributions so as to take into account the local trip distributions (see Stopher and Pointer, 2003). Nevertheless, when one considers that the Adelaide data showed about 30 percent fewer workers per household, an average of 50% more non-car-owning households, and about 25% lower average cars per household, it is remarkable that the simulations for Adelaide did as well as they did. This seems to indicate that the simulation procedure used here is capable of adapting to quite a wide range of differences in socio-demographics and transport supply situations.

Recent work by Kothuri (2002) on Bayesian updating of the simulations suggests that this may be a simple method by which the results can be improved. This would be the next logical step in this procedure of simulation. At the same time, it is important to note that statistical significance of differences between the simulated and original survey results is neither a sufficient nor a necessary condition for acceptance of the simulation procedure. These results are inputs to descriptions of the region, obtained through data expansion and weighting procedures, and to models that describe and forecast travel behaviour. To determine the real acceptability of the simulations, it would be necessary to undertake both of these activities with each of the simulated and actual data and compare the results.

Table 8.12 Trip Length By Purpose from the AHTS and both Simulations

Time Period	Source	Home-work	Home-school	Home-College	Home-shopping	Home-other	Other-work	Other-other	Total
1-5 mins.	AHTS	35,117	47,799	2,519	140,204	332,623	82,825	196,382	837,468
	Simulation	63,507	46,887	6,912	134,451	346,426	92,101	198,311	888,596
	Adjusted	59,766	32,498	2,545	120,438	392,966	136,085	188,669	932,967
6-10 mins.	AHTS	56,982	42,754	2,953	112,690	303,886	76,792	147,015	743,072
	Simulation	79,865	43,704	13,921	112,052	271,386	65,016	162,370	748,313
	Adjusted	75,160	30,292	5,125	100,373	307,845	96,066	154,475	769,336
11-15 mins.	AHTS	71,283	22,715	4,241	54,848	219,354	68,266	99,300	540,007
	Simulation	75,743	31,469	10,680	68,442	179,623	48,150	109,519	523,625
	Adjusted	71,281	21,812	3,932	61,309	203,754	71,145	104,194	537,426
16-20 mins	AHTS	45,188	10,061	2,993	24,335	109,967	44,054	49,651	286,249
	Simulation	61,175	21,817	5,646	41,063	99,519	28,980	63,625	321,825
	Adjusted	57,571	15,122	2,079	36,783	112,889	42,820	60,531	327,795
21-30 mins.	AHTS	96,898	5,574	3,265	20,614	130,446	66,000	59,734	382,532
	Simulation	65,855	25,304	8,975	28,735	85,044	24,865	54,770	293,548
	Adjusted	61,976	17,539	3,304	25,740	96,469	36,740	52,107	293,874
31-45 mins.	AHTS	50,445	2,220	1,582	7,549	59,181	45,414	27,849	194,240
	Simulation	35,424	16,216	4,189	12,676	42,849	12,350	22,909	146,614
	Adjusted	33,337	11,240	1,542	11,355	48,605	18,248	21,795	146,122
46-60 mins.	AHTS	14,900	216	1,142	2,837	20,524	20,592	10,182	70,392
	Simulation	11,204	3,914	816	6,187	10,076	4,057	9,241	45,495
	Adjusted	10,544	2,713	300	5,542	11,430	5,994	8,792	45,315
60+ mins.	AHTS	4,797	475	478	792	17,527	11,707	11,700	47,476
	Simulation	6,348	865	941	2,599	17,232	5,788	11,825	45,598
	Adjusted	5,974	600	346	2,328	19,547	8,552	11,250	48,597
Total	AHTS	375,610	131,814	19,174	363,868	1,193,506	415,650	601,813	3,101,436
	Simulation	399,122	190,175	52,079	406,205	1,052,156	281,307	632,570	3,013,613
	Adjusted	375,610	131,814	19,174	363,868	1,193,506	415,650	601,813	3,101,435

Finally, current research is examining the potential to use the data directly, without estimation of models. The concept here is that the simulation of current data requires the input of socio-demographic data, together with distributions of travel characteristics for different classes of households. Assuming that this approach is valid, then forecasting demographics to the future should permit simulation of future travel patterns, assuming that the underlying distributions of travel characteristics by household class remain stable. It may even be possible to develop a method to predict changes in these distributions that would further improve the prediction of future travel.

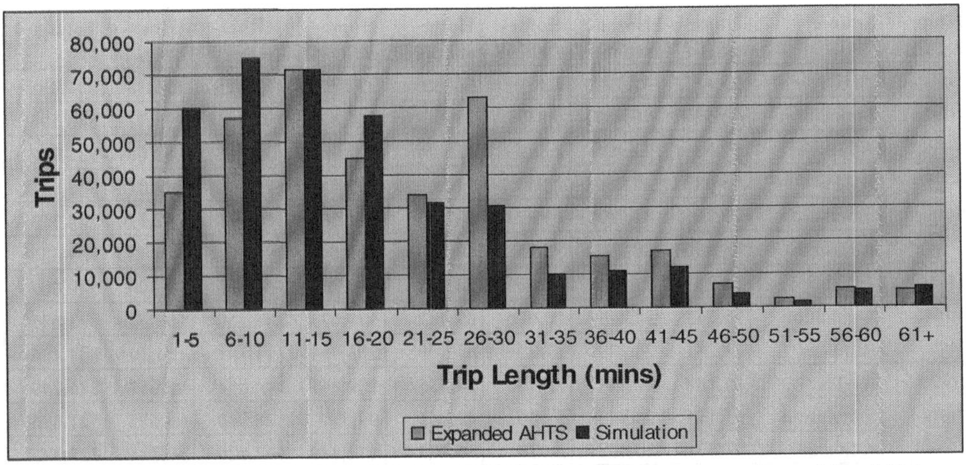

Figure 8.1 Trip Length Frequency Distribution (Adjusted) for Home-Based Work Trips

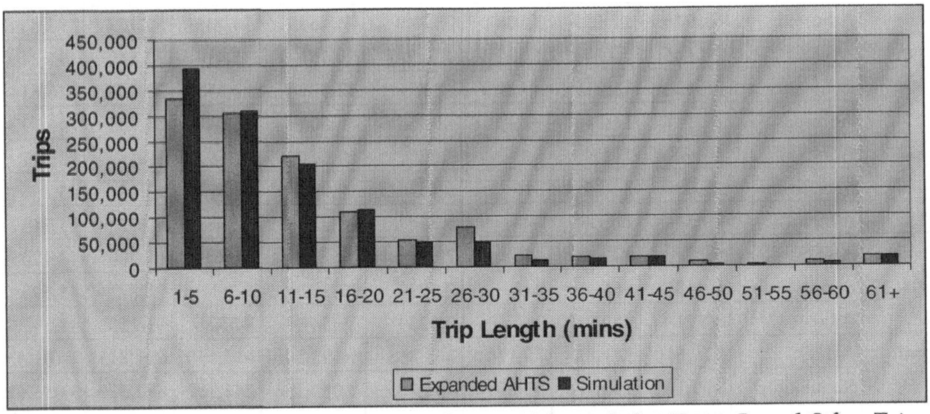

Figure 8.2 Trip Length Frequency Distribution (Adjusted) for Home-Based Other Trips

REFERENCES

Australian Bureau of Statistics Census (1996). 1996 Census Household Sample File (HSF).

Cambridge Systematics, Inc. (1996). *Travel Survey Manual*, prepared for U.S. DOT and U.S. EPA, TMIP Program, Washington DC, July 1996.

Chung, J. and K.G. Goulias (1997). Travel demand forecasting using microsimulation: initial results from a case study in Pennsylvania. Paper presented at the 76th Annual Meeting of the Transportation Research Board, Washington DC, 1997.

Greaves, S.P. (1998). *Applications of GIS Technology in Recent Household Travel Survey Methodologies*. Report prepared for the Travel Model Improvement Program, Federal Highway Administration.

Greaves, S.P. (2000). *Simulation Household Travel Survey Data for Metropolitan Areas*. Unpublished Ph.D. Dissertation, Department of Civil and Environmental Engineering, Louisiana State University, Baton Rouge, Louisiana.

Greaves, S.P. and P.R. Stopher (2000). Creating a simulated household travel/activity survey – rationale and feasibility analysis. *Transportation Research Record 1706,* pp. 82-91.

Kalfs, N. and H. van Evert (2003). Non-response and travel surveys. In: *Transport Survey Quality and Innovation* (P.R. Stopher and P.M. Jones, eds.), Chap. 35, pp. 567-585. Pergamon Press, Oxford.

Kothuri, S. (2002). *Bayesian Updating of Simulated Household Travel Survey Data For Small/Medium Metropolitan Areas.* Master's Thesis, Louisiana State University, Baton Rouge, Louisiana, U.S.A.

McNally, M.G. (1997). An activity-based micro-simulation model for travel demand forecasting. In: *Activity-Based Approaches to Travel Analysis* (D.F. Ettema and H.J.P. Timmermans, eds.), Chap. 2. Pergamon Press, Oxford.

SPSS (1998). *Answer Tree Version 1.0 Users Guide and Software*. Statistical Packages for the Social Sciences (SPSS) Inc.

Stopher, P.R., Greaves, S.P. and P. Bullock (2003). Synthesizing household travel survey data: Application to two urban areas. *Transportation Research Record*, forthcoming.

Stopher, P.R. and G. Pointer (2003). Monte Carlo simulation of household travel survey data with Bayesian updating. Paper presented to the Joint ARRB/REAAA Conference, Cairns, Queensland, May.

Transport South Australia of the Department for Transport, Urban Planning and the Arts (1999). Metropolitan Adelaide Household Travel Survey (AHTS).

U.S. Census Bureau Census (2000). Initial Response Rates at http://www.rates.census.gov.

Integrated Land-Use and Transportation Models: Behavioural Foundations
M. Lee-Gosselin and S. Doherty (Editors)

9

GIS, OBJECTS AND INTEGRATED URBAN MODELS

Ronald N. Buliung, Pavlos S. Kanaroglou and Hanna Maoh
School of Geography and Geology, McMaster University, Hamilton, Canada

ABSTRACT

Integrated land use and transport models address complex relationships between urban land use and transportation systems. These models can be used to simulate urban transportation and land use policy effects. The **I**ntegrated **M**odel of **U**rban **LA**nd-Use and **T**ransportation for **E**nvironmental Analysis (IMULATE) has been developed for the City of Hamilton, Ontario, Canada. Historically, simulations using IMULATE required expertise in the management of complex data structures and transportation modelling software. The literature suggests this was a common occurrence in early operational models. This paper reports on a recent effort to embed IMULATE within a component driven GIS-based environment.

An object-oriented approach has been used to construct a software environment that can accommodate decision support tools of various types. The approach encourages the re-use of components by allowing developers or experts to embed data and models from various sources. In addition, the environment is extensible and enables rapid implementation of changes to the appearance of the interface and the functionality of underlying components. The approach makes use of spatial software components called MapObjects. The use of "spatial components" in development enables implementation of a specific subset of GIS functions that are relevant to a particular modelling exercise. In addition, spatial and non-spatial data can be manipulated, managed and analysed using familiar GIS and windows controls. This eliminates overhead associated with an alternative coupling with comprehensive GIS software. The result is an environment that is extensible, powerful, and facilitates ease of use of embedded decision support tools.

The paper begins with a review of the historical development of links between integrated urban models and GIS. The review ends with a discussion of the convergence of integrated urban modelling, activity-based research and GIS with a specific focus on the role of GIS as a tool to improve the behavioural component of contemporary models. Following this, the advantages distributed to non-expert and expert users are discussed in sections that address the development environment in general terms and user interaction with the working environment/interface. Toward the end of the paper, consideration is given to future directions that can potentially capitalise on the extensibility and re-use properties of the software environment.

INTRODUCTION

Integrated land use and transport models address the relationship between urban land use and transportation systems. Embedded within the formal structure of these models is the recognition that land use influences the properties of transport infrastructure and travel behaviour which recursively affect spatial patterns of urban land use. These models can be used to simulate transport and land use policy effects over time. Several detailed reviews of existing operational models and applications have been conducted (Wegener, 1994; Southworth, 1995; Miller and Hunt, 1998; Wegener and Fürst, 1999; Torrens, 2000a; Kanaroglou and Scott, 2002). Reviewers consistently point to the potential of GIS and in later work, object-oriented technologies for improving the integrated urban modelling experience from both modeller and user perspectives. This paper addresses progress in the development of synergies between integrated urban models and GIS. We report on the experience of embedding the **I**ntegrated **M**odel of **U**rban **LA**nd-Use and **T**ransportation for **E**nvironmental Analysis (IMULATE) within a component driven, object-oriented, GIS-based development environment. We emphasize that as the behavioural foundation of integrated urban models improves over time such a development environment will improve the ability of modellers to incorporate and study the behaviour of modelled actors. Additionally, the provision of a map-driven interface will facilitate efficient practitioner exploration of urban change, through time, by minimizing interaction with underlying complex data structures and transactions that occur between behavioural units of interest.

The paper has five sections. In the first section the origin of integrated urban models (IUMs) and parallel developments in GIS and activity-based research are considered. A sample of linked IUM-GIS applications are examined using a framework based on developmental linkages between systems. The convergence of IUMs, GIS and activity-based research is considered toward the end of the review with attention given to the use of GIS for behavioural analysis. A general description of IMULATE and its submodels are given in the second section of the paper. The third section of the paper focuses on the development environment. Links between IUM components and the GIS-based graphical user interface are discussed with reference to their re-use and extensibility properties. The fourth section addresses the GIS-based user interface directly. Attention is given to the substantive impetus for developing the interface and benefits of the approach. Ongoing research and conclusions are addressed in the fifth and final section of the paper.

INTEGRATED URBAN MODELS AND GIS

The focus of this section is the history of integrated urban modelling, parallel developments in GIS and activity-based research and examples of their convergence that appeared during the 1990s. The review is conducted from a developmental perspective with attention given to the manner in which integrated urban models have been linked with GIS. In the final section of the review, a number of reflections are offered with respect to lessons learned from the work of the last decade and the trajectory of the field particularly where GIS and related technologies potentially improve behavioural content.

Origins

The origin of computer models of urban land use and transportation can be traced back to the 1950's with early operational examples emerging in the 1960's (Wegener, 1994). The models, generally referred to as Large Scale Urban Models (LSUMs), and efforts that have followed, have been the focus of considerable debate in the literature (Lee, 1973; Batty, 1994; Harris, 1994; Klosterman, 1994; Lee, 1994). In the remainder of the discussion the term Integrated Urban Models (IUMs) will replace Large-Scale Urban Models. Parallel to the development of computer based urban models was the emergence of early examples of Geographic Information Systems (GIS). The early efforts of Coppock (1962) and Roger Tomlinson's work with the Canadian Geographic Information System (CGIS) are generally accepted to represent the origin of modern GIS. A limiting factor for both early GIS and IUMs was computational power, a limitation that has now largely been overcome (Batty, 1994; Harris, 1994). What is interesting about this early period of development is that the two evolved largely independent of one another. Potential synergies between GIS and IUMs were not explored with any real vigor until the early to mid-1990s.

Early Efforts

A limited number of applications emerged during the 1990s that demonstrate the potential of GIS as part of the urban modelling framework. This work was likely a partial response to advances in computing, the emergence of relatively low cost commercial GIS software and literature (Simkowitz, 1988; Simkowitz, 1990; Prastacos, 1991) identifying possible connections between transportation applications and GIS. Most applications linked either land use *or* transportation models with GIS. The California Urban Futures Model (CUFM) (Landis, 1994; Landis, 1995; Landis and Zhang, 1998) is generally cited as one of the first urban models to make use of GIS technology (Wegener, 1994). Within the CUFM framework Arc/Info GIS acts as a spatial database management system for the supply of urban land that is available for development. Yeh and Li (1998) provide another example using Remote Sensing and GIS to develop a sustainable land development model that explicitly accounts for agricultural land loss. Their approach, while effective, seems less automated than the CUFM, requiring expertise in Arc/Info and a number of its analytical extensions.

Several other software development and applied research initiatives were undertaken during the 1990s that represent the convergence of GIS and urban modelling. Prastacos (1991) cites two approaches to software development that can be used to describe these early efforts. The first approach involves the development of links between GIS and transport modelling software such as EMME/2 or TRANPLAN. The second approach focuses on embedding transport or land use modelling within a GIS. GIS/Trans, Ltd. (1997) provides an example of the first approach in developing links between TRANPLAN and Arc/Info GIS. As a second example, the Planning Department of Portland Oregon's regional government (METRO) has developed a software link between Arc/Info GIS and EMME/2. Planners use the system to study air quality, congestion and other transport issues. The evolution and application of the TransCAD transportation GIS software environment (Simkowitz, 1988; Simkowitz, 1990) serves as an example of Prastacos' second approach. In general, the advantages of developing links of either sort include reduced maintenance of network databases, the extension of modelling capabilities and enhanced visualisation and mapping of transport related data. Research initiatives and the "practice" of transportation planning and programming continue to make use of these kinds of software developments. The evaluation of adjustments to transport supply, transit service and transit routing represent common applications (Gallimore *et al.*, 1992; O'Neill *et al.*, 1992; Hartgen *et al.*, 1993). Other issues that have received attention include urban air quality, freeway congestion, transit use and urban sprawl (GIS/Trans, 1997). These and other applications are the focus of a series of case studies in the use of GIS in Transportation Planning published through the Travel Model Improvement Program (TMIP) (U.S. Department of Transportation, 1998a).

Encompassing and Modular Frameworks

While the integrated modelling field continued to grow (Wegener, 1994) few IUM-GIS applications could be found during the early to mid-1990s. Examples include links between the Projective Optimization Land Use Information System (POLIS) and TransCAD (Prastacos, 1991; Prastacos and Diamandakis, 2000), and research at McMaster University in Hamilton, Canada building a joint IUM-GIS system that also used TransCAD's capabilities. A full description of the Hamilton system can be found in Anderson *et al.* (1994) with applications to mobile source emissions modelling reported in Anderson *et al.* (1996) and Kanaroglou *et al.* (2000). Toward the end of the decade, You and Kim (1999) had produced the Land Use and Transportation Modeling System with GIS (LUTGIS). The SPARTACUS project resulted in the development of a link between Echenique's MEPLAN model and a raster GIS module (Commission of the European Communities, 1998). Johnston and de la Barra (2000) had linked their long-standing TRANUS model with the California Urban Futures Model (CUFM) and Waddell (1998) explored the potential of GIS component software as a means to integrate spatial data viewing into an IUM.

The remainder of the review will focus on an evaluation of these examples by considering the process used to link GIS and IUMs. Anselin (2000) provides a useful nomenclature for this "architectural" analysis in discussing *encompassing* or *modular frameworks* for linking GIS and statistical analysis software. An *encompassing framework* is one in which the functionality of the core GIS is simply extended using scripting or macro languages that are native to the GIS software. A *modular framework* is one in which a set of systems, optimized for specific functions (i.e., mapping, user interaction), are linked together Anselin (2000). The integrated models of the 1990s that fall under the encompassing framework approach are reviewed first followed by those corresponding to the modular framework.

Both POLIS and the Hamilton model (later to become known as IMULATE) have been linked with TransCAD. While the initial development of POLIS preceded the diffusion of GIS technologies, effort was extended toward integration of the system with GIS (Prastacos and Diamandakis, 2000). This integration had two forms with the initial trial partially relying on manual exchange of files between POLIS and TransCAD. Later, a more substantial effort was conducted to closely integrate POLIS with TRANSCAD. This involved embedding POLIS as a TransCAD procedure. The development process made use of TransCAD's proprietary scripting language. This integration met with some success but unfortunately the continued evolution of the system could not be sustained (Prastacos and Diamandakis, 2000). Interestingly, while the POLIS/TransCAD coupling took shape, the Hamilton group had also implemented their integrated model as a TransCAD procedure. This development project also relied on TransCAD's Procedure Toolkit and proprietary scripting. At the time, both the work in Hamilton and the POLIS/TransCAD integration represented state-of-the-art links between urban models and GIS. You and Kim's (1999) LUTGIS initiative is useful in that it illustrates the role that GIS technology can have in the practice of developing IUMs. They cite general and network database management, spatial analysis and visualisation capabilities alongside maintenance of network topology as the principle contributions of GIS. The technical challenges of implementing an integrated model with contemporary GIS software are apparent. You and Kim utilize two GIS software programs (ArcView and Arc/Info), TRANPLAN for four-stage modelling and a number of scripting and programming languages to patch everything together. The LUTGIS model corresponds to the encompassing framework because the integration of systems is achieved through a patchwork of scripts, programs and software packages.

Johnston and de la Barra (2000) link TRANUS, a market based integrated urban model, with the California Urban Futures Model (CUFM). The location of TRANUS in the spectrum of urban models is addressed in Wegener (1994) while a detailed description can be found in de la Barra (1989). A two-stage land use allocation procedure is developed. In stage one, TRANUS produces aggregate land use projections for 58 zones. In stage two, CUFM allocates land units (polygons) to specific land use classes within each of the 58 zones until the aggregate zonal totals for each land use class are met. This is essentially equivalent to a rule-based spatial disaggregation with a final result of 272,000 developable land-units each corresponding to a specific land use. CUFM uses Arc/Info as its GIS engine. Although not discussed directly in the paper, experience with these tools dictates that linking them would require file sharing and scripting. In addition, technical limitations of Arc/Info force manual editing of the spatial database. It is for these reasons the TRANUS-CUFM link falls under the encompassing framework.

SPARTACUS (System for Planning and Research in Towns and Cities for Urban Sustainability) is based on the MEPLAN land use and transportation model. Discussion of MEPLAN relative to other models can be found in Wegener (1994) and Wegener and Fürst (1999). There are two aspects of GIS included in SPARTACUS: a Raster module and MEPLUS-GIS. The Raster module calculates spatially disaggregate indicators of emissions, air quality and noise. Zonal and network level environmental and social indicators from MEPLAN are disaggregated to a high-resolution raster (100m x 100m). This kind of link overcomes the mismatch between zone-based data of IUMs and the micro-level spatial framework that is necessary to describe certain environmental and social processes (Spiekermann and Wegener, 2000). Spatial conversion is used to implement environmental submodels that assess land cover, air quality and noise levels. Social impact submodels assess population exposure to air pollution and noise levels estimated from environmental models. The output of this process includes five indicators of urban sustainability and maps of air pollution and noise levels. MEPLUS-GIS is used to manage these results. It is a data management, policy analysis and reporting tool with mapping capabilities and is embedded within the MapInfo GIS software package. It is not entirely clear how MEPLUS-GIS and MapInfo are linked. It is likely that the MapBasic programming language has been used to customize and automate MapInfo functionality. The Raster module has been developed independent of any commercial GIS software. SPARTACUS, while closer to the encompassing framework, falls somewhere in the spectrum between encompassing and modular. It is not completely modular because the overall system is not composed of easily re-usable components that could be applied in a different setting. The Raster module may be an exception but this is difficult to ascertain from the literature. In addition, the MEPLUS-GIS module is coupled to MapInfo in a manner that extends MapInfo capabilities.

The reviewed models fall under the encompassing framework by design, but this is largely a function of available GIS software. From this perspective, an encompassing framework is not a weakness, as each example represents the state-of-the-art at a particular point in time. What has emerged from these initial efforts is a solid basis of understanding GIS/IUM complements from both technical and analytical perspectives. Given these observations and the emergence of component driven GIS development in the latter part of the decade, it is useful to consider potential drawbacks of encompassed systems. First, as Anselin (2000) points out, the encompassing framework can potentially lead to implementation of less efficient computational algorithms. Second, in each case a comprehensive commercial GIS software tool has been used as a GIS engine. These tools generally come with a wide array of functions, and in the case of systems like Arc/Info, considerable installation footprint and system requirements. In most cases, the GIS-IUM applications reviewed are using a small portion of the total capability of each GIS. In other words, integrated models are carrying around significant overhead when linked with comprehensive GIS software. The end result has been operational systems, capable of generating results, that can be cumbersome to deploy, costly and require considerable training both in GIS and the integrated urban modelling software.

Carrying around the overhead of a large GIS is functionally useful but not optimal from installation and system resource perspectives. One could argue that to fit the modular framework an urban model should not be linked with an "all-or-nothing" GIS tool. To optimize development, modellers should be able to select and integrate only necessary GIS components. Two examples of this approach are UrbanSIM and the version of IMULATE discussed in the remainder of the paper. UrbanSIM is a state-of-the-art linked microeconomic urban land use and GIS model that handles the location choice of households and firms. It is not considered a complete IUM because it does not model transportation directly instead relying on input from external transport models. Discussion and comparison of UrbanSIM with other models can be found in Miller and Hunt (1998) and Wegener and Fürst (1999). IMULATE, on the other hand, is a zone-based integrated model that handles both transport and land use. IMULATE will be the focus of the remainder of this paper and so additional detail is not covered in this section.

Both examples use GIS development components called MapObjects. Detailed discussion of MapObjects can be found in ESRI (1996). MapObjects is compliant with the Object-Oriented approach to programming. An overview of the general concepts of object-oriented programming and how they relate to MapObjects can be found in Lombard (1997). The modular structure of the MapObjects development environment enables the inclusion of only necessary GIS functionality in applications. Developers can construct familiar windows-style applications with embedded mapping components. This eliminates the overhead associated with links to full GIS software packages. MapObjects was used to build UrbanSIM's GIS data viewer (Wadell, 1998). IMULATE relies more heavily on MapObjects *data access*, *mapping* and *geometric* object classes for spatial database management, data manipulation and visualisation. In addition, both applications maintain file format compatibility with other common desktop and GIS tools. A developmental distinction can be made between UrbanSIM and IMULATE. UrbanSIM is a land use model that has been extended with GIS development components while IMULATE is an integrated urban model embedded within a general purpose, component driven, GIS management environment. This environment can be extended and re-used as the front, back-end or both for other analytical tools in the future. In this sense IMULATE is closer to the modular framework. As Anselin (2000) suggests, the modular framework can potentially yield a collection of reusable spatial tools that can be combined in the future to tackle other issues.

The "GIS-IUM Experience": Emerging Behavioural Trajectories

By the end of the 1990s it was widely recognized that many of the computing and data obstacles of the past had largely been overcome. Advocates speculated as to the future of integrated urban models in the presence of new and maturing technologies including GIS, object-oriented databases and data visualisation tools (Batty, 1994; Wegener, 1994; Southworth, 1995; Miller *et al.*, 1998; Wegener and Fürst, 1999; Kanaroglou and Scott, 2002).

The experiences of the past suggest that GIS adds value to the urban modelling process. Wegener and Fürst (1999) identify three essential roles for GIS. These are, data organisation and management, spatial disaggregation of land use and transport data to support air quality dispersion and noise propagation models, and simultaneous representation of cities in vector and raster formats to make use of the most efficient algorithms available for either case. Wegener (1994) also had the foresight to suggest that adoption of a microsimulation approach could benefit from the representation of spatially disaggregate microdata stored in a GIS. Wegener and Spiekermann (1996) take these ideas further in their discussion of the IRPUD-2 disaggregate integrated land use and transport model.

On the research front, the 1990s also witnessed the emergence of numerous studies conducted using the human activity approach to travel behaviour analysis. This paradigm represents an alternative to the trip-based approach currently embedded in many integrated models. The activity approach assumes that the demand for travel is the product of the need to participate in activities inside and outside of the home (Jones *et al.*, 1990; Ettema *et al.*, 1993; Bhat and Koppelman, 2000). The 1990s became an era also marked by the collection of new types of travel diary data that could support modelling, in many cases, under the trip or activity paradigms. Some researchers during this period developed theoretical frameworks for understanding activity/travel behaviour (Recker *et al.*, 1986a,b; Gärling *et al.*, 1998) while others developed activity-based forecasting models (Bowman and Ben-Akiva, 1996; U.S. Department of Transportation, 1998b). Statistical modelling tools were also used to explore complex relationships between various dimensions of the activity paradigm (Damm, 1982; Pas, 1996). Both Kitamura (1996) and Pas (1996) review a number of these studies. The overall objective of activity-based research has been the evolution of alternatives to integrated models that are considered lacking in behavioural content and policy sensitivity (Stopher, 1993; Stopher *et al.*, 1996).

The shift from trips to activities is in one sense a shift in scale; aggregate to disaggregate and zone to individual or household. The general role of GIS in this framework remains unchanged only now GIS will be used to organise, model and manage much larger databases. From a behavioural perspective GIS provides an environment within which to analyze and visualise the revealed or predicted activity/travel behaviour of individuals and households (Kwan, 2000; Buliung, 2001). GIS has also been used as a predictive tool to determine the space-time prisms of individuals (Miller, 1991) and as part of computational process models (CPMs) designed to simulate the activity scheduling behaviour of individuals (Golledge *et al.*, 1994). In addition, the emergence of object-oriented GIS provides an opportunity to model activity/travel objects and their relationships in a more intuitive manner that reflects the intrinsic complexity of activity/travel behaviour (Makin, 1997; Claramunt and Thériault, 2001). In general object-oriented GIS can be used to manage objects that have spatial behaviour and provides a framework within which to implement methods that model this behaviour. Another interesting development loosely related to GIS, activity-based research, and microsimulation involves efforts to simulate the micro-level behaviour of "agents" that comprise transport and land use systems (Torrens, 2000a; Torrens, 2000b; Torrens, 2001). These agent-based approaches to urban dynamics and activity/travel behaviour are based in complexity theory and, while very interesting, are still largely in the basic research phase. Most current developments in integrated urban modelling reflect an apparent desire to capitalise on the advantages of object-oriented technologies, their presence in GIS and the basic and applied activity-based research of the 1990s.

INTEGRATED MODEL OF URBAN LAND USE AND TRANSPORTATION FOR ENVIRONMENTAL ANALYSIS (IMULATE)

The **I**ntegrated **M**odel of **U**rban **LA**nd-Use and **T**ransportation for **E**nvironmental Analysis (IMULATE) is an operational integrated urban model originally developed to study urban issues in the City of Hamilton, Ontario, Canada. Using IMULATE, one can assess impacts of land use change and generated travel demand on the environment (Anderson *et al.*, 1996; Kanaroglou *et al.*, 2000). IMULATE has been the focal point of a group of researchers at McMaster University since the early 1990s. Detailed descriptions of the model's development and structure can be found in Anderson *et al.* (1994) and Anderson *et al.* (1996). IMULATE consists of three core modules that handle land use, transport, and mobile emissions (Figure 9.1). The land use and transport modules consist of a number of submodels.

The first submodel, POPMOB, is the land use component managing intraurban migration and place of work assignment of resident workers within 151 census tract zones. The land use model has recently been modified to include an employment location model (EMPLOC) that joins with POPMOB to update place of residence/place of work totals for census tract zones (Maoh *et al.*, 2002). The second submodel, TRANDEM, controls trip generation, distribution and modal split. TRANDEM estimates the number of work, school and discretionary (i.e., shopping, recreational) trips by mode. The TRAFFIC ASSIGNMENT submodel uses a stochastic user equilibrium algorithm to assign interzonal automobile trips (from TRANDEM) to a model of the Hamilton road network.

The bi-directional relationship between urban land use and transportation is modelled using the accessibility approach that is common to other models of this sort (Anderson *et al.*, 1994; Southworth, 1995). Transportation network performance affects the evolution of land use patterns by adjusting the accessibility of places within the city. Inter-zone travel costs estimated using TRANDEM and the TRAFFIC ASSIGNMENT are fed back into the residential and employment models which re-distribute residents and workers within the city. The MOBILE EMISSIONS extension uses link flows and average speeds from the TRAFFIC ASSIGNMENT submodel to estimate CO, HC and NO_x on each link. MOBILE5.C emission factors are matched to estimates of link average speeds and adjusted by link volumes. Average link speeds are used as indicators of roadway congestion. Link speeds lower than those under free flow are considered representative of stop-and-go conditions, commonly associated with congested roadways. Under these conditions tail-pipe emission levels of HC and CO will be higher than under driving cycles associated with higher average speeds.

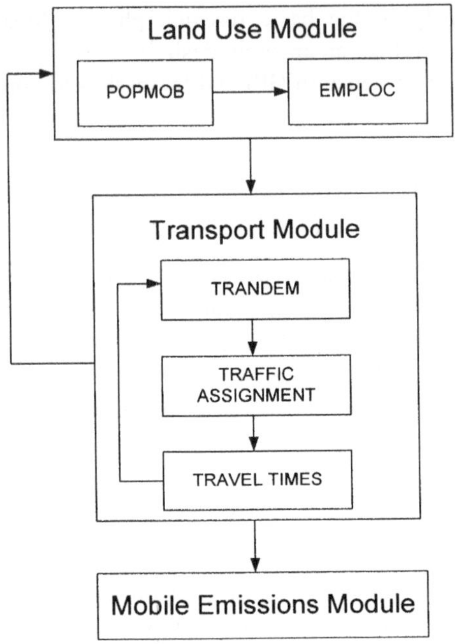

Figure 9.1 IMULATE

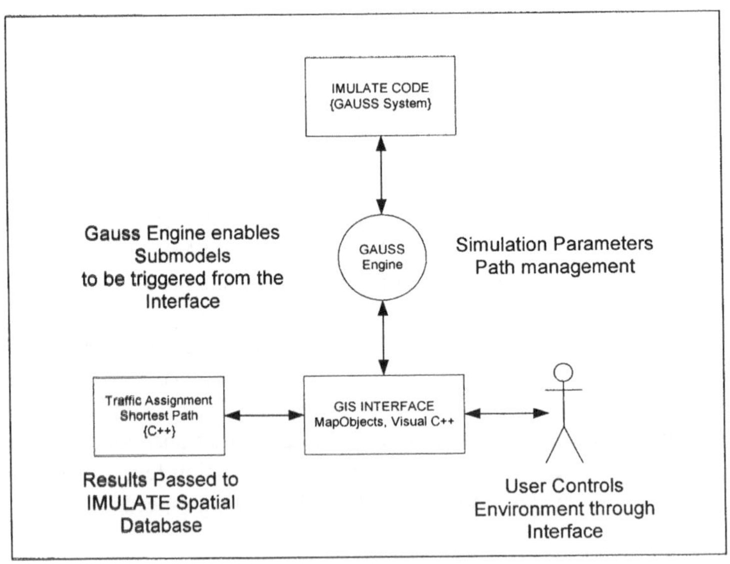

Figure 9.2 Development Environment and Modular Products

DEVELOPMENT ENVIRONMENT

Emerging development tools have facilitated a transition from an encompassing to a modular framework for systems development. This transition will be the focus of this section of the paper with attention given to the suite of development tools used, modular products and benefits of the adopted approach. The relationship between individual modules and an enumeration of the development tools used in their construction is shown in Figure 9.2. This figure provides a graphical representation of the modular framework that has been adopted and should not be confused with a conceptual representation of IMULATE's analytical models.

The first generation of IMULATE was linked with TransCAD (version 2.0) with design guided by the principles of an encompassing framework. Technically, IMULATE was implemented as a procedure using TransCAD's Procedure Toolkit. The toolkit enables the execution of external programs on TransCAD databases (Caliper, 1990). At the time, this represented a state-of-the-art linkage between an integrated urban model and GIS. TransCAD provided spatial and attribute data management, visualisation and mapping, and transportation modelling capabilities. Specifically, TransCAD was used to estimate inter-zone travel times, manage network topology, and map area and link level exogenous and endogenous variables (Anderson *et al.*, 1996). In addition, the traffic assignment portion of IMULATE's four-stage transportation model relied on the TransCAD implementation of a stochastic user equilibrium algorithm (e.g., Sheffi, 1985). Deployment meant a full installation of TransCAD and the integrated modelling software even though a subset of TransCAD capabilities were being used. In addition, users of IMULATE had to become competent users of TransCAD. This made the solution functional but costly, cumbersome, difficult to deploy and relatively challenging to use. As an alternative, we decided to explore independent development of model components historically provided by TransCAD.

The submodels of IMULATE discussed in the previous section have been programmed using the GAUSS Mathematical and Statistical System. GAUSS is a high level programming language based on the principles of matrix algebra. It is widely used by scientists, statisticians and econometricians. Historically, accessing the functionality of GAUSS required expertise in the specifics of the language. With the introduction of the GAUSS Engine (Figure 9.2) non-GAUSS users may access GAUSS functionality through easy-to-use visual applications. The GAUSS Engine is a dynamic library that makes the GAUSS language available to programs written in a number of other languages (e.g., C, C++, VB, Java). This enables developers to embed programs written in GAUSS within applications developed using other contemporary development environments. The result is that legacy software and new analytical models written in GAUSS can be re-used or embedded within new applications. The GAUSS Engine has enabled the research team to capitalise on and extend earlier development efforts as opposed to translating IMULATE's complex mathematical procedures into an alternative programming language. With the IMULATE source code and the GAUSS Engine the research team was able to embed the system within a new graphical user interface that employs state-of-the-art mapping components offerd by MapObjects.

As discussed above, MapObjects are software components that enable developers to incorporate mapping or spatial data handling capabilities into applications (ESRI, 1996). The use of "spatial components" in development allows implementation of a specific subset of relevant spatial tools. Again, spatial and non-spatial data can be manipulated, managed and analysed using familiar GIS and windows controls. This use of components eliminates overhead associated with the unused capabilities of much larger "all-or-nothing" solutions (i.e., ArcInfo Workstation, ArcView, etc.). As Prastacos and Diamandakis (2000) suggest, the GIS services required for urban models are generally less than the range of capabilities offered by most GIS software. In the context of this project, MapObjects facilitates independence from TransCAD specifically where requirements exist for mapping, visualisation and the management of spatial and non-spatial data. In adopting MapObjects, we have established a path for future development that includes object-orientation and the microsimulation approach.

The development tools and system architecture described in this section have allowed the research team to transition IMULATE to a modular framework. Mapping, visualisation and data management are handled by MapObjects components and legacy GAUSS code. In addition, TransCAD's transport modelling capabilities have been replaced with shortest path and stochastic user equilibrium traffic assignment modules recently linked with the rest of the system (Figure 9.2). An alternative to this approach would have involved using the most recent version of TransCAD as an automation server to provide network-modelling routines as services to the core IMULATE application. This would alter the architecture of the system replacing the *Traffic Assignment/Shortest Path* box in the Figure 9.2 with a TransCAD automation server box. As a longer run development objective, we would like to increase the flexibility of our solution by simultaneously supporting our proprietary traffic assignment algorithm and TransCAD services when available. In its present form the existing system supports the Hamilton group's modelling needs for research and teaching.

Advantages offered by the modular approach only materialise in the presence of computationally efficient system modules. Our network modelling routines do not outperform existing routines available in commercial packages such as TransCAD and EMME/2. Nevertheless, the performance of our network tools is more than acceptable. The traffic assignment routine, operating on a network database containing 1500 links, converges after 11 iterations in approximately 77 seconds. This level of performance was achieved running on a workstation with a Xeon 2.0GHz processor. Performance issues are not a limitation of our approach. In addition, our traffic assignment routine provides analysts with additional diagnostic output that is not readily available in TransCAD.

Adoption of the development approach discussed in this section has reduced the overhead associated with the original version of IMULATE and improved the ease with which the model system can be deployed. Data exchange between modules is now completely transparent and "non-expert" users will find it easier to interact with the system using a geographically enabled graphical user interface. The availability of this interface eliminates direct user interaction with complex underlying data structures. These developments represent a significant improvement over the original IMULATE/TransCAD coupling. In addition, researchers can continue to build new analytical tools using GAUSS and the GAUSS Engine and embed these new tools in an interface that can be easily extended to suit other applications.

USER ENVIRONMENT AND INTERFACE

Improving the accessibility of integrated urban modelling software, from a user perspective, has been recommended in several reviews (Southworth, 1995; Miller and Hunt, 1998; Wegener and Fürst, 1999; Torrens, 2000a). It has been suggested that this could be achieved by linking urban models with GIS (Torrens, 2000a). Prastacos and Diamandakis (2000) anticipate that the convergence of GIS and urban models could contribute to increased use of these types of model systems. While the dissemination of integrated models to practitioners is an important objective the appropriate application of these model systems should be simultaneously considered. With that caveat we have provided a framework within which such a link can be constructed so that users and developers benefit.

IMULATE is accessible through an intuitive GIS-based or map-driven graphical user interface. The interface was developed using Visual C++ and MapObjects and consists of a number of standard windows and additional controls that are commonly found in desktop GIS software like ArcView, MapInfo and ArcGIS Desktop (Figure 9.3). In general terms the interface has been designed with the user in mind. We have intentionally limited the complexity of the interface to facilitate focused data exploration and scenario development. Specifically, users are able to view, query and manipulate spatial and attribute data that form the basis of analysis for the integrated urban modelling tools. The set of tools that have been included enable data exploration that can inform the construction of urban change scenarios. These scenarios can be investigated further by exercising the integrated model through time.

Structurally, the interface consists of a *Menu Bar, Layer Control, Spatial Tools* and a *Map Control*. The *Menu Bar* provides access to a number of tools for spatial data viewing, manipulation and simulation. The *Tool Bar* consists of a series of general windows based controls for printing and file management. When the user starts the program the *Layer Control* and *Spatial Tools* (i.e., zoom, pan etc.) automatically appear. As with most windows applications redundancy has been built into the interface. Many of the functions available through custom controls (i.e., Layer Control) are also available through drop down menus in the *Menu Bar*.

The dominant feature of the interface is the *Map Control*. The interested reader is directed to the "ESRI MapObjects Programmer's Reference Guide" for a more thorough discussion of the properties and methods associated with objects discussed in this section. The default property of the *Map Control* is a collection of *Layers* that contain georeferenced vector or raster data. *MapLayer* objects store georeferenced vector data (i.e., shapefiles) and have a number of methods that allow users to access information from them directly (Figure 9.4). *ImageLayer* objects store georeferenced raster information. This implies that in the future orthophotography or satellite imagery could be added as an *ImageLayer* to provide a "real-world" context for vector overlays. The capability to add similarly georeferenced layers into the *Map Control* is currently under development.

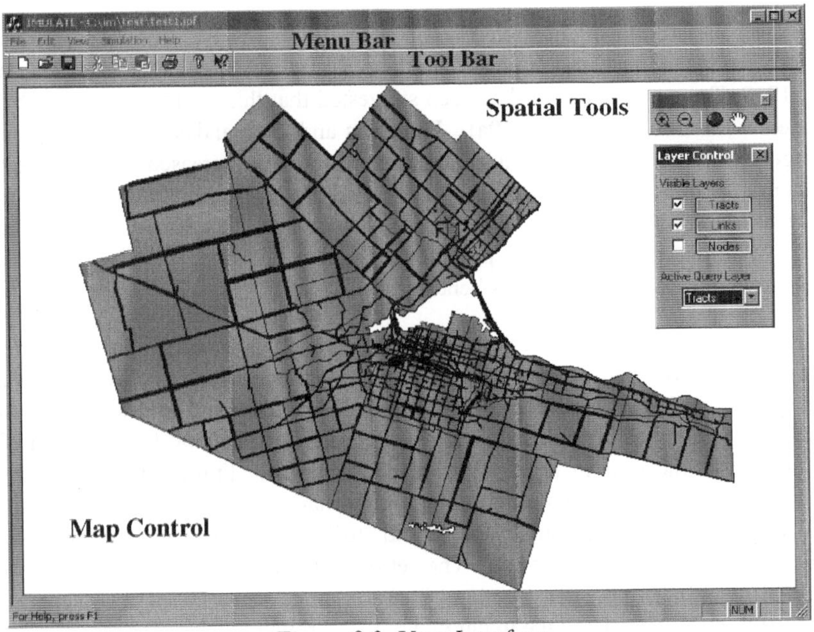

Figure 9.3 User Interface

The display properties (i.e., shading of polygons, line width etc.) of *MapLayers* can be altered using the Layer Control. Additional visualisation capabilities have been programmed for the "Links" layer (Figure 9.5). Estimates of mobile emissions are determined at the link level. Using the advanced link display properties the line width can be scaled to reflect emission values of various mobile source pollutants. This tool allows identification of potential emission "hot spots" within the study area. The tool will be generalised in the future to enable visualisation of any numeric field contained in the link database. In addition, descriptive statistics for mapped variables can be displayed at the bottom of the *Advanced Link Display* dialogue.

At this stage in development IMULATE can be accessed and run through the Simulation dropdown menu of the *Menu* Bar (Figure 9.6). The time period of simulations can be controlled through the interface. The model is not dynamic and operates over discrete five year time periods. Convergence and the number of iterations of the TRANDEM submodel are controlled through the *Simulation Parameter* dialogue as well. Routines have been written that manage the base data and output generated from simulations. In addition, simulation parameters (metadata) are stored in files to maintain a record of model runs. In the future, simulation metadata will be managed by custom controls designed for this purpose.

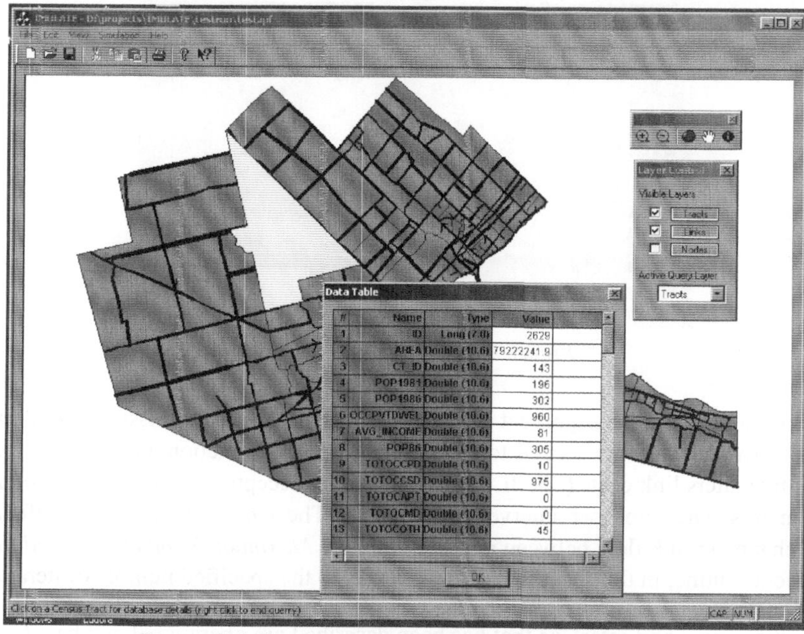

Figure 9.4 Retrieving Information from a MapLayer

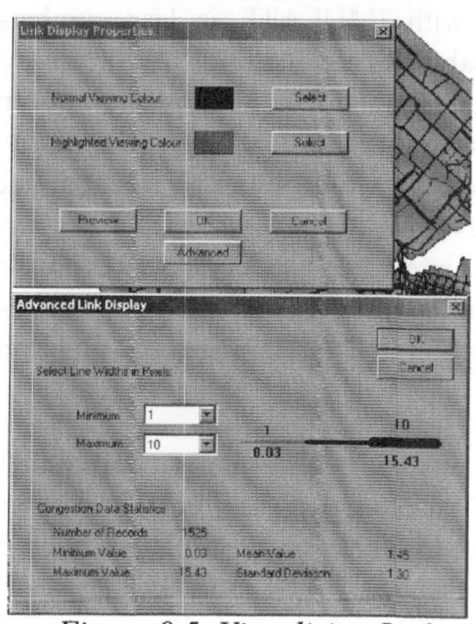

Figure 9.5 Visualising Links

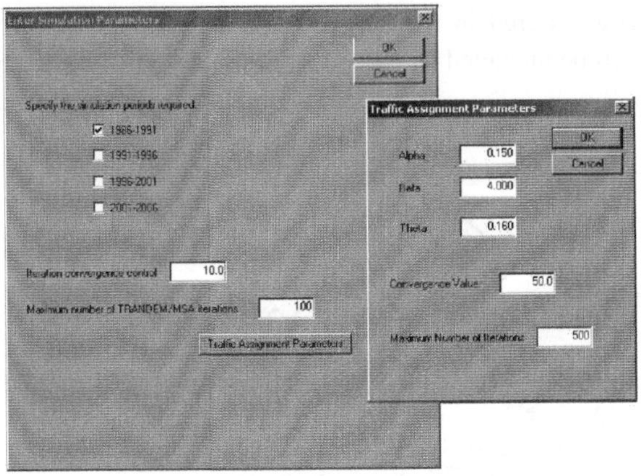

Figure 9.6 Simulation Control

The behaviour of the traffic assignment is managed through the Traffic Assignment Parameters dialogue. *Alpha* and *Beta* are related to the link performance function and *Theta* is a random component that alters link costs (Sheffi, 1985). Generally accepted default values are given for these parameters when no user intervention occurs. The *Convergence Value* refers to the maximum change in link flow between iterations and the *Maximum Number of Iterations* limits the assignment routine, in the absence of convergence, to the specified number of iterations.

The user environment and interface that has been described are operational and beta testing has demonstrated that the new system can reproduce results acquired when IMULATE was linked with TransCAD. Interaction with IMULATE is less cumbersome than in the past and development of additional capabilities is ongoing. With an easier to use integrated urban model the so-called "impedance mismatch" between potential users of integrated urban models (i.e., planners) and developers (i.e., modellers) can potentially be overcome. Impedance mismatch refers to circumstances in which users of systems and developers have difficulty determining the direction of development because each lacks expertise in the other's domain (Booch, 1994). In this context, integrated models have been criticized as being "operational" as opposed to being used in practice (Lee, 1994). Perhaps, once easier to use, end users will be able to gain experience with integrated modelling tools and contribute more directly to the development of such systems.

As integrated urban models evolve behaviourally, requirements for tools that facilitate exploration and understanding of exogenous and endogenous model elements will increase. The GIS-based development and user environment that has been the focus of this paper is open to extension and refinement as the IMULATE model system evolves. The next generation of IMULATE will be developed using a microsimulation approach that capitalises on our recent experiences with GIS-based object-oriented development.

Availability of detailed activity-based microdata and maturing spatial technologies have contributed to research (Frihida *et al.*, 2002; Schönfelder and Axhausen, 2003) focused on development and application of GIS-based tools for studying activity/travel behaviour. Buliung (2004) has developed an environment that facilitates exploration of household activity patterns in space and time. The system was developed using the Geodatabase data model with additional programming of ArcObjects that are part of the ArcGIS software suite (Environmental Systems Research Institute, Redlands CA). The application has been tested using data from the 1994/1995 Portland Metro Household Activity and Travel Behavior Survey. System capabilities include construction of household level, activity-based: space-time paths (Hägerstrand, 1970; Lenntorp, 1978; Kwan, 2000), standard deviational ellipses (Lefever, 1926; Furfey, 1927; Yuill, 1971; Ebdon, 1988), and convex-hull polygons to represent the spatial extent of the point-patterns associated with recorded household activities.

The type of spatial exploration that can be achieved within this environment is shown in Figure 9.7. What we observe are Friday and Saturday activity patterns for a sample household that has two-employed (full or part-time) adults (between the ages 20-59), and a teenage child (age $\geq$ 13). The numbers beside activity geocodes indicate the unique spatial locations visited by household members during survey days. Corresponding participants, activity types and times are given in Table 9.1. We have not included an exhaustive list of at-home activities in this case. From this figure we observe, for this sample household, spatial and temporal differences when comparing weekday with weekend travel behaviour.

The purpose here is not exhaustive study of this example. We are using this particular case to suggest potential directions for future integration of GIS-based behavioural analysis tools and integrated urban models. The incorporation of these types of tools within the current IMULATE environment is a technical exercise that can be undertaken as the predictive model components evolve behaviourally in the presence of new regional micro-data. The development approach that we have adopted is flexible and extensible offering the capacity to incorporate substantively related analytic tools as they become available and as the need arises.

Table 9.1 Sample Household 2-Day Activity Summary

Identifier	Day	Participant(s)	Type	Timing
1	1	Female	Work	5:45 a.m. – 2:45 p.m.
2	1	Male	Work	8:00 a.m. – 5:30 p.m.
3	1	Child	School	8:40 a.m. – 5:30 p.m.
4	1	Female&Child	Religious	6:50 p.m. – 9:30 p.m.
5	1		At-home	
1	2	Male	Work	9:00 a.m. – 3:00 p.m.
2	2	Female&Child	Shopping	2:30 p.m. – 2:45 p.m.
3	2		At-Home	

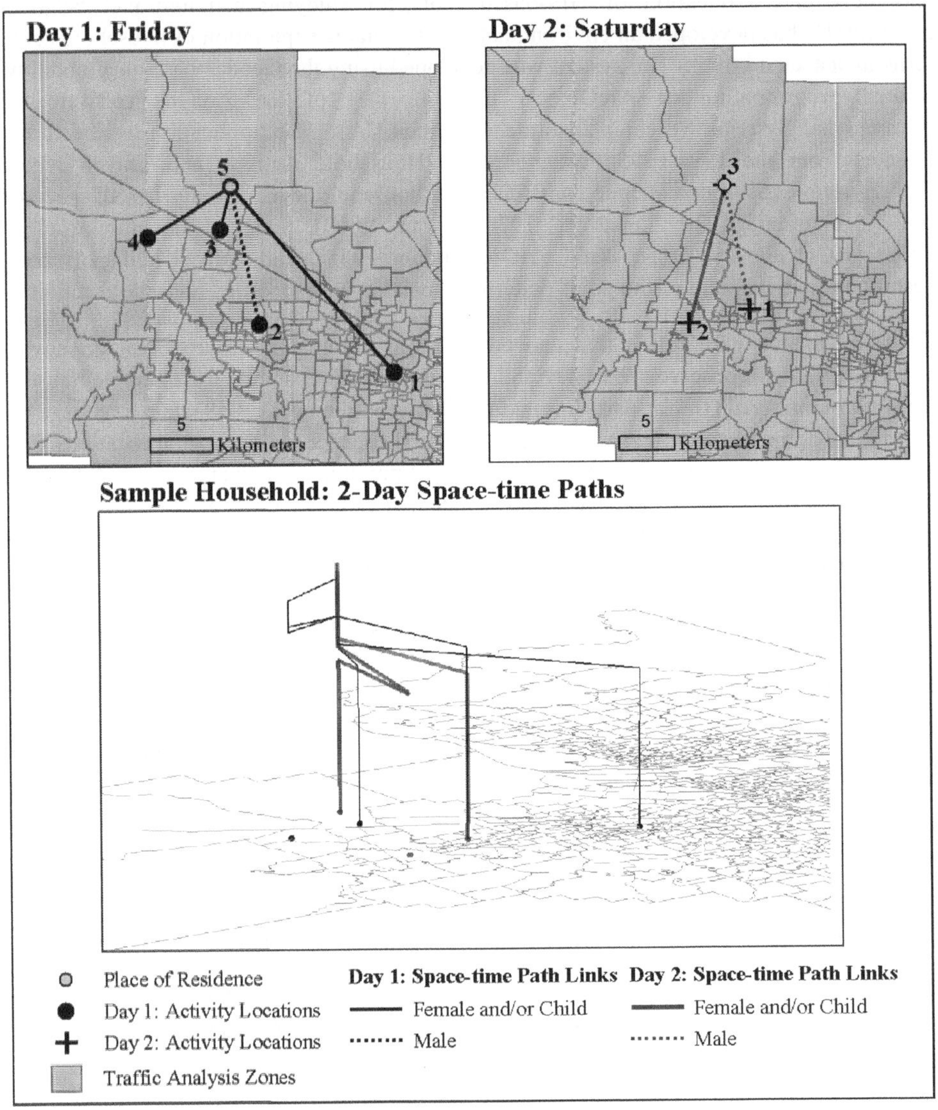

Figure 9.7 Household Level Activity Participation in Space-Time

CONCLUSION

The Geomatics sector has recently witnessed an increase in the availability of spatially enabled software development tools. Tools such as MapObjects allow developers to place controls in applications that permit the manipulation and analysis of spatial data. In addition, data accessed within these applications are treated as spatial objects inheriting properties and methods from powerful generic object classes. Developing "spatial" software using these kinds of tools facilitates incremental and modular development. The implication is that over time the software becomes a "laboratory" with researchers extending functionality and improving the "behaviour" of systems as knowledge grows concerning specific problem domains.

An innovative and cost-effective systems integration approach has been adopted in this project. Legacy code has been re-used and users are now able to communicate with model subsystems using an intuitive GIS-based graphical user interface. The interface has been built using MapObjects and Visual C++. GIS capabilities were built as needed as opposed to a less efficient coupling with a much larger GIS software package. Data management, visualisation and mapping capabilities identified in the literature as essential contributions of GIS to urban modelling have been implemented. The spatial components are not secondary to the integrated modelling exercise. Users will build scenarios directly through the manipulation of spatial objects and other controls that are part of the interface. Standard GIS file types have also been used so that users will be able to bring base data and simulation results into other GIS desktop tools for additional processing. This compatibility can be used to inform the detailed specification of future requirements that could include development of additional analytical capabilities. The adopted development approach facilitates less cumbersome deployment of the system. In addition, users of GIS software will find that spatial and attribute data controls have been implemented in a manner similar to those of other GIS software. Overall, embedding IMULATE with this environment enables efficient exploration of urban change scenarios.

Development of IMULATE is ongoing, with attention being given to the development of controls that can ease the process of investigating transport and land use scenarios. Mapping and visualisation tools are currently being refined as well. Advanced arithmetic operations and query building using attribute fields are also under development. As the GIS components evolve with new requirements so will the potential for re-use in other applications that require similar functionality. The component driven, object-oriented, modular approach is likely to produce multiple returns.

REFERENCES

Anderson, W. P., P. S. Kanaroglou and E. J. Miller (1994). *Integrated Land Use and Transportation Model for Energy and Environmental Analysis: A Report on Design and Implementation*. McMaster University, Hamilton, Canada.

Anderson, W. P., P. S. Kanaroglou, E. J. Miller and R. Buliung (1996). Simulating Automobile Emissions in an Integrated Urban Model. *Transportation Research Record*, **1520**, 71-80.

Anselin, L. (2000). Computing Environments for Spatial Data Analysis. *Journal of Geographical Systems*, **2,** 201-220.

Batty, M. (1994). A Chronicle of Scientific Planning: The Anglo-American Modeling Experience. *Journal of the American Planning Association*, **60,** 7-16.

Bhat, C. R. and F. S. Koppelman (2000). Activity-Based Travel Demand Analysis: History Results and Future Directions. Paper presented at the 79th Annual Meeting of the Transportation Research Board, Washington, D.C.

Booch, G. (1994). Object-Oriented Analysis and Design with Applications. 2nd ed. Benjamin Cummings Publishing Company Inc. Redwood City, CA.

Bowman, J. L. and M. Ben-Akiva (1996). Activity-Based Travel Forecasting. In: Texas Transportation Institute (eds.) Activity-Based Travel Forecasting Conference, June 2-5, 1996. Summary, Recommendations, and Compendium of Papers. Washington, D.C.: Travel Model Improvement Program, U.S. Department of Transportation and U.S. Environmental Protection Agency, pp.3-36.

Buliung, R. N. (2001). Spatiotemporal Patterns of Employment and Non-Work Activities in Portland, Oregon. *ESRI International User Conference Proceedings*. San Diego, California, July, 2001.

Buliung, R. N. (2004). Data Organization and the Design of Measures for Exploration and Analysis of Household Activity/Travel Behaviour. *Unpublished Ph.D. dissertation.* McMaster University, Hamilton, Ontario, Canada.

Caliper Corporation (1990). *TransCAD Transportation Workstation Software: Reference Manual Version 2.0 Caliper Corporation.* Newton, MA.

Claramunt, C. and M. Thériault (2001). A UML-Based Modelling Approach for Analysing Travel Behaviour. Paper presented at Eurosim 2001, Delft, NL.

Commission of the European Communities (1998). SPARTACUS System for Planning and Research in Towns and Cities for Urban Sustainability. EC Environment and Climate Research Programme: Research Theme 4 Human Dimensions and Environmental Change.

Coppock, J. T. (1962). Electronic Data Processing in Geographical Research. *Professional Geographer*, **14,** 1-4.

Damm, D. (1982). Parameters of Activity Behavior for use in Travel Analysis. *Transportation Research A*, **16A,** 135-148.

de la Barra, T. (1989). *Integrated Land Use and Transport Modelling: Decision Chains and Hierarchies.* Cambridge University Press, NY.

Ebdon, D. (1988). *Statistics in Geography* (2nd ed.). Blackwell, Oxford, UK.

ESRI (1996). *Building Applications with MapObjects.* Environmental Systems Research Institute, Redlands CA.

Ettema, D., A. Borgers and H. Timmermans (1993). Simulation Model of Activity Scheduling Behaviour. *Transportation Research Record*, **1413,** 1-11.

Frihida A, D. J. Marceau and M. Thériault (2002). Spatio-Temporal Object-Oriented Data Model for Disaggregate Travel Behavior. *Transactions in GIS,* **6,** 277-294.

Furfey, P. H. (1927). A Note on Lefever's "Standard Deviational Ellipse". *The American Journal of Sociology,* **33,** 94-98.

Gallimore, W. P., D. T. Hartgen and Y. Li (1992). Applications of Geographic Information System-Transportation Analysis Packages in Superregional Transportation Modeling. *Transportation Research Record,* **1364,** 122-130.

Gärling, T., T. Kalen, J. Romanus and M. Selart (1998). Computer Simulation of Household Activity Scheduling. *Environment and Planning A*, **30,** 665-679.

GIS/Trans Ltd. (1997). Transportation Model-GIS Data Integration, Case Study: Southern California Association of Governments (SCAG).

Golledge, R. G., M-P. Kwan and T. Gärling (1994). Computational Process Model of Household Travel Decisions Using a Geographical Information System. *Papers in Regional Science*, **73,** 99-117.

Hägerstrand, T. (1970). What About People in Regional Science? *Papers of the Regional Science Association,* **4,** 7-21.

Harris, B. (1994). The Real Issues Concerning Lee's 'Requiem'. *Journal of the American Planning Association*, **60,** 31-34.

Hartgen, D. T., Y. Li and G. Alexiou (1993). Super-Regional, Very Long Range Transportation Modeling with a Geographic Information System. *Transportation Research Record,* **1413,** 78-87.

Johnston, R. A. and T. de la Barra (2000). Comprehensive Regional Modeling for Long-Range Planning: Linking Integrated Urban Models and Geographic Information Systems. *Transportation Research A*, **34,** 125-136.

Jones, P., F. Koppelman and J. P. Orfeuil (1990). Activity Analysis: State-of-the-Art and Future Directions. In: *Developments in Dynamic and Activity-Based Approaches to Travel Analysis* (P. Jones, ed.), pp. 34-55. Gower, Aldershot, UK.

Kanaroglou, P. S., R. N. Buliung and B. W. Taylor (2000). Mobile Source Emissions in Urban Areas: The Relative Contribution of Trucks, 5th International Conference on Protection and Restoration of the Environment. *Proceedings*, Thassos Island, Greece, July, 2000, Vol. 2, pp.991-998.

Kanaroglou, P. S. and D. M. Scott (2002). Integrated Urban Transportation and Land-Use Models for Policy Analysis. In: *Governing Cities on the Move: Functional and Management Perspectives on Transformations of European Urban Infrastructures* (M. Dijst, I. Thomas and W. Schenkel eds.), pp. 42-72. Ashgate, Aldershot, UK.

Kitamura, R. (1996). Applications of Models of Activity Behavior for Activity Based Demand Forecasting. In: *Activity-Based Travel Forecasting Conference, June 2-5, 1996 Summary, Recommendations, and Compendium of Papers* (Texas Transportation Institute, eds.), pp. 119-150. Washington, D.C.: Travel Model Improvement Program, U.S. Department of Transportation and U.S. Environmental Protection Agency.

Klosterman, R. E. (1994). Large-Scale Urban Models: Retrospect and Prospect. *Journal of the American Planning Association,* **60,** 3-6.

Kwan, M-P. (2000). Interactive Geovisualization of Activity-travel Patterns Using Three-dimensional Geographical Information Systems: A Methodological Exploration with a Large Data Set. *Transportation Research C,* **8C,** 185-203.

Landis, J. D. (1994). The California Urban Futures Model: A New Generation of Metropolitan Simulation Models. *Environment and Planning B,* **21,** 399-420.

Landis, J. D. (1995). Imagining Land Use Futures: Applying the California Urban Futures Model. *Journal of the American Planning Association,* **61,** 438-457.

Landis, J. D. and M. Zhang (1998). The Second Generation of the California Urban Futures Model. Part 1: Method Logic and Theory. *Environment and Planning A,* **30,** 657-666.

Lee, D. B. (1973). Requiem for Large-Scale Models. *Journal of the American Institute of Planners,* **39,** 163-78.

Lee, D. B. (1994). Retrospective on Large-Scale Urban Models. *Journal of the American Planning Association,* **60,** 35-40.

Lefever, D. W. (1926). Measuring Geographic Concentration by Means of the Standard Deviational Ellipse. *The American Journal of Sociology,* **32,** 88-94.

Lenntorp, B. (1978). A Time-Geographic Simulation Model of Individual Activity Programmes. In: *Human Activity and Time Geography* (T. Calrstein, D. Parkes and N. Thrift, eds.), Vol. 2, pp. 162-180. John Wiley and Sons, New York.

Lombard, K. (1997). Application Development with MapObjects and Visual Basic. *ESRI International User Conference Proceedings*. San Diego, California, July, 1997.

Makin, J., R. G. Healey and S. Dowers (1997). Simulation Modelling with Object-Oriented GIS: A Prototype Application to the Time Geography of Shopping Behaviour. *Geographical Systems,* **4,** 397-429.

Maoh, H., P. S. Kanaroglou and R. Buliung (2002). "Modelling the Location of Firms within an Integrated Transport and Land-use Model for Hamilton, Ontario". Paper Presented at International Colloquium: The Behavioural Foundations of Integrated Land-Use and Transportation Models: Assumptions and New Conceptual Frameworks. Quebec City, Canada, June, 2002.

Miller E. J., D. S. Kriger and J. D. Hunt (1998). *Integrated Urban Models for Simulation of Transit and Land-Use Policies*. TCRP Project H-12. University of Toronto Joint Program in Transportation and DELCAN Corporation.

Miller, H. (1991). Modelling Accessibility using Space-Time Prism Concepts with Geographical Information Systems. *International Journal of Geographical Information Systems,* **5,** 287-301.

O'Neill, W. A., R. D. Ramsey and J. Chou (1992). Analysis of Transit Service Areas Using Geographic Information Systems. *Transportation Research Record,* **1364,** 131-138.

Pas, E. I. (1996). Recent Advances in Activity-Based Travel Demand Modeling. In: *Activity-Based Travel Forecasting Conference, June 2-5, 1996 Summary, Recommendations, and Compendium of Papers* (Texas Transportation Institute, eds.), pp. 79-102. Washington, D.C.: Travel Model Improvement Program, U.S. Department of Transportation and U.S. Environmental Protection Agency.

Prastacos, P. (1991). Integrating GIS Technology in Urban Transportation Planning and Modeling. *Transportation Research Record,* **1305,** 123-130.

Prastacos, P. and M. Diamandakis (2000). Applying GIS Technology in Operational Urban Models. In: Spatial Models and GIS: New Potential and New Models (A. S. Fotheringham and M. Wegener, eds.), GISDATA 7, pp.223-234. Taylor and Francis, London, UK.

Recker, W. W., M. G. McNally and G. S. Root (1986a). A Model of Complex Travel Behaviour: Part 1—Theoretical Development. *Transportation Research A*, **20A,** 307-318.

Recker, W. W., M. G. McNally and G. S. Root (1986b). A Model of Complex Travel Behavior: Part II—An Operational Model. *Transportation Research A*, **20A**, 319-330.

Schönfelder, S. and K. W. Axhausen (2003). Activity Spaces: Measures of Social Exclusion? *Transport Policy*, **10**, 273-286.

Sheffi, Y. (1985). *Urban Transportation Networks: Equilibrium Analysis with Mathematical Programming Methods*. Prentice-Hall Inc., New Jersey.

Simkowitz, H. J. (1988). Transportation Applications of Geographical Information Systems. *Computers, Environment and Urban Systems*, **12**, 253-271.

Simkowitz, H. J. (1990). Integrating Geographic Information System Technology and Transportation Models. *Transportation Research Record*, **1271**, 44-47.

Southworth, F. (1995). A Technical Review of Urban Land Use—Transportation Models as Tools for Evaluating Vehicle Travel Reduction Strategies. *Report ORNL-6881*. Oak Ridge (TN): Oak Ridge National Laboratory.

Spiekermann, K. and M. Wegener. (2000). Freedom from the Tyranny of Zones: Towards New GIS-based Spatial Models. In: *Spatial Models and GIS: New Potential and New Models* (S. Fotheringham and M. Wegener, eds.), GISDATA 7, pp.45-61. Taylor and Francis, London, UK.

Stopher, P. R. (1993). Deficiencies of Travel-Forecasting Methods Relative to Mobile Emissions. *Journal of Transportation Engineering*, **119**, 723-741.

Stopher, P. R., D. T. Hartgen and Y. Li (1996). SMART: Simulation Model for Activities, Resources and Travel. *Transportation*, **23**, 293-312.

Torrens, P. M. (2000a). How Land-Use-Transportation Models Work. *Working Paper Series: Centre for Advanced Spatial Analysis*. University College, London, UK.

Torrens, P. M. (2000b). How Cellular Models of Urban Systems Work: (1. Theory). *Working Paper Series: Centre for Advanced Spatial Analysis*. University College, London, UK.

Torrens, P. M. (2001). Can Geocomputation Save Urban Simulation? Throw Some Agents into the Mixture, Simmer, and Wait... *Working Paper Series: Centre for Advanced Spatial Analysis*. University College, London, UK.

U.S. Department of Transportation (1998a). *Transportation Case Studies in GIS*. Available at http://tmip.fhwa.dot.gov/clearinghouse/docs/gis/.

U.S. Department of Transportation (1998b). *A System of Activity-Based Models for Portland*. Oregon. FHWA-PD-99-003.

Waddell, P. (1998). The Oregon Prototype Metropolitan Land Use Model. Presented at ACSE Conference Transportation, Land Use, and Air Quality: Making the Connection. Portland, Oregon May 1998.

Wegener, M. (1994). Operational Urban Models: State of the Art. *Journal of the American Planning Association*, **60**, 17-29.

Wegener, M. and K. Spiekermann (1996). The Potential of Microsimulation for Urban Models. In: *Microsimulation for Urban and Regional Policy Analysis* (G. P. Clarke, ed.), pp.149-163. European Research in Regional Science 6. Pion, London, UK.

Wegener, M. and F. Fürst (1999). *Land-Use Transport Interaction: State of the Art.* Deliverable 2a of the project TRANSLAND of the 4th RTD Framework Programme of the European Commissions: Institut fur Raumplanung (IRPUD), Dortmund.

Yeh, A. G. and X. Li (1998). Sustainable Land Development Model for Rapid Growth Areas using GIS. *International Journal of Geographical Information Science*, **12,** 169-189.

You, J. and T. J. Kim (1999). An Integrated Urban Systems Model with GIS. *Journal of Geographical Systems,* **1,** 305-321.

Yuill, R. S. (1971) The Standard Deviational Ellipse; An Updated Tool for Spatial Description. *Geografiska Annaler. Series B, Human Geography,* **53,** 28-39.

ACKNOWLEDGEMENTS

This paper was supported by GEOIDE, the Network of Centres of Excellence in Geomatics, Canada.

Integrated Land-Use and Transportation Models: Behavioural Foundations
M. Lee-Gosselin and S. Doherty (Editors)

10

A Distributed Computing Approach for Transportation Modelling and Simulation: Application to an Airport System

Cyril Ray, INRIA, Rennes, France
Christophe Claramunt, IRENav - Ecole Navale, France

Abstract

This chapter introduces a modelling and distributed simulation approach for the analysis of transportation data flows. The research is based on a conceptual and logical mapping between a real-world system, in which the interest of study is the modelling of transportation data flows, and a distributed computing environment. The dynamic and static properties of a transportation system are modelled using a hierarchical-based approach at different levels of granularity. The mapping between the application and the computing level is ensured by a graph that represents a structural decomposition of the transportation system on the one hand, and the computing nodes of the distributed system on the other. Our computing platform supports time-driven distributed simulations whose dynamic properties replicate the behaviour of transportation data flows. One of the peculiarities of the platform relies in the concept of migration by group that supports the coordinated displacements of objects between the computers of the distributed system, thus replicating the transportation data flows identified at the application level. Simulation objectives are the analysis of emergent transportation behaviours that result from the non-linear interactions and displacements of mobile objects within the distributed system. The potential of our distributed computing approach is illustrated by a case study that models and simulates passenger flows between the different halls of an airport terminal.

INTRODUCTION

An important objective of modelling and simulation is the characterisation of real-world entities by a set of unique mappings towards abstract objects that can be used to simulate the behaviour of a complex system (Zeigler *et al.*, 2000). A significant trend in modelling is that of simplifying a real-world system to create artificial representations. These virtual worlds act as platforms to develop dynamic representations at either the aggregated or disaggregated simulation levels (also called macro- vs. micro-simulations). The main advantage of disaggregated over aggregated oriented simulations relies upon the fact that they reproduce the behaviour of a represented system at the finest granularity level (Clarke and Holm, 1987; Goodchild, 1998; Miller, 1998; Thériault *et al.*, 2002).

Modelling disaggregated-based applications logically implies a hierarchical approach in which properties of each modelling level are derived from those of the finer levels of granularity (Minsky, 1986). Hierarchical modelling represents powerful and programmable abstractions to deal with complex and interconnected problems (Miller, 2005). Each level of a given hierarchy has its own and specific temporal granularity and spatial scale of analysis, the higher those levels the lower the spatial scale and the speed of evolution.

Disaggregated-based applications constitute a form of distributed and non-linear system made of autonomous entities that behave, interact, evolve and generate global patterns whose analysis, prediction and planning are important objectives of many urban studies. Several natural and artificial systems have been analysed using disaggregated-based modelling principles in the domains of economy, biology, physics, cognitive psychology and transportation (Renders, 1994). In the field of transportation, disaggregated-based models have been widely used for the development of micro-simulation models. Those include origin-destination surveys and analysis in which emergent properties and patterns of a given transportation system are derived from local transportation choices and interactions (Frihida *et al.*, 2002; Buliung *et al.*, 2005; Roorda *et al.*, 2005), and the micro-simulation of business establishments (Khan *et al.*, 2002). However, disaggregated-based simulations still suffer from several drawbacks. Firstly, despite the fact that some common mathematical and modelling principles emerge through their application to different disciplines, it is often difficult to apply and re-use some of the modelling and simulation principles from one discipline to another one. Secondly there is also a gap between the models and paradigms often used in disaggregated-based models and computational capabilities. This is due to the difficulty of mapping modelling concepts used in disaggregated-based approaches towards relevant computing units, this both at the static (data properties) and dynamic levels (data behaviour). It has been also noted that a few efforts have been made so far to integrate ideas drawn from computation to conceptualise, understand and model disaggregated-based models (Torrens, 2001).

Distributed computing environments offer a new avenue for the simulation of dynamic and large geographical systems to be represented and implemented by inter-connecting component models (Bargiela, 2000). They constitute promising platforms to explore the underlying properties and behaviour of large real-world systems. Two essential characteristics define a distributed system: such a system has a number of independent, autonomous, communicating CPUs and the system appears to the users as a single computer (Tanenbaum, 1993). From a modelling point of view, a distributed system is a collection of interacting objects that intercommunicate only using requests sent to well-defined interfaces (Guerraoui and Vinoski, 1997). The advantages of these systems are multiple. First, they potentially allow for a distribution of simulations through local or wide area networks, offering then a form of collaborative computing environment. Secondly, complex and costly computational simulations can be supported by a distribution of data and processes at the appropriate granularity level.

Despite its potential, computing distribution of real-world data and processes still requires the exploration and development of appropriate data communication models and synchronisation mechanisms between different computing nodes (an old problem of distributed systems), a continuous estimation of the uncertainty factor, and the design of interfaces and visualisations adapted to interactive communication between simulation and end-users. This research deals with some of the above issues and introduces a distributed computing environment that acts as a dynamic modelling support for the simulation and coordination of transportation data flows. This distributed computing environment replicates the behaviour of a transportation system, including modelling of real-worlds entities at the object level (i.e., static and dynamic properties), and physical communication rules that replicate transportation data flows. These modelling concepts are applied to an airport case study that acts here as a reference to illustrate the capabilities of our approach. The intention is to apply these concepts to range of transport behaviours.

MODELLING APPROACH

Without loss of generality, the principles of our modelling approach are introduced and illustrated in the context of an airport system that serves as a case study. Airport systems face important modelling and simulation challenges to infrastructure and logistics planners. The evaluation of the gearing between the current terminal organisation, available infrastructure, transportation capacity and future needs are important aspects to consider in airport management and planning. It has been also noted that although simulations of human-driven flows in airport terminals are not always a deterministic process, simulations of such behaviours are very valuable in transportation logistics of planning (Gatersleben and van der Wiej, 1999). This fact is particularly reinforced thanks to the ability of computing systems to report the emergence of high level properties of behaviours and interactions of fine-grained entities (passengers and visitors in the present example). These arise from non-deterministic choices and non-coded behaviours in the computing model.

Several related simulation systems have been proposed so far for the design and planning of airport terminals. Those include micro-simulations for the distribution of aircraft on airport gates (Cheng, 1998), flight scheduling (Teretsch, 1993) and roadway and curbside traffic (Tunasar *et al.*, 1998). Another important aspect in airport systems is the analysis of the distribution of passenger flows within airport terminals (Setti and Hutchinson, 1994; Gatersleben and van der Wiej, 1999). In (Gatersleben and van der Wiej, 1999) a micro-simulation system analyses passenger flows at a lower level of abstraction; modelling units include check-in, boarding, immigration and transfer operations. This has been applied to the internal design and optimisation of passenger holding at Schiphol airport in the Netherlands. In a related work (Setti and Hutchinson, 1994), a fluid approximation model, based on simulation principles used in hydrology, simulates the operations of terminal facilities; a computer implementation has been implemented and applied to the simulation of the operations of a terminal of Toronto airport.

The model presented in this section describe an airport system, Charles de Gaulle Terminal 2 (CDG terminal 2), where passenger flows are distributed through several halls, and in which (flight) passengers and visitors can be considered as mobile entities acting in the airport system. CDG terminal 2 has a hierarchical-based structure that can be decomposed over a distributed system. Passenger flows have the property of being disaggregated but also coordinated by transportation resources. The model is based on several simplifying assumptions. Our aim is not to reproduce exhaustively the complex behaviour of CDG Terminal 2, but rather to provide a flexible modelling platform to approximate and explore performance of different planning strategies. We intend to simulate the distribution of transportation flows, bottlenecks and emergent patterns according to different planning scenarios (e.g. number of halls, flight timetables, transportation resources and flows within the terminal). Local behaviours such as check-in, boarding and immigration controls are not considered by our modelling approach.

Modelling principles

Our modelling approach relies on a hierarchical and a graph-based analysis of the structure and behaviour of a given transportation system, preferably distributed (Figure 10.1). At each level of the hierarchical organisation, entities and their characteristics are identified. At the finest level of the hierarchy, objects are the basic modelling units. Those objects constitute the disaggregated and main dynamic component of the model. Although these objects behave according to their initial properties and displacement objectives, their actions and displacements are controlled, at a higher level of the system, by the modelling unit in which they are currently located. This facilitates their distribution and monitoring in a computing environment. Time is distributed and synchronized across the different levels of the hierarchy. This guarantees a timely management of object interactions and displacements within the transportation system.

The resulting hierarchical and graph-based model is mapped towards a computational representation. This computational representation is supported by a middleware, so-called

Atlas (Ray and Claramunt, 2002), that supports logical and physical management of a computing network and mobile objects on top of a distributed system. This distributed system is modelled as a bi-directional connected graph where the set of nodes represents the computing elements, and the set of edges communication links and data flows. Simulations allow for an interactive exploration of emergent patterns, constraints and performances of the system represented, according to different graph configurations. An important property of this modelling and distributed simulation approach, over a single computer solution, is the replication of the logical structure of the given real-world system at the physical computing level. This implies that each computer works at the finest granularity level. The system reports its global state using bottom-up behaviour by merging the local states of its computing elements, where the complex system is viewed as a set of interacting objects.

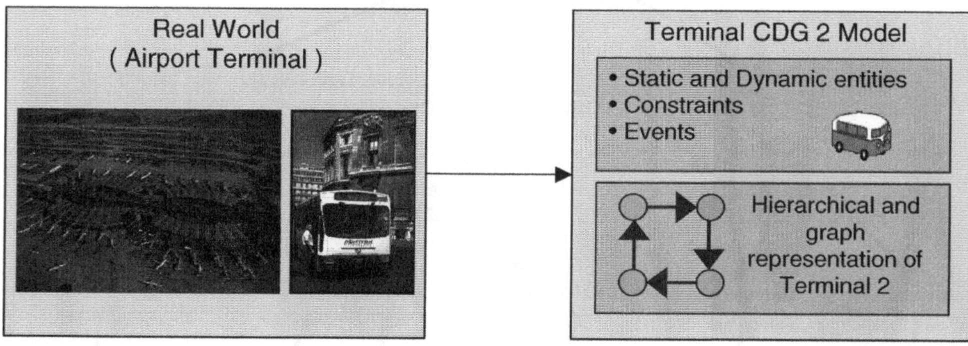

Figure 10.1 Modelling principles

Hierarchical modelling

CDG terminal 2 is made of six halls (A to F). It is modelled as a hierarchical system made of several levels of granularity where people displacements are represented at the appropriate granularity level (Figure 10.2).

This hierarchical modelling method highlights the partition of the system as spatial (aggregated) components that can be projected over a distributed computing environment. In Figure 10.2 those components are parts of the second level, meanings halls. These spatial properties can be implicitly modelled as a connected and directed graph where each node represents a hall, and edges connections between these halls. The dynamism of the system is given by flight timetables according to different schedules, connections constraints throughout the terminal, and people origin and destinations within the terminal (e.g. hall of origin, hall of departure).

People flows between the different halls of Paris CDG terminal 2 are modelled using a bus metaphor that performs a cyclic and regular tour between the halls of CDG terminal 2. This bus takes/drops people at their origin/destination hall (i.e., bus stops linked to the halls). People flows between the different halls of the terminal are made of flight passenger and visitor flows. Distances between the nodes of the graph are represented as temporal parameters that constrain people displacements between those nodes. A second displacement constraint controls those displacements: bus timetables and capacities. Basic modelling elements are as follows:

- Flight passengers and visitors (objects to migrate);
- Halls and their interconnections (modelled by the graph structure);
- Bus and bus stops;
- Flights (arrivals, departures).

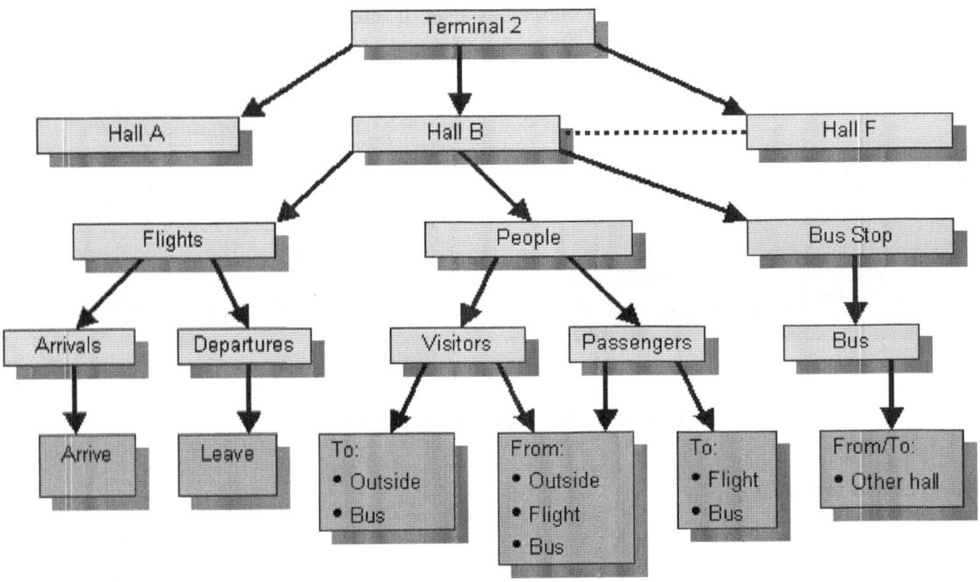

Figure 10.2 Hierarchical model

Functional modelling

During a simulation, each hall regularly generates the arrivals of people according to figures derived from CDG Terminal 2 timetables. A similar pattern is defined for people deletion, according to flight departures, and using a random departure time given to each visitor generated in the system. People states inside the terminal depend on their origin and destination, and on their type (visitors or flight passengers). Flight passengers are constrained by their flight departure time while visitors stay inside the terminal until their visiting time is over. The outside world denotes the external part of the terminal. The events that support the modelling of passenger and visitor displacements during a simulation are as follows:

- From outside to a hall;
- From a flight to a hall;

- From a hall to outside;
- From a hall to a flight;
- From a hall to another hall.

Passenger and visitor activities in the terminal are illustrated by an activity diagram (Figure 10.3). In the terminal, passengers and visitors behave (in terms of mobility) according to their arriving and their destination hall. Within a hall, visitors and passengers either wait for a flight or a bus or simply wait before leaving the terminal. When a person gets into a bus, he stays into the bus until that bus reaches the destination hall.

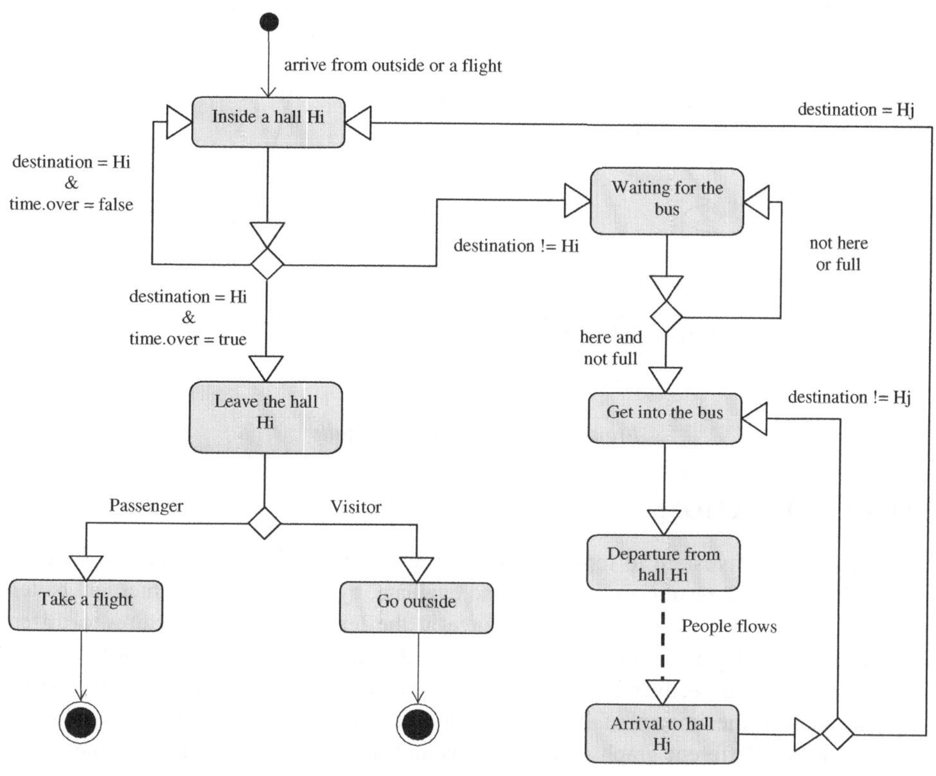

Figure 10.3 People life cycle

At the computational and simulation levels, each hall (i.e., computing node) manages its flights according to their arrival and departure times. This leads to a regular generation and deletion of visitors and passengers. Flights timetables are initialised and monitored per hall according to different scenarios and on a 24-hour basis. Each 24-hour simulation reflects a usual temporal distribution of flights at CDG 2 Terminal.

Graph modelling

The graph retained for the modelling of the terminal is composed by four nodes which represent the initial configuration of CDG terminal 2 (halls recently created are not considered). Those halls (i.e., modelled as nodes) are connected by a bus metaphor that distributes flight passengers and visitors, according to their flight timetable constraints and arrival and/or destination halls. The edges of the graph represent the path followed by the bus. The bus performs a cyclic path from hall B to D, C, A to B as illustrated in Figure 10.4. Bus displacement times replicate observed bus timetables.

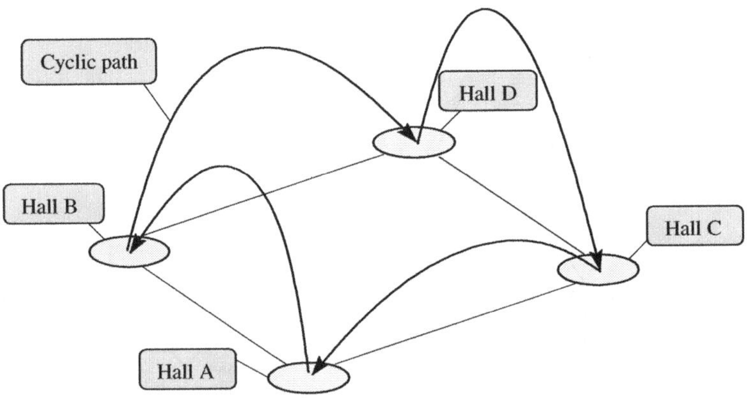

Figure 10.4 Path between halls

Airport Simulations

The simulation objectives are to represent and simulate people flows and patterns within the terminal. The expected results are to derive and evaluate, by the simulation, emergent patterns, presence of bottlenecks and overall performance at the terminal level according to different hall configurations, flight timetables and estimated displacement times between the terminal halls. The simulation objectives are not only to evaluate the final state of the simulated system, but also to analyse the impact of initial conditions on that final state and the evolution of successive states. Different graph configurations and transportation schemes of the terminal halls are distributed on a computing environment and evaluated to analyse how different requirements can be met.

Simulation principles

The computational representation of CDG 2 Terminal model is supported by Atlas, a distributed platform for synchronous, parallel, discrete event simulation (Righter and Walrand, 1989). The Atlas middleware is a Java-based software developed on top of a cluster made of a controller and several computing nodes that together form a local area network. The objective behind the design and the development of the Atlas platform is to offer a reusable computing environment for distributed simulations of transportation system. Atlas supports the initialisation of airport system modelling units at the different levels of the hierarchy, together with the dynamic properties of the system. Each computing node simulates a well-defined part of the transportation system. Atlas is composed by complementary modules:

- a communication module ensures physical communications between the computing nodes and the controller;
- a migration per-group module (Round Trip Bus protocol: RTB) allows object displacements between the computing nodes (Ray, 2001);
- an object management module offers basic functions for object initialisation, updates and deletion within the distributed system;
- a module defines a global distributed time which is synchronised across the computing nodes, ensures a time-driven progress of the simulation, and supports time-stamping of simulation events;
- an application–oriented interface module controls and executes the graphic interface that monitors a given simulation and;
- a monitoring module controls Atlas states and manages several internal indicators that reports on the evolution of a given simulation.

An Atlas simulation is composed by several computing multithreaded processes distributed and physically located on each node (Figure 10.5). Each computer represents a hall implemented by a Java agent. The entities represented as Java objects (e.g. passengers, flights) are managed and monitored by the hall/computer where they are located. Atlas initial parameters include the logical structure of the graph (i.e., nodes and edges), bus transportation timetables and capacities and temporal constraints (e.g. displacement times).

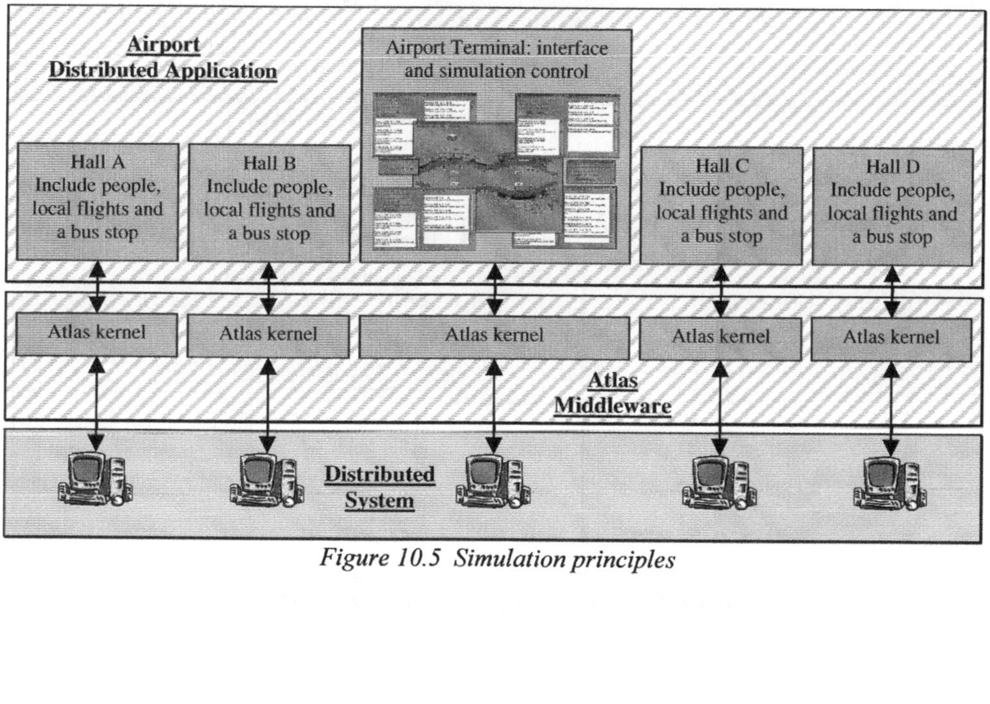

Figure 10.5 Simulation principles

One of the peculiarities of our approach relies in the parallelism made between the graph structure of the terminal, and that of the distributed computing local area network (i.e., each computer models a hall). The per-group migration protocol (RTB) replicates and controls bus cyclic displacements, thus ensuring at the computing level people displacements between the halls according to the system constraints.

The simulation interface offers a visual control of the simulation at different time-steps (Figure 10.6). The Atlas controller manages and displays a graphic interface that regularly reports on the simulation states. Each local change at the node level is transmitted to the controller (e.g. arrival of a new passenger at a terminal hall). Simulation time is given by the controller, using a mapping between airport and physical times. The Atlas airport interface proposes several interaction levels whose objectives are to trigger different simulations, report on simulation states, and analyse bottlenecks and emergent properties. The airport interface developed integrates several monitoring components. The main panel displays a graphic representation of CDG terminal 2, and reports successive bus locations and displacements. Additional event-oriented panels display information on bus states, hall states, and airport time. Atlas also offers additional facilities to export simulation results to graphic packages in order to generate graphs, charts and radviz plots.

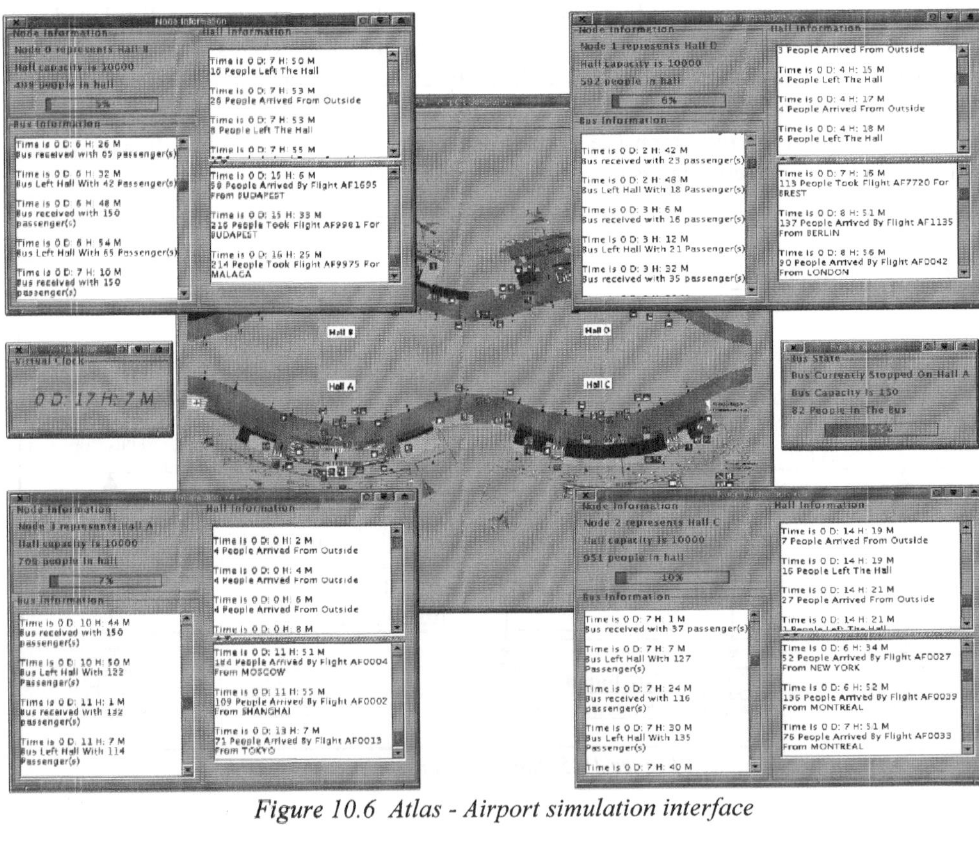

Figure 10.6 Atlas - Airport simulation interface

Simulation experiments

Let us introduce three examples of simulation initialised according to different configurations: balanced vs. unbalanced distribution of flights across a four-hall terminal, alternative three-hall terminal with a well balanced distribution of flights. Those simulations illustrate how different initial configurations behave and have an impact on overall terminal performances.

Measuring and evaluating those simulations require an agreement on some performance indicators. An indicator of interest is given by estimating the number/percentage of passengers that miss their connecting flights for each hall. Simulation should also address balanced versus unbalanced performances and possible bottlenecks. This is based on the assumption that a successful terminal design minimises unsuccessful passenger connections. Simulation results are evaluated at different timestamps using the Atlas simulation interface.

Airport simulations are parameterised using initial conditions defined for each scenario. The following scenarios include some constant values such as the hall capacity (10 000), number of people per hall at initialisation time (50), number of arrival flights per hall (16 flights) and flight capacity (200 passengers). The number of flight departures per hall is a variable (between 5 and 24 flights) defined homogeneously, or not, according to simulation scenario objectives. Let us remark that these simulations are calibrated using flight departures only, as this gives sufficient modularity and flexibility. All airport simulations cover a 24-hour period, which is the acceptable level of analysis for the evaluation of terminal behaviour. The RTB protocol is currently parameterised with one bus. A bus displacement between two nodes is triggered either after a given estimated delay or when its capacity is reached (set to 150 people). Finally bus travel times are modulated according to day-night cycles.

Four balanced-halls scenario

A first simulation scenario models a four-node simulation with a well-balanced scheduling of flights across the four nodes and 18 flight departures per hall. Figures 10.7 to 10.10 represent the halls behaviour. Each chart evaluates incoming and outgoing objects according to their origin and destination. It contains two kinds of indicator: number of objects per hall as a function of time, and several object counters from simulation starting time (e.g. total number of people who missed their flight per hall). In this scenario the numbers of people that enter/leave each hall from/to outside is relatively well-balanced across the halls.

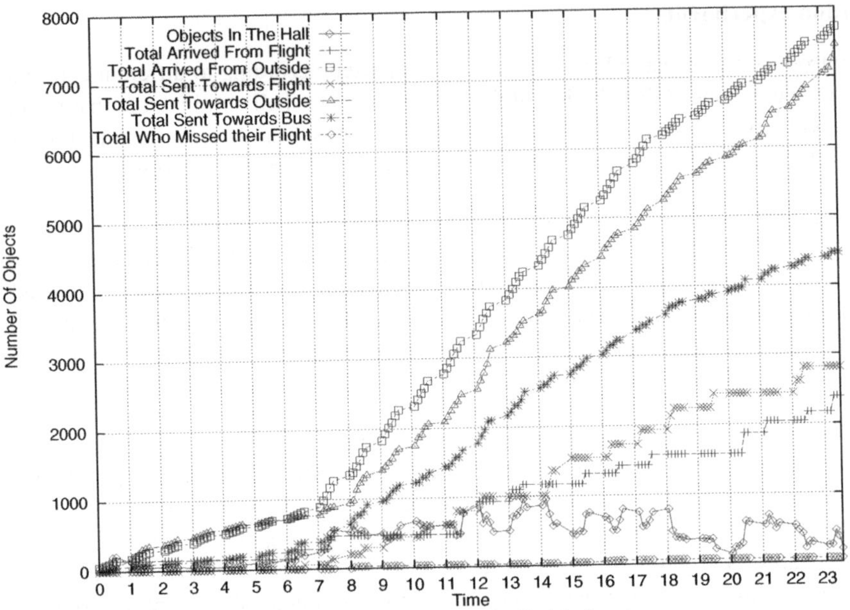

Figure 10.7 Scenario 1 - hall A behaviour

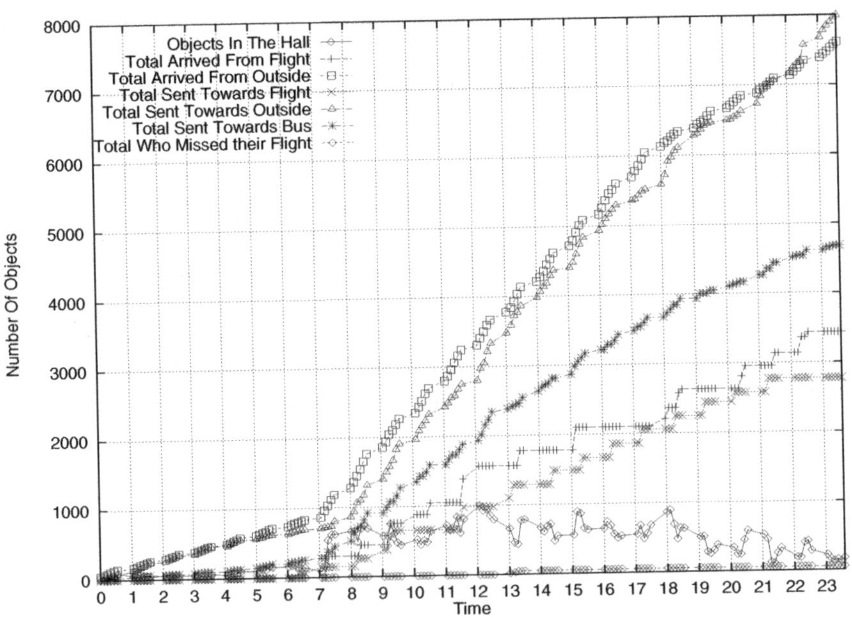

Figure 10.8 Scenario 1 - hall B behaviour

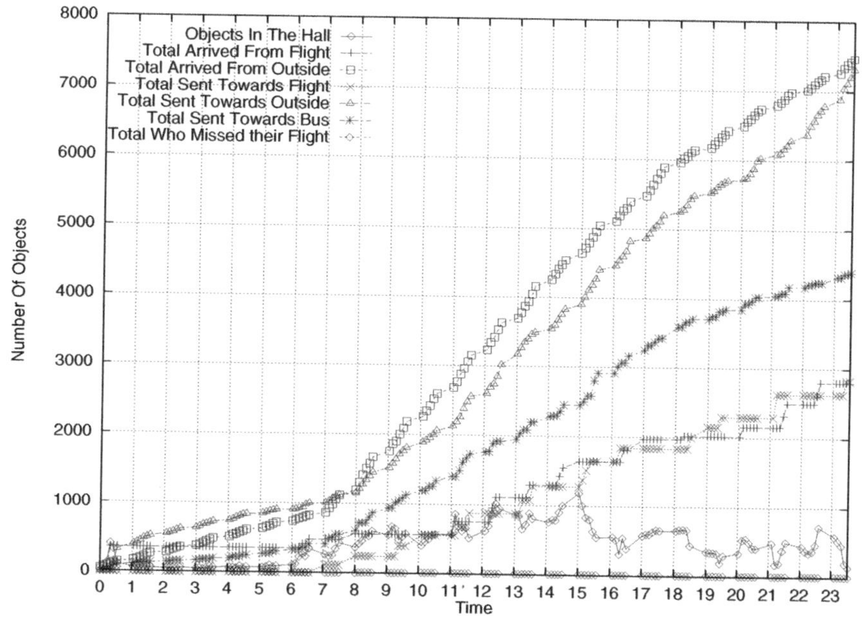

Figure 10.9 Scenario 1 - hall C behaviour

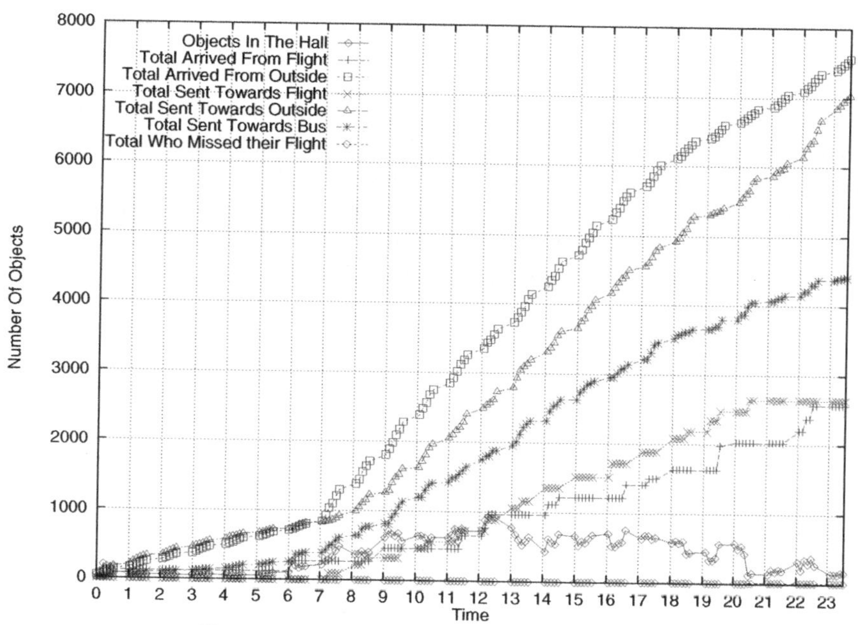

Figure 10.10 Scenario 1 - hall D behaviour

Figure 10.11 gives a comparative analysis on each hall performance by giving the evolution of the percentage of people who missed their flight regarding to the total number of passengers who planned to take a flight.

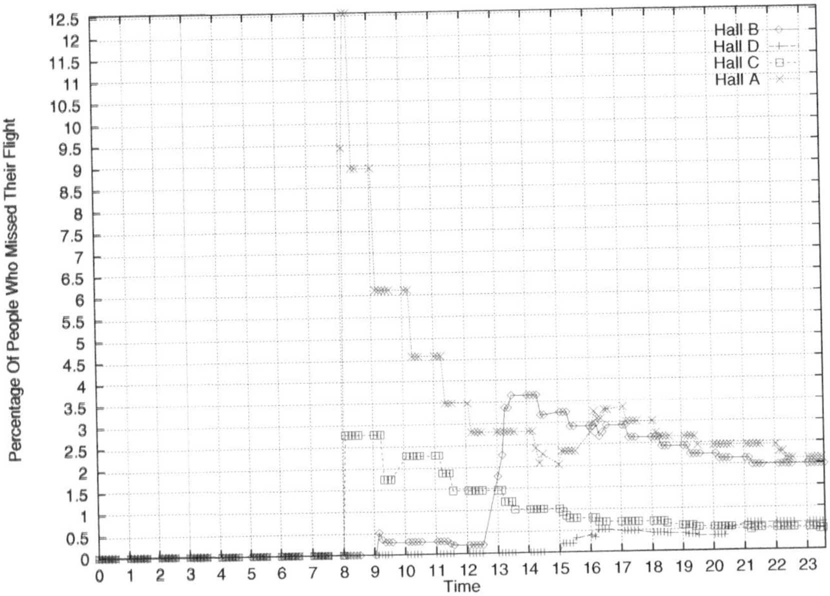

Figure 10.11 Scenario 1 - halls performance

This simulation reaches a relative equilibrium at the end of the day: the percentage of people that missed their flight varies between 0,5% and 2%. The peak in hall A (Figure 10.11) can be explained by the fact that at 8am a few people have taken a flight in that hall (only one flight at 7:30 am), so a relative small number of flights missed has a great impact on that performance indicator. This pattern is reinforced by the fact that this period is a relatively busy time with several flights arriving causing a transportation bottleneck.

Four unbalanced-halls scenario

A second simulation analyses the halls behaviour of a four-node terminal, but with an unbalanced scheduling of flights across the four nodes with 22 flight departures for halls B, D, C and 5 flight departures only for hall A. As the numbers of people generated and arrival flights are still equals in each hall, displacement requests from hall A to other halls increase. This results in a higher demand on displacements from hall A and affects connections to other hall flights.

Figure 10.12 shows the evolution of the percentage of people who missed their flight in relation to the total number of people expected to take a flight. This simulation gives some interesting results as the unbalanced flights scheduling is reported at the performance level. Hall B, D and C flights overload, and connection problems from hall A affect their performance. One can remark that hall B performs better than hall C and D. This reflects the clockwise direction of the bus cycle. Conversely hall A performs relatively well as there is no flight missed in this hall.

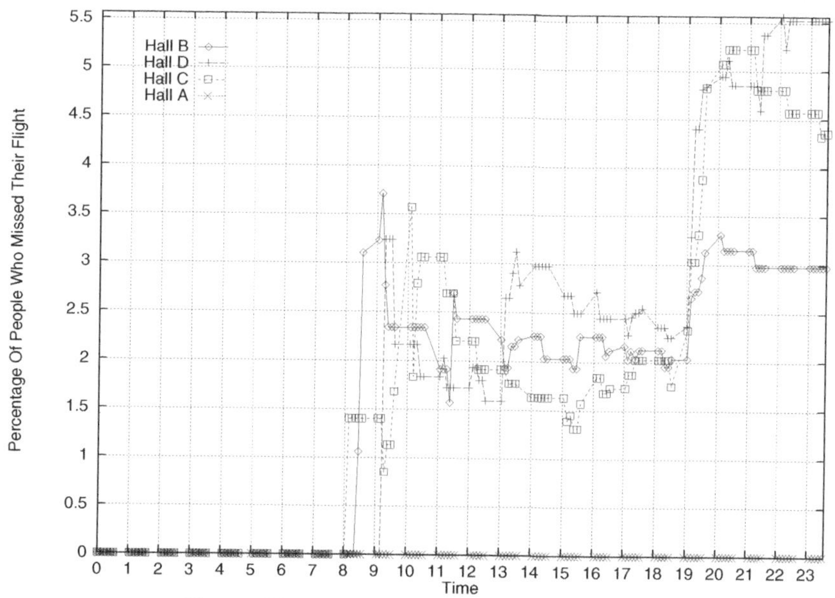

Figure 10.12 Scenario 2 - halls performance

Three balanced-halls scenario

A third simulation replicates the four-node simulation flight and passenger terminal figures but with a three-node terminal configuration. Daily numbers of flights and passenger flows are equal to those of the four-node simulations. This simulation presents the interest of evaluating how the terminal reacts when a hall becomes unavailable. The simulation is initialised using a well-balanced scheduling of flights across the different halls (24 flight departures per hall). At the hall level, people generation and deletion flows, as well as the numbers of incoming and outgoing passengers increase so does the computational pressure on each hall. This reflects an increase of internal passenger loads in the three halls.

Figure 10.13 illustrates the evolution of the percentage of people who missed their flight in function of the total number of people who were expected to take a flight in this hall. In the same way the numbers of people per hall increase to reach a peak at the end of the day. This scenario illustrates the fact that reducing the number of halls of the terminal implies to at least augment transportation capabilities (under the same initial conditions) in order to ensure satisfactory connections.

The configuration of this scenario has the same characteristics as in the previous four hall simulations in terms of people figures, bus capabilities and scheduling, number of flights and flight connections. The scenario generates a situation where each hall has more requests for displacements and people flows to manage (although the number or requests for displacement proportionally decrease in each hall, the augmentation of people in each hall leads to a quantitative increase of displacement flows within the terminal). This leads to larger waiting queues in each hall, unsuccessful connections and overall poor performance of this scenario.

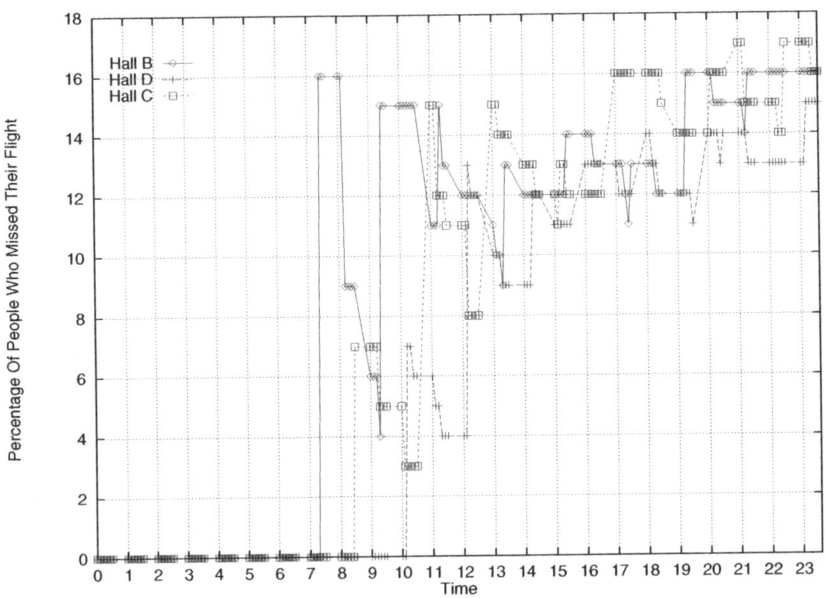

Figure 10.13 Scenario 3 - halls performance

A matrix of passengers' flows presents respective simulation performances at different time-stamps (Figure 10.14). Each graph gives, for each hall, percentages of passengers that took/missed their flight, and total numbers of passengers that took a flight. Passenger flows in the terminal are displayed in relative values. The different scenarios presented evolve to reach equilibrium at the end of the day. One can remark the well-balanced simulation states of scenario 1. Conversely, the last state of scenario 2 denotes an unbalanced configuration and its effect on hall performances. One can also note the impact of the increase of flights in scenario 3 on hall performances. These graphs are generated on demand for any time-stamp of a given simulation; they provide an intuitive interface for the analysis of the distribution of hall performances.

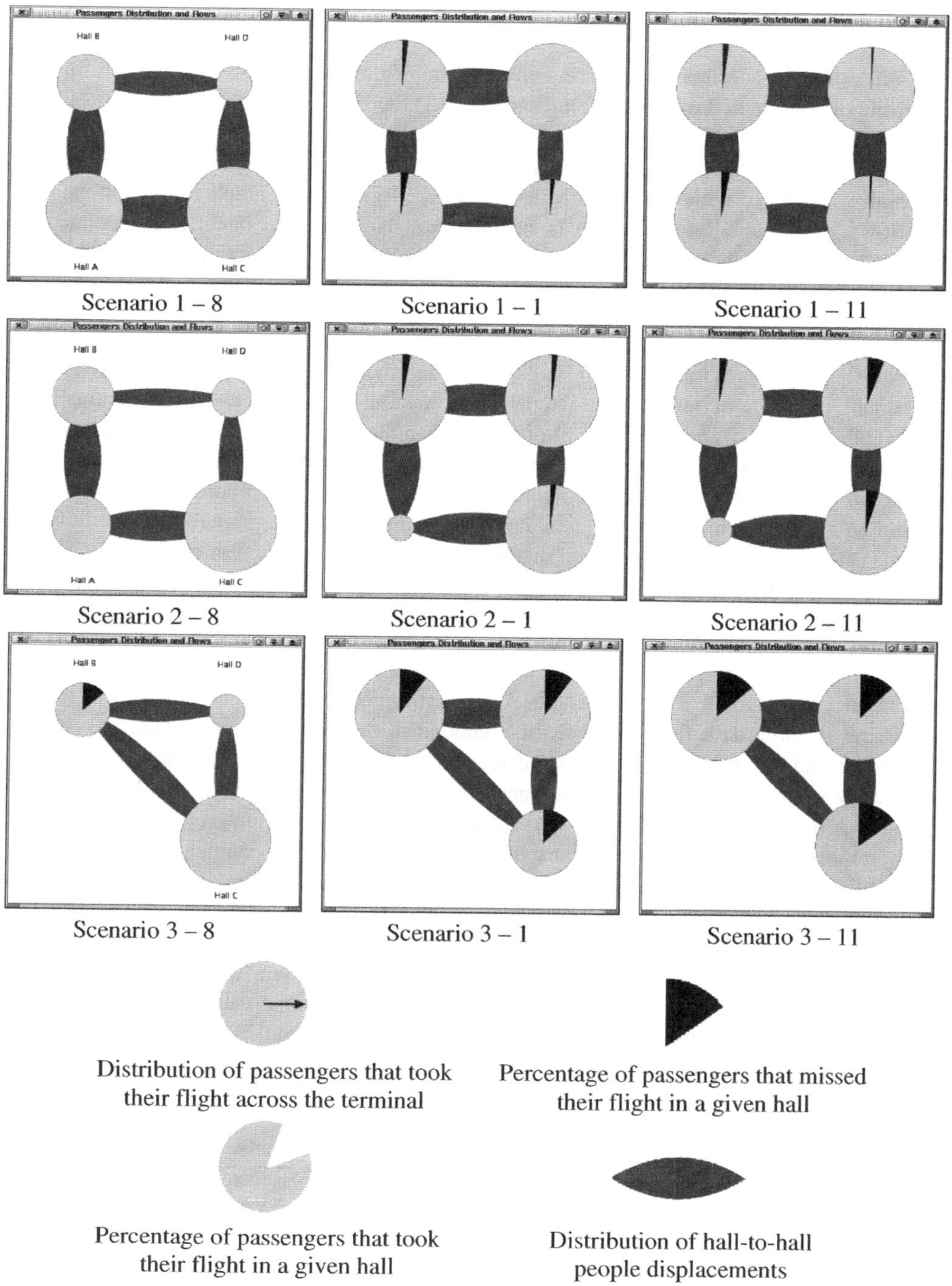

Figure 10.14 Passengers' distribution and flow

Discussion

This airport system, designed as a case study, illustrates the potential of the approach for the modelling of coordinated transportation data flows. The proposed model replicates data flows that can be logically structured as a graph. A modelling objective relies on defining user-defined constraints that model the application semantics, particularly how data flows are temporally distributed within the graph. These flows reflect different graph configurations and data flows. The modelling approach is flexible enough to be used in different application contexts as far as a given transportation system can be mapped to a graph, and as the user's objective is to understand data flow patterns at the global level from the aggregation of local behaviours.

The Atlas platform provides an interactive simulation environment that supports analysis and assessment of different terminal structures and timetabling strategies. These simulations illustrate the flexibility and suitability of our simulation platform for exploring, visualising and evaluating different terminal configuration performances in preliminary design studies. The simulation platform is flexible as the number of nodes, transportation constraints and flights scheduling are user-defined. These simulations are supported by the modelling capabilities of the Atlas system, and the RTB protocol that supports physical migration of computing objects. Atlas capabilities have been illustrated by a reference/demonstrative physical middleware made of several nodes and a cyclic migrating protocol. This physical configuration can be extended by increasing the number of computing nodes as necessary.

Performance evaluation shows that the Atlas distributed environment is adapted to the processing of large data flows, it is flexible enough to model larger transportation systems. The model and simulations developed are far from being complete with respect to the real complexity of the distribution of passenger flows within CDG terminal 2. However they provide a first level of efficient and flexible computational analysis to the terminal configuration, flight timetables and transportation flows within the terminal. Comparisons of our distributed simulations to single computer simulations under similar conditions show that a mono-processor execution (i.e., one computer) quickly reaches its limits in terms of memory and processing capacity (Ray, 2003). Under the conditions of the scenarios presented in the case study, a mono-machine execution almost reaches its maximum computational capacity. Thus it is straightforward to conclude that for larger simulation involving huge number of objects, a distributed computing approach is required. This has the advantage not only to support large simulations, but also to improve dramatically computational performances, that is, the execution time of the simulations.

So far, the model presented includes a one-to-one node association between the application and the distributed computing environment, and a single bus line. Recent developments of the Atlas platform integrate a RTB multi-graph protocol that extends the modelling capabilities by supporting several bus lines (Ray, 2003). This protocol works with many migration groups schemas, where each migration group represents a bus line. Displacements are optimised according to an adapted Dijkstra's shortest path algorithm which manages bus line identities, distance between nodes and penalties for bus connections. The Atlas simulation platform can be extended to an indefinite number of nodes, the limitations being the number of processing units available.

RELATED WORK

The distributed computing of transportation systems leads to several modelling and implementation challenges. At the representation level, there is a need for a model adapted to the complexity of the transportation application. At the computational level, the distributed environment should be a faithful representation of that model. The former is the scope of transportation modelling studies. The latter implies identifying a relationship between the partition of the distributed computing and the transportation application, which is the main aim of our research. It is useful to briefly review distributed systems that attempt to partition and simulate the behaviour of large set of objects acting in geographical systems.

In (Merchant *et al.*, 1995) a distributed computing environment replicates the underlying structure of geographical application using a quadtree decomposition. Simulated objects support migration, these migrations being controlled by the line segments presenting the represented space or directed by a move in a specific direction. In the domain of traffic systems, the Cybele prototype provides a distributed multi-agent system (Erol *et al.*, 1999). This agent-based simulator supports load-balancing over a large set of heterogeneous computers, agent migration, communication and interaction. Each computing node models a geographical region (i.e., partition of the traffic system) at the micro level. Partitioning space to improve computational time has been also applied to ecological simulations. In (Maniatty *et al.*, 1993), a spatially explicit model of epidemics behaves like a probabilistic cellular automaton in which discrete cells partition space, making the simulation suitable for distributed and parallel implementation. In comparable work, a parallel simulation system has been developed for landscape modelling (Maxwell, 1997). Each processor is assigned a different partition of space, each region of space being structured as a set of cells. Interprocessor communications are based on neighbouring relationships. This allows a cellular automata landscape model to be efficiently executed over a relatively high resolution. The computing environment is also designed as an open environment for ecological and economic modelling.

These distributed systems share several characteristics: they use either the graph properties of a distributed system or a partition of space to simulate a geographical phenomenon. Propagation of objects is asynchronous, objects are autonomous in the sense that their behaviours are independent and triggered at the local level. Our distributed computing approach, thanks to the migration per-group protocol, simulates the properties of a graph-based system under some similar principles. The modelling and simulation approach considered offers a flexible and scalable distributed platform, support for the development of customised applications thanks to its multi-layer architecture. This is an important advantage compared to other distributed and multi-agent approaches that are often mono-application oriented.

SUMMARY

The aim of the research presented in this chapter is to explore and validate the interest of distributed systems for the modelling and simulation of transportation data flows. We introduce a hierarchical and graph-based modelling approach, on top of a computational framework, that represents a first step towards the development of a computing architecture flexible enough to be applied to different application areas. Relying on distributed simulation principles, our approach includes several features particularly adapted for the simulation of transportation system from the modelling, distribution and simulation of data flows. The computational implementation includes several functional levels from the logical representation of data structures and properties at different levels of granularity, to the control, monitoring and analysis of data flows within the represented transportation system. The Atlas platform is based on a time-driven engine that ensures a synchronous distributed simulation and maximizes simulation performance. A user-oriented interface supports the parameterisation, execution, control, monitoring and visualisation of the simulations. This gives the user an interactive environment to observe and analyse the behaviour, performance, bottlenecks and emergent patterns of the represented system.

A case study, that models and simulates passenger flows within the different halls of an airport terminal, illustrates the potential of our modelling and distributed simulation approach. Simulation experiments show how such a terminal behaves at the global level in function of different initial conditions. The simulation platform presented is designed as application independent so it can be re-used with some minor adaptations to replicate the static and dynamic properties of transportation systems. This application is a first step towards the development of a modular and flexible distributed computing environment oriented to the simulation of transportation systems where the object of study is the simulation of data flows.

On-going work concerns the adaptation, validation and calibration of our modelling platform in the context of transportation systems. We plan to extend the modelling concepts currently identified by our approach to enhance interaction during the simulation process and better scalability. A modelling extension currently under study is the representation of collaborative agents that coordinate their actions within the system. Another challenging issue to consider is the dynamic reconfiguration of the graph structure during a simulation. This should support the modelling of critical situations and dynamic reconfigurations of the transportation system. Scalability implies to model large transportation systems and to validate our approach in the context of real systems.

REFERENCES

Bargiela, A. (2000). Strategic directions in simulation and modelling. *UK Computing Research Strategy Meeting*, The Nottingham Trent University, Nottingham, UK.

Buliung, R., P. Kanaroglou and H. Moah. (2005). GIS, Objects and Integrated Urban Models. *In this volume*.

Cheng, Y. (1998). A rule-based reactive model for the simulation of aircraft on airport gates. *Knowledge-Based Systems,* **10**(4), pp. 225-236.

Clarke, M. and E. Holm (1987). Microsimulation methods in spatial analysis and planning. *Geografiska Annaler*, **69B**, pp. 145-164.

Erol, K., R. Levy and J. Wentworth (1999). Application of Agent Technology to Traffic Simulation. Technical report. *U.S. Department of Transportation*. McLean, Virginia, USA.

Frihida, A., D.J. Marceau and M. Thériault (2002). Spatio-temporal object-oriented data model for disaggregate travel behavior. *Transactions in GIS,* **6**(3), pp. 277-294, Blackwell Publishing.

Gatersleben, M.R. and S. van der Wiej (1999). Analysis and simulation of passenger flows in an airport terminal. *Proceedings of the 1999 Winter Simulation conference*. Hilton Squaw Peak Resort, Phoenix, Arizona, USA, pp. 1226-1231.

Goodchild, M.F. (1998). Geographical information systems and disaggregate transportation modelling. *Geographical Systems*, **5**(1-2), pp. 19-44.

Guerraoui, R. and S.M. Vinoski (1997). Guest editorial. *Distributed Systems Engineering*, **4**(3), pp.129-130.

Khan, A.S., J.E. Abraham and J.D. Hunt (2002). A system for microsimulating business establishments: Analysis, design and results. *Geoide International Colloquium on the Behavioural Foundations of Integrated Land-Use and Transportation Models*. Québec, Canada.

Maniatty, W., B.K. Szymanski and T. Caraco (1993). Implementation and performance of parallel ecological simulations. *Proceedings of IFIP WG10.3 International Conference on Applications in Parallel and Distributed Computing*. Amsterdam, The Netherlands.

Maxwell, T. (1997). An open geographic modeling environment. *Simulation Journal,* **68**(3), pp. 175-185.

Merchant, F., L.F. Bic, P.M. Borst, M. Corbin, M.B. Dillencourt, M. Fukuda and P.S. Sapaty (1995). Simulating autonomous objects in a spatial database using WAVE. *Proceedings of the 9th European Simulation Multiconference (ESM95)*. Prague, Czech Republic, pp. 768-773.

Miller, E.J. (1998). Emerging themes and research frontiers in GIS and activity-based travel demand forecasting. *Geographical Systems*, **5**(1-2), pp. 189-198.

Miller, E.J. (2005). Propositions for modelling household decision-making. *In this volume*.

Minsky, M. (1986). *The society of mind.* Simon and Shuster, New York.

Nagel, K. (2002). Distributed intelligence in large scale traffic simulations on parallel computers. Proceedings of Collective Cognition: Mathematical Foundations of Distributed Intelligence. Santa Fe, New Mexico, USA

Ray, C. (2001). Round Trip Bus: A protocol for migrating groups of objects. *11th Ph.D. Workshop on Object Oriented Systems, 16th European Conference on Object-Oriented Programming (ECOOP 2001)*. Budapest, Hungary.

Ray, C. and C. Claramunt (2002). Atlas: A distributed system for the simulation of large-scale systems. *Proceedings of 10th ACM International Symposium On Advances In Geographic Information Systems*. McLean, Virginia, USA, pp. 155-162, ACM Press.

Ray, C. (2003). Atlas, une plate-forme pour la modélisation et la simulation de systèmes désagrégés. Ph.D. Thesis. Université de Rennes 1, France.

Renders, J.-M. (1994). *Algorithmes génétiques et réseaux de neurones*. Hermès Science. Paris, France.

Righter, R. and J.C. Walrand (1989). Distributed Simulation of Discrete Event Systems. *Proceedings of the IEEE,* **77**(1), pp. 99-113.

Roorda, M., S. Doherty and E. Miller (2005). Operationalising household activity scheduling models: Addressing assumptions and the use of new sources of behavioural data. *In this volume*.

Setti, J.R. and B.G. Hutchinson (1994). Passenger-terminal simulation model. *Journal of Transportation Engineering,* **120**(4), pp. 517-535.

Tanenbaum, A.S. (1993). Distributed Systems Anno 1992: What We Have Learned so Far? *Distributed Systems Engineering,* **1**(1), pp. 3-10.

Teretsch, D. (1993). Scheduling flights at hub airports. *Transportation Research, Part B,* **27**(2), pp. 133-150.

Thériault, M., C. Claramunt, A.-M. Séguin and P.Y. Villeneuve (2002). A spatio-temporal model for lifelines analysis. *Proceedings of the 10th International Symposium on Spatial Data Handling* (D. Richardson and P. van Oosterom, eds.), pp. 433-450. Springer-Verlag.

Torrens, P.M. (2001). *Can geocomputation save urban simulation ? Throw some agents into the mixture, simmer, and wait...* Technical Report 32, University College, London, UK.

Tunasar, C., G.G. Bender and H. Young. (1998). Modelling curbside vehicular traffic at airports. *Proceedings of the 1998 Winter Simulation Conference*. Piscataway, New Jersey, USA, pp. 1113-1117.

Zeigler, B.P., H. Praehofer and T.G. Kim (2000). *Theory of modeling and simulation*, second edition. Academic Press, New York.

Integrated Land-Use and Transportation Models: Behavioural Foundations
M. Lee-Gosselin and S. Doherty (Editors)

11

Design and Implementation of PECAS: A Generalised System for Allocating Economic Production, Exchange and Consumption Quantities

J.D. Hunt and J.E. Abraham, University of Calgary, Canada

Introduction

PECAS is a recently-developed, generalised approach for simulating spatial economic systems. It is designed to provide a simulation of the land use component of land use transport interactive modelling systems.

PECAS stands for Production, Exchange and Consumption Allocation System. It uses an aggregate, quasi-dynamic equilibrium structure with separate flows of exchanges (including goods, services, labour and space) going from production to consumption based on variable technical coefficients and market clearing with exchange prices. Flows of exchanges from production to exchange zones and from exchange zones to consumption are allocated using nested logit models according to exchange prices and transport (dis)utilities. These flows are converted to transport demands that are loaded to networks in order to determine congested travel disutilities. Exchange prices determined for space inform the calculation of changes in space thereby simulating developer actions. The system is run for each year being simulated, with the travel disutilities and changes in space for one year influencing the flows of exchanges in the next year.

PECAS is a generalisation of the spatial input-output modelling approach used in the MEPLAN (Hunt and Simmonds, 1993) and TRANUS (de la Barra, 1998) land use transport modelling systems. PECAS also includes elements of the land use components of 'connected' land use transport modelling systems such as DELTA (Simmonds, 1996). As such, under certain simplifying assumptions, PECAS reduces to these other systems.

PECAS or its components are currently being applied in the development of state-wide transportation land use modelling systems for Ohio and Oregon, in the development of an urban land use model for Sacramento and in the anticipated development of urban land use models of Calgary and Edmonton in Canada. It is also being used as the basis for a recommended design for a model of the Los Angeles Region.

PECAS STRUCTURE

PECAS has two component modules. One is the space development module, which represents the actions of developers in the provision of space (land and floorspace) where activities can locate, including the new development, demolition and re-development that occurs from one point in time to the next. The other is the activity allocation module, which represents how activities locate within the space provided by developers and how these activities interact with each other at a given point in time.

Treatment of Space

The area covered by PECAS is organised into a set of land use zones. Activities locate in these zones and interact between them. Ideally these zones are the same as the transport zones used in the transport model being run with PECAS, or at least are aggregations of whole numbers of adjacent transport zones. The connectivity among the land use zones is then based on the representation provided by the transport model, where the transport model uses its networks to establish congested network times and costs and associated interchange disutilities that PECAS uses in its consideration of the interactions between land use zones.

Currently it is recommended that a maximum of about 750 such land use zones be used because of the 2 gigabyte memory address space limitations of most current operating systems.

Treatment of Time

PECAS works through time in a series of discrete, fixed steps from one point in time to the next, with the activity allocation module running at each point in time and the space development module considering the length of the time step. The steps can be of any fixed duration, but 1 year time steps are recommended since they allow an appropriately quick response of land developers in the land development module to the prices established in the activity allocation module.

Figure 11.1 illustrates the interactions among the modules and with other models when the two PECAS modules are run as part of an integrated land use transport modelling system. The transport model used to calculate the congested travel times and disutilities is normally run for each year (after PECAS completes its run for that year) or, if travel conditions are relatively stable, the transport model can be run less often to save computation time.

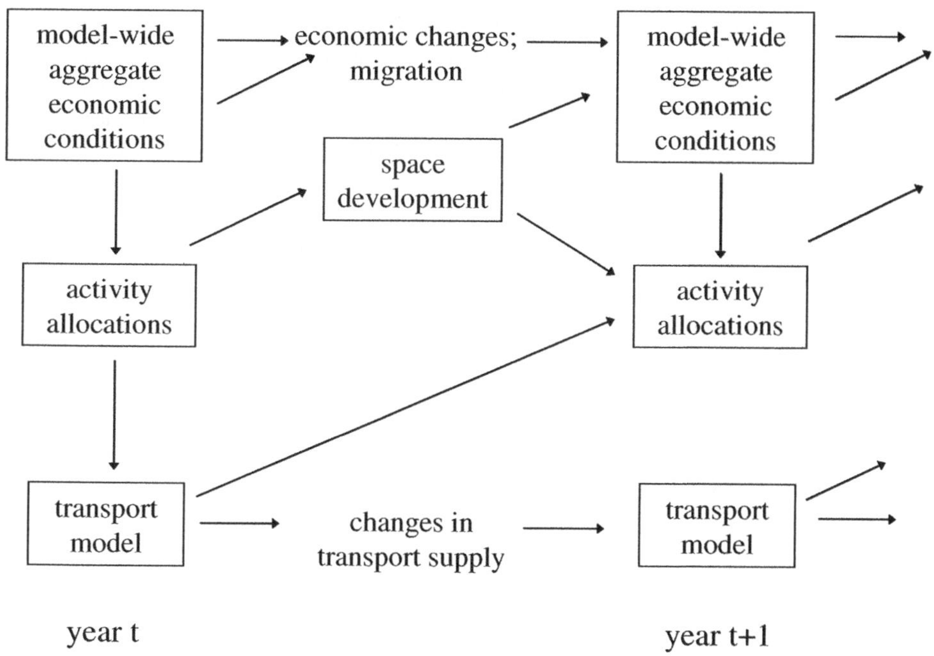

Figure 11.1 Interactions among modules simulating temporal dynamics

ACTIVITY ALLOCATION MODULE

The PECAS activity allocation module is an aggregate representation based on spatially-disaggregated forms of extended make and use input-output tables, with variable technical coefficients and with households included – making the tables 'social accounting tables' rather than just input-output tables. Categories of activity produce 'commodities' as indicated in the make table and consume these 'commodities' as indicated in the use table, where these 'commodities' include goods and services, labour and space (land and/or floorspace). The movement of these commodities from where they are produced (in zones) to where they are consumed (in zones) is the economic basis for travel and transport in the land use model. The travel conditions – distances, costs, times and associated disutilities by mode – for the movements of these commodities form the basis of the influence of the transportation system on the attractiveness of zones as locations for different activities.

For each zone there are two 'lowest-level' logit allocation models for each category of commodity considered in the model. The first is an allocation of the quantities purchased among various 'exchange locations' where they are sold to other activities. The second is an allocation of the quantities sold among the various exchange locations where they are bought by other activities. The utility of each alternative in these models is influenced by the price at the exchange location, and the characteristics of transporting the commodity to or from the exchange location. The composite utility values from these two 'lowest-level' logit models are called the 'buying utility' and the 'selling utility' for the commodity in the zone. They are used as the transportation-related inputs into the 'next-higher-level' logit allocation process for allocating the production and consumption of commodities by activities. From there, via the resulting utilities of production and consumption, they feed into the location utilities used at the 'highest-level' to allocate categories of industries and households to the land use zones.

This spatial allocation system with both buying and selling allocations giving rise to exchanges from production to consumption locations in zones is illustrated in Figure 11.2.

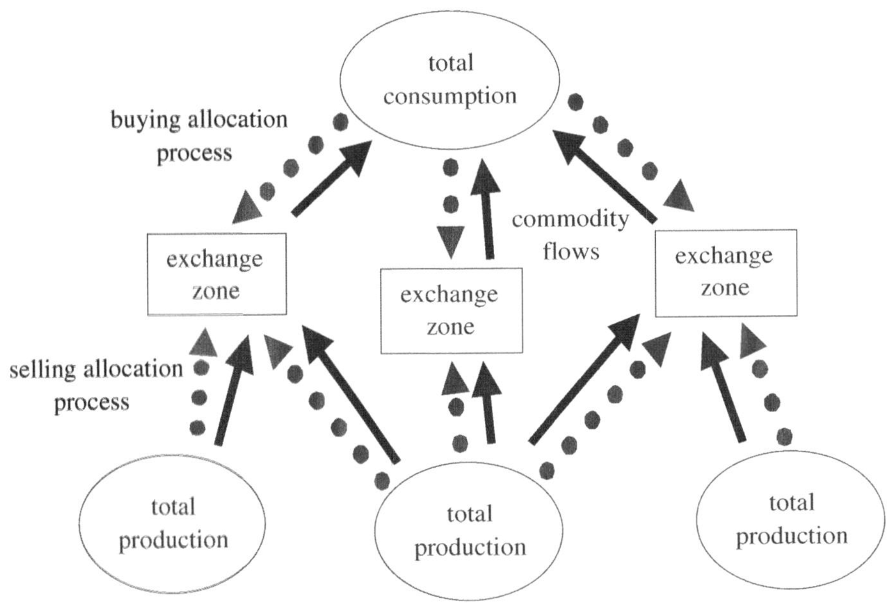

Figure 11.2 Buying and selling allocations giving rise to commodity flows from production zone to consumption zone via exchange location

The exchange locations are location-specific markets for commodities, where sellers sell commodities to buyers. Prices are established at exchange locations so that the quantity bought equals the quantity sold – thus the spatial allocation procedure assumes a short-run market equilibrium in commodities.

For simplicity and realism, commodities can be assigned to be either purchased from the exchange location in the zone they are consumed (e.g. for labour, this occurs in the employment zone, where the labour is exchanged and the price is set) or purchased in the exchange location in the zone they are produced (e.g. for retail goods this is at the retail establishment, where the goods are exchanged and the price is set). In this case for each commodity in each zone, one of the logit allocation models (either the buying or the selling) consists of only one alternative.

Imports and exports also enter the system at exchange locations. The commodity import and export quantities at each exchange location are determined as functions of the exchange price, so that as prices rise more imports are attracted and as prices fall more exports are produced. These import and export functions connect the model economy to the rest of the world.

Quantity Definitions and Categories

Figure 11.3 provides a summary of the make and use tables and the categories of activities and commodity categories used within them, showing the specific types of interactions being represented. Commodity production is represented in the make table at the top of Figure 11.1; commodity consumption is represented in the use table at the bottom; commodities 'flow' down the columns from where they are produced to where they are consumed. The general nature of the exchange location for each commodity category and the treatment of imports and exports are also indicated.

Table 11.1 provides an example listing of the activity categories that could be used in the activity allocation module. In general, standard industrial sector categories are used as the starting point for the representation of the business and government components of the economy, with these categories split according to the type of employment and associated space consumed. In general, a given industrial sector is split into blue-collar and white-collar components and associated 'shopfloor' and 'office' uses in this way so that the allocation process in the model can consider the very different behaviour of these two components regarding location, labour use and space type use.

Different types of household are included as different activity categories. Labour is included as a commodity that is produced by households and consumed by economic production activities, with labour in different occupation categories treated as different commodity categories.

Different types of floorspace are also included as commodities in the use table in that they are consumed by activities, but the amount of floorspace available for consumption in each zone is fixed in any one time period being considered with the activity allocation module. It is the space development module running between time periods that adjusts the amount of floorspace available in zones, as described below. Floorspace is a 'non-transportable commodity' in that it must be exchanged and consumed in the zone where it is produced and thus located.

	commodities								
	goods	internal management services	services	education services	retail services	domestic services	labor	non-residential space	residential space
producing activities (making)									
industry	goods production by industry	internal management services production by industry	services production by industry	education services production by industry	retail services production by industry				
government	goods production by government		services production by government	education services production by government					
households						domestic services production by households	labor production by households		
exports	yes		yes		yes		yes		
exchange locations	at production	at consumption	at production	at production	at production	at consumption	at consumption	non-transportable	non-transportable
imports	yes	yes	yes		yes		yes		
consuming activities (using)									
industries	goods consumption by industry	internal management services consumption by industry	services consumption by industry	education services consumption by industry			labor consumption by industry	space consumption by industry	
government	goods consumption by government		services consumption by government	education services consumption by government			labor consumption by government	space consumption by government	
households	goods consumption by households-direct		services consumption by households	education services consumption by households	goods consumption by households-via retail	domestic services consumption by households			space consumption by households

Figure 11.3 Summary of categories of activities and commodities in make and use tables

Table 11.1 Potential activity categories for activity allocation module

Industry	Government
Agriculture in Office Floorspace	Government in Office Space
Agriculture on Agricultural Land	Government in Government Support Land
Light Industry in Office Floorspace	
Light Industry in Light Industrial Floorspace	**Households**
Heavy Industry in Office Floorspace	Low Income Households
Heavy Industry on Heavy Industrial Land	Medium Income Households
Wholesale in Office Floorspace	High Income Households
Wholesale in Warehouse Floorspace	
Retail in Office Floorspace	
Retail in Retail Floorspace	
Transportation Handling in Office Floorspace	
Transportation Handling in Depot Floorspace	
Other Services in Office Floorspace	
Other Services in Light Industrial Floorspace	
Other Services in Retail Floorspace	
Gradeschool Education in Office Floorspace	
Gradeschool Education on Gradeschool Land	
Post-Secondary Education in Office Floorspace	
Post-Secondary Education on Insititutional Land	

Water, energy, and air or water borne emissions can also be included as commodities, which allows the model to also determine aggregate demands or impacts arising from activities.

Activity Allocation

The model-wide total quantity of production activity for each activity category is input to the activity allocation module. This amount is allocated among the land use zones using a logit allocation as follows:

$$W_{a,z} \quad = \quad TW_a \cdot (\exp (\lambda_a \cdot LU_{a,z}) / \Sigma_{z \in Z} (\exp (\lambda_a \cdot LU_{a,z}))) \qquad (01)$$

with:

$$LU_{a,z} \quad = \quad \alpha_{size} \cdot \ln [Size_{a,z}] + \alpha_{inertia} \cdot PrevW_{a,z} + Constant_{a,z} + \Sigma_{v \in V} (\alpha_{a,v} \cdot X_{v,z}) + \alpha_{prod} \cdot CUProd_{a,z} + \alpha_{cons} \cdot CUCons_{a,z} \qquad (02)$$

where:

z = index representing land use zones;
Z = the set of all land use zones;
a = index representing activity categories;
c = index representing commodity categories;
$W_{a,z}$ = quantity of activity a in zone z;
TW_a = model-wide total quantity of activity a;
$LU_{a,z}$ = location utility for a unit of activity a in zone z;
$Size_{a,z}$ = representation of relative size of zone z for activity a, indicating the *a priori* expected share of activity a for zone z;
$PrevW_{a,z}$ = the proportion of model-wide quantity of activity a in zone z in the previous time period;
v = index representing 'other' zonal attributes;
V = the set of all 'other' zonal attributes;
$X_{v,z}$ = one of the 'other' zonal attributes;
$CUProd_{a,z}$ = composite utility associated with production by activity a in zone z, defined in equation 07 and discussed below;
$CUCons_{a,z}$ = composite utility associated with consumption by activity a in zone z, defined in equation 08 and discussed below;
α_{size} = utility function coefficient for the sensitivity to size;
$\alpha_{inertia}$ = utility function coefficient for the sensitivity to the previous proportion of activity a in zone z, representing inertia in allocation of activity a;
$\alpha_{a,v}$ = utility function coefficient for the sensitivity to the 'other' zonal attribute $X_{v,z}$;
$Constant_{a,z}$ = utility function alternative specific constant for zone z for allocation of activity a;
α_{prod} = utility function coefficient for the sensitivity to composite utility associated with production;
α_{cons} = utility function coefficient for the sensitivity to composite utility associated with consumption;
λ_a = utility function dispersion parameter for allocation of activity a.

The 'other' zonal attributes vary depending on the production activity being allocated. For residential activities in particular, these 'other' attributes include representations of various amenities relevant to housing location choice, including school quality, general noise levels, air quality, open space density, municipal taxation levels and possibly zonal-level income distributions and racial compositions.

Production and Consumption Allocation

The make and use technical coefficients indicating the rates at which activities produce and consume commodities are elastic, representing opportunities for substitution (between inputs in consumption and between by-products in production), as follows:

$$M_{c,a,z} = MMin_{c,a} + MDisc_{c,a} \cdot (\exp(\lambda_{m,a} \cdot UIProd_{c,a,z}) / \Sigma_{c \in C} \exp(\lambda_{m,a} \cdot UIProd_{c,a,z})) \quad (03)$$

with:

$$UIProd_{c,a,z} = MDisc_{c,a} \cdot CUSell_{c,a,z} \quad (04)$$

and

$$U_{c,a,z} = UMin_{c,a} + UDisc_{c,a} \cdot (\exp(\lambda_{u,a} \cdot UICons_{c,a,z}) / \Sigma_{c \in C} \exp(\lambda_{u,a} \cdot UICons_{c,a,z})) \quad (05)$$

with:

$$UICons_{c,a,z} = UDisc_{c,a} \cdot CUBuy_{c,a,z} \quad (06)$$

where:

$M_{c,a,z}$ = make technical coefficient for production of commodity c by activity a in zone z, indicating the quantity of commodity c produced per unit of activity a

$U_{c,a,z}$ = use technical coefficient for consumption of commodity c by activity a in zone z, indicating the quantity of commodity c consumed per unit of activity a;

$UIProd_{c,a,z}$ = utility of producing the potential discretionary amount of commodity c associated with a unit of activity a in zone z

$UICons_{c,a,z}$ = utility of consuming the potential discretionary amount of commodity c associated with a unit of activity a in zone z

$MMin_{c,a}$ = minimum amount of commodity c made per unit of activity a

$MDisc_{c,a}$ = potential discretionary amount of commodity c made per unit of activity a

$UMin_{c,a}$ = minimum amount of commodity c used per unit of activity a

$UDisc_{c,a}$ = potential discretionary amount of commodity c used per unit of activity a

$CUSell_{c,a,z}$ = composite utility associated with selling a unit of commodity c produced by activity a in zone z;

$CUBuy_{c,a,z}$ = composite utility associated with buying a unit of commodity c consumed by activity a in zone z;

$\lambda_{m,a}$ = utility function dispersion parameter for allocation of by-product substitutes made by activity a;

$\lambda_{u,a}$ = utility function dispersion parameter for allocation of input substitutes used by activity a.

Then the composite utilities associated with production and consumption by activity a in zone z, as used in equation 02 above, are as follows:

$$CUProd_{a,z} = (\Sigma_{c\in C} MMin_{c,a} \cdot CUSell_{c,a,z}) + (1/\lambda_{m,a}) \cdot \ln [\, \Sigma_{c\in C} \exp(\lambda_{m,a} \cdot UIProd_{c,a,z})] \quad (07)$$

and

$$CUCons_{a,z} = (\Sigma_{c\in C} UMin_{c,a} \cdot CUBuy_{c,a,z}) + (1/\lambda_{u,a}) \cdot \ln [\, \Sigma_{c\in C} \exp(\lambda_{u,a} \cdot UICons_{c,a,z})] \quad (08)$$

These composite utilities combine the accessibilities for buying and selling individual commodities consistently with the production and consumption options for the activity. The location utility (determined in equation 02) of an activity that consumes a large amount of a commodity will be strongly influenced by the composite utility of buying that commodity, and the location utility of an activity that produces a large amount will be strongly influenced by the composite utility of selling. As such, the last two terms in equation 02 provide overall indications of the accessibility of the zone for the activity consistent with the technology of the production process for the activity.

To allow consistency with earlier spatial input-output representations in MEPLAN (Hunt and Simmonds, 1993) and TRANUS (de la Barra, 1998), a form of elastic consumption and production using independent exponential functions for each commodity is also supported by the PECAS software. This alternate form does not have the same consistency with random utility theory as the form described here.

Buying and Selling Allocation

The total quantities of commodities produced and consumed by each activity in each zone are calculated using the technical coefficients as follows:

$$TP_{c,z} = \Sigma_{a\in A} M_{c,a,z} \cdot W_{a,z} \quad (09)$$

and

$$TC_{c,z} = \Sigma_{a\in A} U_{c,a,z} \cdot W_{a,z} \quad (10)$$

where:

$TP_{c,z}$ = total quantity of commodity c produced in zone z;
$TC_{c,z}$ = total quantity of commodity c consumed in zone z.

The total quantity of each commodity produced by production activities in a given zone is allocated among exchange locations (where it is sold) as follows:

$$S_{c,z,k} = TP_{c,z} \cdot (\exp(\lambda_{s,c} \cdot SU_{c,z,k}) / \Sigma_{k\in K} \exp(\lambda_{s,c} \cdot SU_{c,z,k})) \quad (11)$$

with:

$$SU_{c,z,k} = \delta_{size,s} \cdot \ln [Size_{s,c,k}] + \delta_{price,s} \cdot Price_{c,k} + \delta_{tran} \cdot Tran_{c,z,k} \tag{12}$$

where:

k	=	index representing exchange locations (land use zones);
K	=	the set of all exchange locations;
$S_{c,z,k}$	=	quantity of commodity c produced in zone z that is allocated to (go to) exchange location k;
$SU_{c,z,k}$	=	utility for selling to exchange location k a unit of commodity c produced in zone z;
$Size_{s,c,k}$	=	representation of relative size of exchange location k for selling commodity c, indicating the *a priori* expected share of commodity c sold in exchange location k;
$Price_{c,k}$	=	the unit exchange price for commodity c in exchange location k;
$Tran_{c,z,k}$	=	the utility for transporting a unit of commodity c from zone z to exchange location k;
$\delta_{size,s}$	=	utility function coefficient for the sensitivity to size when selling;
$\delta_{price,s}$	=	utility function coefficient for the sensitivity to price when selling;
δ_{tran}	=	utility function coefficient for the sensitivity to transport utility;
$\lambda_{s,c}$	=	utility function dispersion parameter for allocation of selling of commodity c.

In the case of selling, the coefficient $\delta_{price,s}$ is positive.

The selling composite utility for commodity c (independent of the producing activity) is determined consistent with equations 11 and 12, as follows:

$$CUSell_{c,z} = (1/\lambda_{s,c}) \cdot \ln (\Sigma_{k \in K} \exp (\lambda_{s,c} \cdot SU_{c,z,k})) \tag{13}$$

The selling composite utility for commodity c when produced by activity a in particular, as used in equations 04 and 07, is then established as follows:

$$CUSell_{c,a,z} = \phi_{s,c,a} \cdot CUSell_{c,z} + USellRef_{c,a} \tag{14}$$

where:

$\phi_{s,c,a}$	=	factor adjustment to the composite utility of selling commodity c by activity a;
$USellRef_{c,a}$	=	offset adjustment to the composite utility of selling commodity c by activity a.

The scaling and offset adjustments $\phi_{s,c,a}$ and $USellRef_{c,a}$ have been found useful in calibration to adjust (usually remove by setting $\phi_{s,c,a}$ to zero) the effect of individual buying and selling composite utilities on production utilities and thus on location utilities.

The allocation of quantities consumed to the exchange zones where they are purchased is analogous to the allocation of quantities produced as described above. Specifically, the total quantity of each commodity consumed by a production activity in a given zone is allocated among exchange locations (where it is bought) as follows:

$$B_{c,z,k} = TC_{c,z} \cdot (\exp (\lambda_{b,c} \cdot BU_{c,z,k}) / \Sigma_{k \in K} \exp (\lambda_{b,c} \cdot BU_{c,z,k})) \quad (15)$$

with:

$$BU_{c,z,k} = \delta_{size,b} \cdot \ln [Size_{b,c,k}] + \delta_{price,b} \cdot Price_{c,k} + \delta_{tran} \cdot Tran_{c,k,z} \quad (16)$$

where:

$B_{c,z,k}$ = quantity of commodity c consumed in zone z that is allocated to (come from) exchange location k;
$BU_{c,z,k}$ = utility for buying from exchange location k a unit of commodity c consumed in zone z;
$Size_{b,c,k}$ = representation of relative size of exchange location k for buying commodity c, indicating the *a priori* expected share of commodity c bought in exchange location k;
$Tran_{c,k,z}$ = the utility for transporting a unit of commodity c from exchange location k to zone z;
$\delta_{size,b}$ = utility function coefficient for the sensitivity to size when buying;
$\delta_{price,b}$ = utility function coefficient for the sensitivity to price when buying;

δ_{tran} = utility function coefficient for the sensitivity to transport utility;
$\lambda_{b,c}$ = utility function dispersion parameter for allocation of buying of commodity c.

In the case of buying, the coefficient $\delta_{price,b}$ is negative.

The buying composite utility for commodity c (independent of the consuming activity) is determined consistent with equations 15 and 16, as follows:

$$CUBuy_{c,z} = (1/\lambda_{b,c}) \cdot \ln (\Sigma_{k \in K} \exp (\lambda_{b,c} \cdot BU_{c,z,k})) \quad (17)$$

The buying composite utility for commodity c when consumed by activity a in particular, as used in equations 06 and 08, is then established as follows:

$$CUBuy_{c,a,z} = \phi_{b,c,a} \cdot CUBuy_{c,z} + UBuyRef_{c,a} \quad (18)$$

where:

$\phi_{b,c,a}$ = factor adjustment to the composite utility of buying commodity c by activity a;
$UBuyRef_{c,a}$ = offset adjustment to the composite utility of buying commodity c by activity a.

Imports and Exports

The quantities of imports and exports for a given commodity in a given exchange zone are determined using:

$$Q_{c,i,k} = QRef_{c,i} + \gamma_{c,i} \cdot ([G_i\text{-}1]/[G_i\text{+}1]) + \mu_{c,i} \cdot (Price_{c,k} - PriceRef_{c,i}) \quad (19)$$

and

$$Q_{c,e,k} = QRef_{c,e} + \gamma_{c,e} \cdot ([G_e\text{-}1]/[G_e\text{+}1]) + \mu_{c,e} \cdot (Price_{c,k} - PriceRef_{c,e}) \quad (20)$$

with:

$$G_i = \exp (\eta_{c,i} \cdot (Price_{c,k} - PriceRef_{c,i})) \quad (21)$$

and

$$G_e = \exp (\eta_{c,e} \cdot (Price_{c,k} - PriceRef_{c,e})) \quad (22)$$

where:

$Q_{c,i,k}$ = quantity of commodity c imported to exchange location k;
$Q_{c,e,k}$ = quantity of commodity c exported from exchange location k;
$QRef_{c,i}$ = quantity of commodity c imported to exchange location k when the unit exchange price for commodity c in exchange zone k is at its import reference level $PriceRef_{c,i}$;
$PriceRef_{c,i}$ = reference price per unit for import of commodity c;
$QRef_{c,e}$ = quantity of commodity c exported from exchange location k when the unit exchange price for commodity c in exchange zone k is at its export reference level $PriceRef_{c,e}$;
$PriceRef_{c,e}$ = reference price per unit for export of commodity c;
$\gamma_{c,i}$ = function coefficient for the rate of increase in imports of commodity c for exponent term;
$\mu_{c,i}$ = function coefficient for the rate of increase in imports of commodity c for linear term;
$\eta_{c,i}$ = function coefficient for sensitivity to difference in exchange price for commodity c concerning increase in imports of commodity c for exponent term;
$\gamma_{c,e}$ = function coefficient for the rate of increase in exports of commodity c for exponent term;
$\mu_{c,e}$ = function coefficient for the rate of increase in exports of commodity c for linear term;
$\eta_{c,e}$ = function coefficient for sensitivity to difference in exchange price for commodity c concerning increase in exports of commodity c for exponent term.

In the case of imports, the coefficient $\gamma_{c,i}$ is positive and the coefficient $\mu_{c,i}$ is positive provided $\eta_{c,i}$ is positive. In the case of exports, the coefficient $\gamma_{c,e}$ is negative and the coefficient $\mu_{c,e}$ is negative provided $\eta_{c,e}$ is negative.

Mathematical Solution

The exchange zones simulate markets where the aggregate supply (which is the sum of the selling allocations together with the quantity of imports, both elastic with respect to the exchange price) meets the aggregate demand (which is the sum of the buying allocations together with the quantity of exports). The module solves for the equilibrium solution, adjusting the exchange prices in the exchange locations until all the markets clear.

The search algorithm calculates the partial derivatives of the total excess demand (excess of demand by buyers and exporters over supply by sellers and importers) in each exchange zone with respect to the price in that exchange zone. The derivatives with respect to other prices in other zones or for other commodities are assumed to be zero, and a price change is calculated. A step adjustment factor is applied to the step to speed and aid convergence. If a step results in a lower aggregate sum-of-squares excess demand, then the step adjustment factor is increased slightly for the next iteration. If a step results in a higher aggregate sum-of-squares excess demand, then the step is abandoned, the step adjustment factor is reduced substantially, and a new and smaller step is calculated to replace the abandoned one.

Model Outputs

The model provides consistent activity allocations by activity category by zone, commodity flow quantities from production zone to consumption zone via exchange zone, imports and exports by exchange zone and exchange prices by commodity by exchange zone.

It also provides the basis for assessing the changes in model-wide consumer surplus by production activity. In particular, the model-wide location composite utility for a given activity category is determined consistent with equations 01 and 02 as follows:

$$CLU_a = (1/\lambda_a) \cdot \ln\left(\Sigma_{z \in Z} \exp(\lambda_a \cdot LU_{a,z}) \right) \tag{23}$$

where:

CLU_a = model-wide location composite for a unit of activity a

Changes in CLU_a for a given activity category can be used to calculate measures of the changes in consumer benefit for the activity category arising with the policy actions input to the model. These measures can help assess quality of life and business attractiveness for different household and industrial categories as represented.

Calibration

The data to calibrate the activity allocation module include:

- input-output (social accounting) make and use tables for base year, including import and export quantities;
- information or assumptions on elasticity of technical coefficients in social-accounting tables for the base year, including floorspace consumption rates;
- spatial distribution of employment by industry and occupation in base year and if possible for one or more other years after the base year;
- labour force participation rates by household type and by occupation in the base year;
- trip length distributions by commodity in the base year;
- indications of the average price for each commodity, the overall variation in the price around the region, and some indication of where the prices are higher and where they are lower.

The calibration of the activity allocation module is challenging because the system is so highly inter-related (much like the real world) and because the production activity quantities for the industrial sectors are not generally available as observed values by zone. The base year employment in each zone is calibrated to match aggregate observed employment values at least approximately. This is accomplished by using observed values of employment by zones as 'targets' in an automatic calibration procedure, which adjusts alternative specific constants for each activity category in each zone. If the data describing the amount of floorspace of each type in each zone is consistent with the data describing the amount of employment of each type in each zone (and consistent with the representation of the amount of floorspace consumed per employee by industry) then the calibration procedure will be able to match the observed employment patterns exactly. An exact match with spatial employment data is not necessarily appropriate, however, as it may mean a large mismatch with other target data, such as commodity price data. There will be certain data inconsistencies that need to be accounted for, and it may be that in some cases the best way to account for inconsistencies is to put less faith in the employment distribution data.

Space Development Module

The PECAS space development module is an aggregate representation based on a set of logit allocation models. These logit models determine the changes in aggregate quantities of space (including land and floorspace) in each category in each land use zone occurring in the period from one point in time to the next.

Each non-zero quantity of space in each zone in each of the following categories is considered separately:

- each separate category of existing floorspace and developed land;
- serviced vacant land;
- unserviced vacant land;
- derelict.

When one of these non-zero quantities of space is considered, a logit model is used to determine the proportions of the quantity to allocate to each of the following discrete alternatives (with some restrictions as appropriate):

- remaining in the same space category;
- becoming derelict;
- transforming into each of the other floorspace and developed land categories.

Quantity Definitions and Categories

Figure 11.4 provides an example listing of the space categories that could be used, together with a summary of the allowable transitions among categories. Each space type tends to map to a fairly narrow set of activity categories by setting to zero most of the use coefficients, $UMin_{c,a}$ and $UDisc_{c,a}$, for the consumption of floorspace type c by activity a.

Mathematical Structure and Processes

In a given period from one point in time 't' to the next 't+1' – between runs of the activity allocation module – in each land use zone the quantity of space in each category is allocated among allowable categories as follows:

$$R_{j,h,z} = TR_{j,z} \cdot (\exp(\lambda_{r,j} \cdot RU_{j,h,z}) / \Sigma_{h \in H(j)} \exp(\lambda_{r,j} \cdot RU_{j,h,z})) \quad (24)$$

with:

$$RU_{j,h,z} = \phi_{price,r} \cdot Price_{h,z} + \phi_{price,t} \cdot TCost_{j,h,z} + Constant_{j,h} + \phi_{regprice} \cdot (RegPrice_h - RegPRef_h) + \phi_{cap} \cdot (ER_{h,z} / Cap_{h,z})^{\theta cap} \quad (25)$$

where:

j, h	=	indices representing space categories;
$H(j)$	=	the set of all new space categories allowable for existing space category j;
$R_{j,h,z}$	=	quantity of space transitioning from space category j (at time 't') to space category h (at time 't+1') in zone z;
$TR_{j,z}$	=	total quantity of space category j (at time 't') in zone z;
$RU_{j,h,z}$	=	transition utility for space in category j (at time 't') changing into space in category h (at time 't+1) in zone z;
$Price_{h,z}$	=	unit price for space category h (at time 't') in zone z;
$TCost_{j,h,z}$	=	unit cost for transforming space from category j (at time 't') to category h (at time 't+1') in zone z, including relevant demolition, clean-up, construction and servicing costs;
$RegPrice_h$	=	model-wide price for space category h (at time 't'), calculated as the weighted average of zonal prices for space category h
$RegPRef_h$	=	reference level for model-wide price of space category h;
$ER_{h,z}$	=	transitional total quantity of space category h in zone z;
$Cap_{h,z}$	=	capacity for space category h in zone z;
$\phi_{price,r}$	=	utility function coefficient for the sensitivity to space price;

$\phi_{price,t}$ = utility function coefficient for the sensitivity to cost of transforming space;
$\phi_{regprice}$ = utility function coefficient for the sensitivity to region-wide price for space;
$\phi_{density}$ = utility function coefficient for the sensitivity to model-wide density of use of space relative to the relevant reference density;
ϕ_{cap} = utility function coefficient for the sensitivity to the available capacity term;
θ_{cap} = function power coefficient for the available capacity term;
$Constant_{j,h}$ = utility function alternative specific constant for transition of space from category j (at time 't') to category h (at time 't+1');
$\lambda_{r,j}$ = utility function dispersion parameter for transition of space category j.

This is done for each space category, starting with vacant serviced land, then vacant unserviced land, and then working down the list of other categories in order, such as listed in Figure 11.4, but considering derelict space last. As each space category j is considered, the amount for $TR_{j,z}$ at time 't' is allocated among the $R_{j,h,z}$ and then the amount for $ER_{h,z}$ for each space category h is updated so that the impacts on the available capacity are as current as possible before considering the next space category. The $ER_{h,z}$ amounts at the end of this process, after all the $TR_{j,z}$ at time 't' have been considered, become the $TR_{j,z}$ at time 't+1'.

All of $\phi_{price,r}$, $\phi_{density}$ and θ_{cap} are positive and both $\phi_{price,t}$ and ϕ_{cap} are negative.

The model-wide price term is included in order to represent the influence of a range of development opportunities available for developers and the associated influence of area-wide vacancy rates on transitions. This can provide some dampening of a more rapid response (in particular to prices) in particular zones. Thus, the capability for space development to 'overreact', resulting in boom and bust cycles in available space, as sometimes occurs in reality, can be calibrated as part of the model development.

In general there is a lot of inertia in the system over a 1-year time step, where large portions of space remain in the same category. This is implemented in PECAS using large positive values for the alternative specific constants, $Constant_{j,h}$, in the cases where h=j, that is, where the space is remaining in the same category.

In the Oregon and Sacramento implementations of PECAS, the total amount of space in each zone in each category is further divided into zoning schemes. A zoning scheme causes a modification to the $Constant_{j,h}$ for transitions when $j \neq h$. This allows the capacity information to be based on an overlay of an existing use layer in a GIS with a future permitted use GIS layer. When this mechanism is used, ϕ_{cap} is usually set to zero and $Cap_{h,z}$ is irrelevant because the capacity is expressed directly as the supply of land in different zoning schemes.

Space Category (transforming from) \ Space Category (transforming to)	vacant unserviced land	vacant serviced land	derelict land and/or floorspace	single-family detached residential	multi-family residential	office floorspace	retail floorspace	warehouse floorspace	hotel floorspace	hospital space	transportation depot floorspace	light industrial floorspace	heavy industrial land	gradeschool land	post-secondary institutional land	government support land	agricultural land
vacant unserviced land	X	X		X	X	X	X	X	X	X	X	X	X	X	X	X	X
vacant serviced land		X		X	X	X	X	X	X	X	X	X	X	X	X	X	
derelict land and/or floorspace			X	X	X	X	X	X	X	X	X	X	X	X	X	X	X
single-family detached residential			X	X	X	X	X	X	X	X	X	X	X	X	X	X	
multi-family residential			X	X	X	X	X	X	X	X	X	X	X	X	X	X	
office floorspace			X	X	X	X	X	X	X	X	X	X	X	X	X	X	
retail floorspace			X	X	X	X	X	X	X	X	X	X	X	X	X	X	
warehouse floorspace			X	X	X	X	X	X	X	X	X	X	X	X	X	X	
hotel floorspace			X	X	X	X	X	X	X	X	X	X	X	X	X	X	
hospital space			X	X	X	X	X	X	X	X	X	X	X	X	X	X	
transportation depot floorspace			X	X	X	X	X	X	X	X	X	X	X	X	X	X	
light industrial floorspace			X	X	X	X	X	X	X	X	X	X	X	X	X	X	
heavy industrial land			X			X	X	X	X	X	X	X	X				
gradeschool land														X		X	
post-secondary institutional land															X	X	
government support land														X	X	X	
agricultural land		X		X	X	X	X	X	X	X	X	X	X	X	X	X	X
undevelopable and protected land																	

Figure 11.4 Space categories and allowable transitions

Calibration

The data to calibrate the space development module includes changes in the quantities of space by category occurring over the time from the base year to another known year (about two or three years after the base year) in at least 200 or so of the land use zones, together with indications of the zoning regulations, construction costs and quantities of unserviced and serviced vacant land in these zones in the base year.

Such data concerning changes over a two-to-three year period may not be readily and immediately available from existing sources in a given application. But they can be collected for a sufficient number of zones with only a relatively modest effort by drawing from and/or coordinating with relevant sources, including municipal tax rolls, development applications records and real estate transaction information, at regular intervals over a period of about three years. It the meantime, while waiting for such data, the model can be developed with data on just construction costs, assuming that developers ignore sunk costs for land and future prices.

DISCUSSION

PECAS provides a comprehensive, generalised structure for developing land use models. It represents an ambitious attempt to include many of the relevant components of the land use system, including those related to the transport system. The intention is to provide a 'tool kit' for the development of the land use side of land use transport interaction models that can be suitably policy-responsive in practical application.

PECAS or its components are being used in a number of practical modelling applications. A statewide model of Ohio is being developed using the full PECAS structure (PB Ohio *et al.*, 1999). The Generation 2 modelling work in Oregon is using many of the elements of the PECAS activity allocation module as part of a larger system where various micro-simulation treatments are applied with the results of the activity allocation acting as inputs (Hunt, 2001). The full PECAS structure has been used as the basis for a model specification for the Los Angeles Region. A land use model for the Sacramento Region that is just starting to be developed is based in part on the PECAS activity allocation module (DKS *et al.*, 2001). The intention is to use the full PECAS structure in the development of urban-level land use transport interaction models of both Edmonton and Calgary in Canada.

In these various applications, the PECAS structure and its components provide a framework not only for representing the system elements and how they behave and interact, but also for identifying data needs in each case and for assessing and sharing experience and 'know-how' among the different cases. As indicated, the entire PECAS framework is not used in entirety in each case. Rather, in some cases components of the PECAS framework are replaced by other treatments. In these cases the larger PECAS framework still acts to define these components and their interactions with other components.

The development of PECAS has benefited greatly from previous experience with the MEPLAN (Hunt and Simmonds, 1993) and TRANUS (de la Barra, 1998) frameworks, both of which use an 'integrated' structure based on the spatial disaggregation of input-output tables to link land use with transport. It has also benefited from consideration of connected modelling systems such as those based on DELTA (Simmonds, 1996) and UrbanSim (Waddell, 2002), which use a 'connected' structure where model-wide quantities of activities are allocated in space according to the distribution of relevant accessibilities.

PECAS is a generalisation that incorporates both of these 'integrated' and 'connected' structures. If its make and use input-output tables are replaced by the corresponding reduced square table and the implied multipliers are used to determine the activity totals to be allocated incrementally, then the system becomes much like the MEPLAN and TRANUS systems -

without explicit representation of either commodity flows or commodity-specific accessibilities. In PECAS, the use of accessibilities that are aggregations of composite utilities from buying and selling allocations of commodities used and made by activities, with weights based on the corresponding technology coefficients, represents a re-interpretation of the vector of accessibilities in 'connected' structures. If the buying and selling allocation processes in PECAS are removed, and the corresponding aggregations of composite utilities from these are replaced by a more ad-hoc vector of accessibilities, then the system becomes much like the 'connected' systems as used in DELTA and UrbanSim. The incorporation of these two approaches into a single framework in PECAS is seen to be a step forward that allows the benefits of each approach to be realised as appropriate.

The PECAS activity allocation module can be viewed as the aggregate outcome of a joint choice model of location, production options, consumption options, and destinations for buying and selling by households, business establishments, and other institutions. Hence it is consistent with a micro-simulation of business establishments and of households. Further, if prices are updated over time in response to excess demand (instead of adjusted to force excess demand to zero in each time period) the PECAS activity allocation module provides a dynamic representation of the economy consistent with Walrasian economics (Abraham and Hunt, 2002). Thus PECAS is seen as a practical stepping stone towards explicitly dynamic aggregate models and/or dynamic micro-simulation models.

One of the big, challenging tasks in the practical application of PECAS is model calibration. There are many coefficients whose values have to be determined, and the interactions among the model elements that influence and are influenced by these coefficients are complex. Semi-automated processes have been developed for different sets of coefficients, largely in previous work using just certain elements of PECAS. Overall, a Bayesian sequential estimation procedure is used for the entire system. In the first stage, statistical methods are used with observations of actual behaviour to the extent possible, with the activity allocation module and the space development module treated separately. In such processes, hypotheses are made as to the size and structure of error and likelihood functions are maximized in order to establish parameter values that maximize the probabilities the component models in the modules assign to the behaviour indicated in the observed data. Then the land use model, as the combination of the component activity allocation and space development modules, along with the transport model, is calibrated as an entire system: In particular, the full set of modules are run in combination and the outputs compared with a selection of weighted observed data to provide an overall goodness-of-fit measure. Specific parameters within the modules are adjusted, and the model re-run to determine the effect of these adjustments on the overall goodness-of-fit. This provides information used in a search procedure that establishes the values for the set of parameters found to be more directly related to the overall model fit. When this search has converged on a set of parameters, the remaining lack-of-fit is explored interactively using a weight sensitivity matrix. This may lead to small changes in the details of the model design and specification, so that the entire modelling system can provide a more accurate representation (Abraham, 2000).

The other big, challenging task in the practical application of PECAS is the collection and/or development of the data required for calibration and for subsequent operation of the model. In all practical applications thus far there has had to be considerable work done in data development and synthesis. The view taken is that it is better to work from a more comprehensive modelling framework, and seek to develop data as required, rather than to limit the model form consistent with the (readily?) available data. It is also felt that a consistent treatment within a comprehensive framework is even more important when data are not so readily available. In cases where complete data are not immediately available, some form of synthetic data is used to develop a 'placeholder', so that the full system representation with PECAS is still available and the entire model is working. It is anticipated that there will be opportunities to collect and further enhance the data and adjust the model accordingly after a working model is available and, hopefully, has demonstrated its usefulness by being applied to inform policy development.

Previous experience in modelling of this sort has taught that the working versions of these models are constantly being updated and modified in order to include enhanced representation seen to be required for particular applications. The use of a consistent, generalised structure as provided by PECAS, where further categories and representations can be added and interpreted, is very helpful in this regard.

With the PECAS framework defined as outlined above, the experiences being gained in the practical modelling that is underway can be more completely interpreted and shared, not just among the specific applications of PECAS, but also more widely among other practical land use transport modelling applications elsewhere. It is anticipated that such experiences will be reported in this light in the future.

REFERENCES

Abraham, J.E. and J.D. Hunt (2002). Spatial market representations: concepts and application to integrated planning models. Presented at the 49th annual North American Meetings of the Regional Science Association International, San Juan, Puerto Rico, November.

Abraham, J.E. (2000). *Parameter Estimation in Urban Models: Theory and Application to the MEPLAN Model of Sacramento*. PhD Thesis, University of Calgary Department of Civil Engineering, Calgary AB, Canada.

de la Barra, T. (1998). The mathematical and algorithmic structure of TRANUS. Unpublished paper available at http://www.modelistica.com.

DKS, Mark Bradley Research and Consulting, HBA Specto and Bowman JL (2001). Land Use and Transport Modeling Design Report. Prepared for Sacramento Area Council of Governments, Sacramento CA, USA.

Hunt, J.D. (2001). The 2nd generation Oregon statewide land use transport interaction model. *Proceedings of the 7th International Conference on Computers in Urban Planning and Urban Management*, Honolulu HI, USA, July.

Hunt, J.D. and D.C. Simmonds (1993). Theory and application of an integrated land-use and transport modelling framework. *Environment and Planning*, **20B**, 221-244.

PB Ohio and Hunt Analytics (1999). Ohio Statewide Travel Demand Model Final Report, PID 17803. Prepared for Ohio Department of Transportation Office of Technical Services, Columbus OH, USA.

Simmonds, D.C. (1996). *DELTA Model Design.* David Simmonds Consultancy, Cambridge UK.

Waddell, P.A. (2002). UrbanSim: Modeling urban development for land use, transportation and environmental planning. *American Planning Association J.*, **68**, 297-314.

ACKNOWLEDGEMENTS

The development of the framework described here has benefited from experience gained in various practical modelling projects, including those sponsored by The Ohio Department of Transportation, The Oregon Department of Transportation, The Southern California Association of Governments, The Sacramento Area Council of Governments, The City of Calgary and The City of Edmonton. Notwithstanding, any errors and omissions are the sole responsibility of the authors.

The preparation of this paper has been supported in part by funding from the Natural Sciences and Engineering Research Council of Canada, and the Social Sciences and Humanities Research Council of Canada via a MCRI Grant.

Integrated Land-Use and Transportation Models: Behavioural Foundations
M. Lee-Gosselin and S. Doherty (Editors)

12

HYBRID CHOICE MODELS WITH LOGIT KERNEL: APPLICABILITY TO LARGE SCALE MODELS[1]

Denis Bolduc, Université Laval, Québec, Canada
Moshe Ben-Akiva, Massachusetts Institute of Technology, USA
Joan Walker, Boston University, USA
Alain Michaud, Ministère des Finances, Ottawa, Canada

ABSTRACT

Traditionally, an economic analysis of micro-level data views individual agents (household members, firms) as entities that optimize in order to take actions or decisions that are best for them. The widely used discrete choice models represent the decision process through a simple conceptual framework depicted as a black box for which alternative attributes and individual characteristics are inputs and where the observed choice corresponds to the output of the system. Its simplicity has left the door open to critics arguing that it often neglects important idiosyncratic aspects and that it cannot deal with decisions that look irrational. The Hybrid Choice Model's conceptual framework possesses many ingredients to enhance the behavioural

[1] Acknowledgements: We would like to thank the Social Sciences and Humanities Research Council of Canada (SSHRC) and the Fonds FCAR for their financial contributions.

representation of the choice process and therefore address the problems raised above. As a direct result, the choice model specification is improved and it gains in predictive power.

So far, the applications of hybrid choice models have been limited to small scale models with two to three alternatives. This paper focuses on the major issue that arises from applying hybrid choice models with logit kernel to large scale problems. It concerns probability simulation of large dimensional integrals that arise from the inclusion of numerous attitudes and perceptions in models with large sets of potentially inter related choices. We suggest simulation-driven Bayesian and classical approaches to the econometric estimation of models with large number of dimensions and flexible choice model formulations. Econometrically, the estimation of hybrid choice models can be extremely involved, but the progress in computer technologies now permits us to address those complicated problems.

INTRODUCTION

The Random Utility Model (RUM) framework provides a conceptual setting meant to give a reliable representation of the process leading to a decision. Assuming that the economic agents make their consumption decisions by maximizing a given indicator of satisfaction, standard discrete choice models view the decision process as a black box for which alternative attributes and individual characteristics are inputs. The observed choice made by the individual corresponds to the output of this system. The availability of micro-economic databases (about households or individuals) in fields like transportation, finance or labour economics has permitted us to submit these models to empirical evidence. With traditional models, the economic agent decisions depend on observable individual characteristics, observable attributes of the alternatives and on an error term which takes into account aspects not explicitly incorporated in the model. Up until recently, discrete choice modellers have used rather simple conceptual frameworks. Conceptual frameworks that are too simple might force the models to depict decisions that appear to be irrational within a rational based RUM setting. Using the suggestions in McFadden (1999), Ben-Akiva, McFadden *et al.* (2002) and Walker and Ben-Akiva (2002), we propose a very general conceptual framework that possesses many ingredients to deal with the problems raised above. The more general model framework, termed *hybrid choice model* (HCM), allows one to incorporate many ingredients, such as latent (unobserved) psychometric variables, in order to enhance the behavioural representation of the decision process. Ben-Akiva, McFadden *et al.* (2002) view the HCM framework as a *practical predictive model* that goes beyond the RUM model in its narrowest formulation and that draws into the analysis the elements of cognitive process identified as important to the choice process, including strong dependence on history and context, perception formation, heterogeneity and latent constraints. The whole process allows one to improve the flexibility and the predictive power of discrete choice modelling tools.

So far, the applications of the HCM have been limited to small scale models with few alternatives. Ben-Akiva and Boccara (1987) constitutes one of the earliest contributions.

Morikawa, Ben-Akiva and McFadden (2002) show how some sub-models of the HCM framework can be estimated using a maximum likelihood approach. Walker and Ben-Akiva (2002) integrate other important enrichments and show, with the help of numerous examples, how they can be implemented in practice. The extensions considered include: flexible disturbances; latent variables; latent classes and combined revealed and stated preferences. For most examples, although only binary choices were involved, the number of latent variables used justified the use of a simulation-based maximum likelihood estimation approach because of the presence of multidimensional integrals. The HCM setting has never been implemented in more realistic situations with large number of latent variables and large number of possibly inter-related choices. The econometric complexity comes from the fact that each latent variable introduced in the model and each choice adds one level of integration to the functions required to calculate the choice probabilities and/or some other useful moments needed for estimation and statistical inference.

This paper is concerned with applying HCM to problems with a large number of latent variables and/or large choice sets. We focus on the most important issue that arises from applying hybrid choice models with logit kernel (mixed logit) to such large problems. It concerns simulation of high dimensional integrals that arise from the inclusion of numerous attitudes and perceptions in models with large sets of potentially inter related choices. After introducing the notation, we present and discuss classical and Bayesian estimation techniques. The first technique replaces the choice probabilities in a conventional maximum likelihood estimation setting with an efficient choice probability simulator. The second approach uses data augmentation to implement a Markov Chain Monte Carlo (MCMC) sampler, now extremely popular in Bayesian analysis. It is a natural extension of the work considered in Bolduc, Fortin and Gordon (1997). As stated in Train (2002), Bayesian tools are often used by classical modellers to produce asymptotically equivalent "classical" estimators. Econometrically, the estimation of hybrid choice models can be extremely involved, but the progress in computer technologies now permits us to address those complicated problems. The paper gives only the technical details regarding the implementation of the two approaches. Empirical evidence on the applicability of the techniques proposed and parameter estimability considerations are provided in a forthcoming separate paper by Bolduc and Godbout (2004).

THE HYBRID CHOICE MODEL

Traditionally, discrete choice models represent the economic agent decisions as being functions of the observable individual characteristics, the observable attributes of the alternatives and an error term to take into account aspects not explicitly included in the model. Up until recently, discrete choice modellers have used rather simple conceptual frameworks. The basic RUM is shown on the central vertical axis of Figure 12.1[2], in which observable explanatory variables (such as characteristics of the decision maker and attributes of the alternatives) are shown to

[2] Adapted from Walker and Ben-Akiva (2002)

influence the decision process (utility maximizing), which then leads to the revealed choices. In the traditional quantitative choice models, the emphasis has been on the systematic, invariant features of choice behaviour, which are assumed to dominate the decision-making process. The HCM conceptual framework builds on the basic RUM core in several directions. The first extension considered in Walker and Ben-Akiva (2002) concerns flexible disturbances. They suggest using a logit kernel (mixed logit) formulation because this model was shown (McFadden and Train, 2000) to approximate any type of RUM model. This flexible formulation is very useful as it easily handles heterogeneity through random parameters. It also allows for flexible interdependencies across the alternatives (Ben-Akiva *et al.*, 2001).

Adding the right hand side of Figure 12.1 (the Latent Variable Model) to this flexible formulation leads to the Integrated Choice and Latent Variable (ICLV) model initially suggested in Ben-Akiva *et al.* (1999). To include latent variables makes it possible to better explain behaviour by explicitly representing the effects of latent constructs such as attitudes and perceptions on the decision process. The ICLV setting might lead to model specifications that are quite difficult to estimate in practice when the number of latent variables is large and when the choice model uses a flexible structure in the context of a large number of choices. The last two extensions considered in Walker and Ben-Akiva (2002) concern latent class and the combination of revealed (RP) and stated preferences (SP).

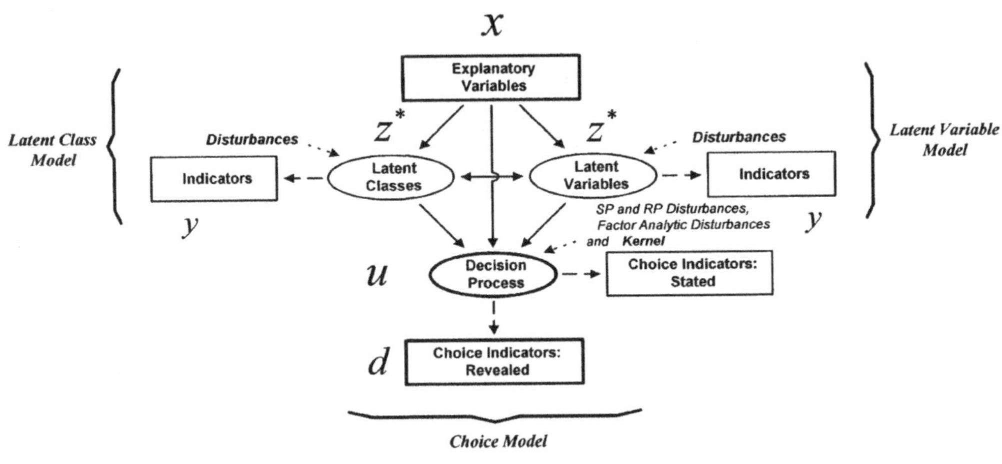

Figure 12.1 Hybrid Choice Model

The latent class dimension replaces the choice probability of a given alternative i with a weighted sum of the conditional probabilities of alternative i given that the subject belongs to a particular market segment s, where each weight corresponds to the probability that the subject belongs to that segment. Under the assumption that the trade-off relationships among major attributes are common to both revealed and stated preferences, to combine them improves the quality of the estimation and of the forecasts. Morikawa *et al.* (2002) give a complete treatment on the combination of RP/SP data. From a methodological standpoint, those two extensions do not add significant methodological complexities on top of those present for the ICLV model. To simplify the notation and to limit the difficulties, the present paper is restricted to the implementability of the ICLV model in large scale situations.

THE INTEGRATED CHOICE AND LATENT VARIABLE (ICLV) MODEL

The ICLV model can potentially be fairly realistic because it explicitly describes how *perceptions* and *attitudes* affect preferences. The *perception* variables measure the capacity of an individual to represent and evaluate the characteristics of different alternatives while the *attitude* variables measure the importance that an individual assigns to the features of the different alternatives. Conceptually, the *preference* variables measure the level of satisfaction that each alternative brings to the individual. The directionality of relationships between the different latent variables, the indicator variables and the explanatory variables allowed in the HCM framework are detailed in Figure 12.1. We use a variable u to denote a utility measure that gets transformed into an explicit choice indicator d. Factors that affect the utility are observed explanatory variables x and latent (unobserved) explanatory variables z^*. Arrows indicate the direction of relationships among the variables. Latent variables (represented by ovals) are linked to observed measures (rectangles) that include indicators (which represent manifestations of the latent constructs) and explanatory variables. In terms of modelling, the latent variables are viewed as structural variables which are related to other variables using a structural latent variable model framework of which LISREL (Linear Structural Relation) (Jöreskog and Sörbom, 1984) is a special case. An obvious example in transportation demand modelling is a specification including comfort as an explanatory variable to explain mode choice. Indicators could be overall rating of the available modes or their perceived reliability of the arrival time, for instances.

Empirically, this problem can be analyzed with an approach that combines LISREL and a conventional discrete choice model. The LISREL formulation incorporates a factor analysis sub-model, which links latent variables to answers to questions of a survey (the *indicators*). The complete model is composed of a group of structural equations (*structural model*), which describe the latent variables in terms of observable exogenous variables and a group of measurement relationships (*measurement model*) linking latent variables to the indicators. Estimation of the parameters in the full system can be done sequentially or jointly. Sequential estimation provides consistency while joint estimation adds efficiency. Details regarding the implementation and the estimation of ICLV model with binary choice model and few latent variables are provided in Morikawa *et al.* (2002).

We extend their model formulation and notation to handle cases with multinomial choices. Ben-Akiva, Walker *et al.* (2002), also tried to handle cases with multinomial choices. The present paper (i) further generalises the techniques, improving the notation and equations to better handle large problems, and (ii) derives and presents the equations and the approach necessary for MCMC estimation, which is critical for large problems. After introducing the notation, the structural model (latent variables and utility process) and the measurement model (indicator and decision process) will be described separately.

Notation

Before we describe the different relationships involved in the ICLV model with J choices, let us establish some notation that we use later on. The major objective of the paper is to set up the relationships that need to be used in implementing the estimation. Both simulated maximum likelihood and the Bayesian implementation will use at some point one of those defined objects. A simple example is provided to help in understanding the notation.

$x_{in} =$	$(K \times 1)$ vector of socio-economic characteristics of individual n and attributes of alternative i.
$X_n =$	$(J \times K)$ specification matrix formed by stacking the J vectors $x_{in}^{'}$.
$z_{in}^{*} =$	$(L \times 1)$ vector of latent variables concerning individual n as it applies to alternative i.
$z_n^{*} =$	$(JL \times 1)$ vector that vertically stacks the J vectors z_{in}^{*}.
$Z_n^{*s} =$	$(J \times P)$ specification matrix formed by appropriately using the elements in z_{in}^{*}, in the structural model.
$Z_n^{*m} =$	$(JR \times T)$ specification matrix formed by appropriately using the elements in z_{in}^{*}, in the measurement model.
$y_{in} =$	$(R \times 1)$ vector of indicators of latent variables of individual n as it applies to alternative i.
$y_n =$	$(JR \times 1)$ vector of indicators of latent variables of individual n formed by stacking the J vectors y_{in}.
$u_{in} =$	Utility of alternative i as perceived by individual n.
$v_n =$	$(J \times 1)$ vector of error terms associated with utility terms u_{in}, $i = 1,...,J$.
$w_n =$	$(M \times 1)$ vector of explanatory variables explaining (causing) the latent variables.
$W_n =$	$(JL \times Q)$ specification matrix containing the elements in w_n. See example for details.
$d_{in} =$	Decision variable which is equal to 1 if alternative i is chosen by individual n, and 0 otherwise.
$d_n =$	$(J \times 1)$ vector that contains the d_{in}'s.
$a, B, C, \Lambda =$	Matrices and vectors (small letters) of unknown model parameters.
$b, c, \lambda =$	Free parameters in B, C and Λ matrices, respectively.
$\gamma =$	The joint set of parameters in the choice model $\gamma = (a^{'}, c^{'})^{'}$.

$\Theta, \Psi, \Sigma =$	Matrices of nuisance parameters.
$\delta =$	The full set of parameters in the system.
$\zeta, \varepsilon =$	Error terms in structural and measurement equation models, respectively.
ξ	Error term modelling the correlations among the utilities.

Estimability of the parameters is not discussed in this paper. The full system involves many parameters. The specific constraints imposed have a direct impact on the identifiability of the remaining free parameters. Those issues are also considered in Bolduc and Godbout (2004).

Algebraically, the decision process of an economic agent that applies to a multinomial situation with J choices can be represented using a general framework composed of structural and measurement relationships. All the relationships required to set up an econometric estimation of the ICLV model are now presented.

Structural Model

Using a matrix notation, the structural model used to describe the inter-relationships that exist between latent variables and the observable explanatory variables is assumed to be as follows:

$$\underset{(JL\times 1)}{z_n^*} = \underset{(JL\times M)}{B} \; \underset{(M\times 1)}{w_n} + \underset{(JL\times 1)}{\zeta_n}, \qquad \zeta_n \sim MVN(0,\Psi), \quad \text{or equivalently as} \tag{1}$$

$$\underset{(JL\times 1)}{z_n^*} = \underset{(JL\times Q)}{W_n} \; \underset{(Q\times 1)}{b} + \underset{(JL\times 1)}{\zeta_n}, \qquad \zeta_n \sim MVN(0,\Psi), \tag{2}$$

and

$$\underset{(J\times 1)}{u_n} = \underset{(J\times K)}{X_n} \; \underset{(K\times 1)}{a} + \underset{(J\times JL)}{C} \; \underset{(JL\times 1)}{z_n^*} + \underset{(J\times 1)}{v_n}, \quad \text{or equivalently as} \tag{3}$$

$$\underset{(J\times 1)}{u_n} = \underset{(J\times K)}{X_n} \; \underset{(K\times 1)}{a} + \underset{(J\times P)}{Z_n^{*s}} \; \underset{(P\times 1)}{c} + \underset{(J\times 1)}{v_n}. \tag{4}$$

According to our notation, a is a vector of K free parameters, b is a vector of the Q free parameters in the B matrix, and c is a vector of the P free parameters in the C matrix. It is important to make sure that the different notations for the structural model is clear since the Bayesian implementation will alternate between the two notations. Which notation to use depends on whether we want the specification to be linear with respect to the vector of variables or the vector of parameters. As we will see, the Bayesian implementation alternates between both forms of the specification. Let us consider a simple trinomial example which involves two latent variables.

Example 1 *Let z^*_{in1} denote, for example, comfort of mode i as perceived by individual n and let z^*_{in2} represent convenience. Using first the notation in equation (4), where we assume that all latent variable effects are alternative specific, and where X_n has a very simple structure with a single generic x_{in} variable with two alternative specific dummies, we would then get:*

$$\begin{pmatrix} u_{1n} \\ u_{2n} \\ u_{3n} \end{pmatrix} = \begin{pmatrix} 1 & 0 & x_{1n} \\ 0 & 1 & x_{2n} \\ 0 & 0 & x_{3n} \end{pmatrix} \begin{pmatrix} a_1 \\ a_2 \\ a_3 \end{pmatrix} + \begin{pmatrix} z^*_{1n1} & z^*_{1n2} & 0 & 0 & 0 & 0 \\ 0 & 0 & z^*_{2n1} & z^*_{2n2} & 0 & 0 \\ 0 & 0 & 0 & 0 & z^*_{3n1} & z^*_{3n2} \end{pmatrix} \begin{pmatrix} c_1 \\ c_2 \\ c_3 \\ c_4 \\ c_5 \\ c_6 \end{pmatrix} + \begin{pmatrix} v_{1n} \\ v_{2n} \\ v_{3n} \end{pmatrix}.$$

Using the alternative notation in equation (3), the same utility model is written as:

$$\begin{pmatrix} u_{1n} \\ u_{2n} \\ u_{3n} \end{pmatrix} = \begin{pmatrix} 1 & 0 & x_{1n} \\ 0 & 1 & x_{2n} \\ 0 & 0 & x_{3n} \end{pmatrix} \begin{pmatrix} a_1 \\ a_2 \\ a_3 \end{pmatrix} + \begin{pmatrix} c_1 & c_2 & 0 & 0 & 0 & 0 \\ 0 & 0 & c_3 & c_4 & 0 & 0 \\ 0 & 0 & 0 & 0 & c_5 & c_6 \end{pmatrix} \begin{pmatrix} z^*_{1n1} \\ z^*_{1n2} \\ z^*_{2n1} \\ z^*_{2n2} \\ z^*_{3n1} \\ z^*_{3n2} \end{pmatrix} + \begin{pmatrix} v_{1n} \\ v_{2n} \\ v_{3n} \end{pmatrix}.$$

In the same way, the latent variable model with two w_n explanatory variables and a pre-specified structure with generic effects across alternatives 2 and 3 could equivalently be written as:

$$\begin{pmatrix} z^*_{1n1} \\ z^*_{1n2} \\ z^*_{2n1} \\ z^*_{2n2} \\ z^*_{3n1} \\ z^*_{3n2} \end{pmatrix} = \begin{pmatrix} b_1 & 0 \\ b_2 & b_3 \\ b_4 & b_5 \\ b_6 & b_7 \\ b_4 & b_5 \\ b_6 & b_7 \end{pmatrix} \begin{pmatrix} w_{1n} \\ w_{2n} \end{pmatrix} + \begin{pmatrix} \zeta_{1n1} \\ \zeta_{1n2} \\ \zeta_{2n1} \\ \zeta_{2n2} \\ \zeta_{3n1} \\ \zeta_{3n2} \end{pmatrix},$$

or as:

$$\begin{pmatrix} z^*_{1n1} \\ z^*_{1n2} \\ z^*_{2n1} \\ z^*_{2n2} \\ z^*_{3n1} \\ z^*_{3n2} \end{pmatrix} = \begin{pmatrix} w_{1n} & 0 & 0 & 0 & 0 & 0 & 0 \\ 0 & w_{1n} & w_{2n} & 0 & 0 & 0 & 0 \\ 0 & 0 & 0 & w_{1n} & w_{2n} & 0 & 0 \\ 0 & 0 & 0 & 0 & 0 & w_{1n} & w_{2n} \\ 0 & 0 & 0 & w_{1n} & w_{2n} & 0 & 0 \\ 0 & 0 & 0 & 0 & 0 & w_{1n} & w_{2n} \end{pmatrix} \begin{pmatrix} b_1 \\ b_2 \\ b_3 \\ b_4 \\ b_5 \\ b_6 \\ b_7 \end{pmatrix} + \begin{pmatrix} \zeta_{1n1} \\ \zeta_{1n2} \\ \zeta_{2n1} \\ \zeta_{2n2} \\ \zeta_{3n1} \\ \zeta_{3n2} \end{pmatrix}.$$

To impose generic effects on the latent variables is straightforward. For example assuming that both z^*_{in1} and z^*_{in2} are generic for alternatives 2 and 3 would lead to:

$$\begin{pmatrix} u_{1n} \\ u_{2n} \\ u_{3n} \end{pmatrix} = \begin{pmatrix} 1 & 0 & x_{1n} \\ 0 & 1 & x_{2n} \\ 0 & 0 & x_{3n} \end{pmatrix} \begin{pmatrix} a_1 \\ a_2 \\ a_3 \end{pmatrix} + \begin{pmatrix} z^*_{1n1} & z^*_{1n2} & 0 & 0 \\ 0 & 0 & z^*_{2n1} & z^*_{2n2} \\ 0 & 0 & z^*_{3n1} & z^*_{3n2} \end{pmatrix} \begin{pmatrix} c_1 \\ c_2 \\ c_3 \\ c_4 \end{pmatrix} + \begin{pmatrix} v_{1n} \\ v_{2n} \\ v_{3n} \end{pmatrix}, \text{ and}$$

$$\begin{pmatrix} u_{1n} \\ u_{2n} \\ u_{3n} \end{pmatrix} = \begin{pmatrix} 1 & 0 & x_{1n} \\ 0 & 1 & x_{2n} \\ 0 & 0 & x_{3n} \end{pmatrix} \begin{pmatrix} a_1 \\ a_2 \\ a_3 \end{pmatrix} + \begin{pmatrix} c_1 & c_2 & 0 & 0 & 0 & 0 \\ 0 & 0 & c_3 & c_4 & 0 & 0 \\ 0 & 0 & 0 & 0 & c_3 & c_4 \end{pmatrix} \begin{pmatrix} z^*_{1n1} \\ z^*_{1n2} \\ z^*_{2n1} \\ z^*_{2n2} \\ z^*_{3n1} \\ z^*_{3n2} \end{pmatrix} + \begin{pmatrix} v_{1n} \\ v_{2n} \\ v_{3n} \end{pmatrix}.$$

We can note that in equation (1), the latent variables z^*_n are function of observable variables w_n and of a JL-dimensional normally distributed random error vector ζ_n with mean zero and variance covariance matrix Ψ. For identification, since latent variables are unobserved, variances should all be set to one. All kinds of covariances can be modelled in theory. Covariance across alternatives and/or across latent variables could thus be handled. A simple form would force the matrix to have a block structure. The weights that w_n has on a particular z^*_{inl} is given by the corresponding row of matrix B. As for the C matrix, the B matrix could contain 0's to handle exclusion restrictions (some w_{mn} terms could be assumed not to affect a particular latent variable). The free parameters in B are contained in a Q - dimensional joint vector b.

Distributional assumptions regarding v_n directly determine the kernel of the discrete choice model that can be used. If v_n is i.i.d. Gumbel distributed, then the discrete choice model conditional on z_n^* is of type Logit, while v_n being i.i.d. normal leads to an i.i.d.-Probit model. To assume that the joint vector of parameters or a subset of parameters is (jointly) randomly distributed or that the v_n is composed of two error terms, one from a joint distribution and one from an i.i.d. Gumbel distribution leads to a Logit Kernel (mixed logit) model. See Ben-Akiva *et al.* (2001), Bhat (2001) or McFadden and Train (2000) for detailed discussions about this flexible discrete choice model formulation.

We now assume that v_n in equations (3) and (4) is composed of two terms:

$$v_n = \xi_n + e_n, \tag{5}$$

where ξ_n is joint multivariate normal of mean 0 and variance covariance matrix Σ and where e_n is i.i.d. Gumbel. Then, the choice probability of alternative i given z_n^* and ξ_n has the following form:

$$P_n(i \mid z_n^*, \xi_n) = \frac{\exp(X_{in}a + C_i z_n^* + \xi_{in})}{\sum_{j=1}^{J} \exp(X_{jn}a + C_j z_n^* + \xi_{jn})}, \tag{6}$$

where X_{hn} and C_h denotes row h of X_n and C, respectively. Assuming that z^* and ξ are mutually distributed independently, the unconditional choice probability of alternative i is

$$P_n(i) = \int_\xi \int_{z^*} \frac{\exp(X_{in}a + C_i z^* + \xi)}{\sum_{j=1}^{J} \exp(X_{jn}a + C_j z^* + \xi)} \mathrm{n}(z^*; Bw_n, \Psi)\, \mathrm{n}(\xi; 0, \Sigma) dz^* d\xi, \tag{7}$$

where $\mathrm{n}(e; h, H)$ denotes the density of a multivariate normal distribution with mean h and variance covariance matrix H. Note that this is an integral of dimension $J(L+1)$. In a moderate size model with say: $J = 5$ alternatives and $L = 4$ latent variables, this integral is of dimension 25. Clearly, simulation would be required. Because equation (7) has the form of an expectation, a simulator based on an empirical mean taken over draws from normal variates naturally arises. This corresponds to the logit kernel simulator which can be computed as follows:

$$\tilde{P}_n(i) = \frac{1}{R} \sum_{r=1}^{R} \frac{\exp(X_{in}a + C_i z_n^{*r} + \xi_{in}^r)}{\sum_{j=1}^{J} \exp(X_{jn}a + C_j z_n^{*r} + \xi_{jn}^r)}, \tag{8}$$

where z_n^{*r} and ξ_n^r denote draw r from their respective distributions for individual n. This simulator is known to be unbiased, consistent and smooth with respect to the unknown parameters. A simulation based estimation technique would replace the true probabilities in equation (7) with simulation computed as in (8). Bolduc *et al.* (1997) used such a logit kernel approach in an 18-dimensional inter-related spatial choice problem.

Measurement Model

We now consider the measurement dimension of the ICLV model framework. Starting with a given alternative i, the measurement equation that links the $(R \times 1)$ vector of indicators y_{in} and the latent variables associated with the ith alternative may be written as:

$$\underset{(R\times 1)}{y_{in}} = \underset{(R\times L)}{\Lambda_i}\,\underset{(L\times 1)}{z^*_{in}} + \underset{(R\times 1)}{\varepsilon_{in}}, \quad \varepsilon_{in} \sim MVN(0, \Theta_i). \tag{9}$$

In the example considered so far, y_{in} could contain relaxation and reliability of mode i as indicators for *comfort* while relaxation and ease could be indicators for *convenience*. The variance covariance matrix Θ_i could be diagonal or full. Because y_{in} is observed, all terms are estimable in theory. However, our experience with linear systems of equations containing latent variables show that identification is not trivial. Applying a similar relationship to each alternative and vertically concatenating those relationships lead to:

$$\underset{(JR\times 1)}{y_n} = \underset{(JR\times JL)}{\Lambda}\,\underset{(JL\times 1)}{z^*_n} + \underset{(JR\times 1)}{\varepsilon_n}, \quad \varepsilon_n \sim MVN(0, \Theta \equiv diag(\Theta_i)), \tag{10}$$

where Λ is a block diagonal matrix with the $\Lambda_i, i = 1,...,J$ matrices on the main diagonal. This can equivalently be written as:

$$\underset{(JR\times 1)}{y_n} = \underset{(JR\times T)}{Z^{*m}_n}\,\underset{(T\times 1)}{\lambda} + \underset{(JR\times 1)}{\varepsilon_n}, \quad \varepsilon_n \sim MVN(0, \Theta \equiv diag(\Theta_i)), \tag{11}$$

where λ is a $(T \times 1)$ vector containing the free parameters in Λ. The term $\Theta \equiv diag(\Theta_i)$ corresponds to a block diagonal matrix with J distinct $(R \times R)$ $\Theta_i, i = 1,...,J$ matrices along the main diagonal.

Example 2 *continuation of example 1:*

Using relaxation y_{in1} and reliability y_{in2} of mode i as indicators for comfort (z^*_{in1}) and using relaxation y_{in1} and ease y_{in3} as indicators for convenience (z^*_{in2}) in our trinomial example, and making the effect to be generic across modes 2 and 3, we would get:

$$\begin{pmatrix} y_{1n1} \\ y_{1n2} \\ y_{1n3} \\ y_{2n1} \\ y_{2n2} \\ y_{2n3} \\ y_{3n1} \\ y_{3n2} \\ y_{3n3} \end{pmatrix} = \begin{pmatrix} \lambda_1 & \lambda_2 & 0 & 0 & 0 & 0 \\ \lambda_3 & 0 & 0 & 0 & 0 & 0 \\ 0 & \lambda_4 & 0 & 0 & 0 & 0 \\ 0 & 0 & \lambda_5 & \lambda_6 & 0 & 0 \\ 0 & 0 & \lambda_7 & 0 & 0 & 0 \\ 0 & 0 & 0 & \lambda_8 & 0 & 0 \\ 0 & 0 & 0 & 0 & \lambda_5 & \lambda_6 \\ 0 & 0 & 0 & 0 & \lambda_7 & 0 \\ 0 & 0 & 0 & 0 & 0 & \lambda_8 \end{pmatrix} \begin{pmatrix} z^*_{1n1} \\ z^*_{1n2} \\ z^*_{2n1} \\ z^*_{2n2} \\ z^*_{3n1} \\ z^*_{3n2} \end{pmatrix} + \begin{pmatrix} \varepsilon_{1n1} \\ \varepsilon_{1n2} \\ \varepsilon_{1n3} \\ \varepsilon_{2n1} \\ \varepsilon_{2n2} \\ \varepsilon_{2n3} \\ \varepsilon_{3n1} \\ \varepsilon_{3n2} \\ \varepsilon_{3n3} \end{pmatrix},$$

or when written using the notation in equation (11) as:

$$\begin{pmatrix} y_{1n1} \\ y_{1n2} \\ y_{1n3} \\ y_{2n1} \\ y_{2n2} \\ y_{2n3} \\ y_{3n1} \\ y_{3n2} \\ y_{3n3} \end{pmatrix} = \begin{pmatrix} z^*_{1n1} & z^*_{1n2} & 0 & 0 & 0 & 0 & 0 & 0 \\ 0 & 0 & z^*_{1n1} & 0 & 0 & 0 & 0 & 0 \\ 0 & 0 & 0 & z^*_{1n2} & 0 & 0 & 0 & 0 \\ 0 & 0 & 0 & 0 & z^*_{2n1} & z^*_{2n2} & 0 & 0 \\ 0 & 0 & 0 & 0 & 0 & 0 & z^*_{2n1} & 0 \\ 0 & 0 & 0 & 0 & 0 & 0 & 0 & z^*_{2n2} \\ 0 & 0 & 0 & 0 & z^*_{3n1} & z^*_{3n2} & 0 & 0 \\ 0 & 0 & 0 & 0 & 0 & 0 & z^*_{3n1} & 0 \\ 0 & 0 & 0 & 0 & 0 & 0 & 0 & z^*_{3n2} \end{pmatrix} \begin{pmatrix} \lambda_1 \\ \lambda_2 \\ \lambda_3 \\ \lambda_4 \\ \lambda_5 \\ \lambda_6 \\ \lambda_7 \\ \lambda_8 \end{pmatrix} + \begin{pmatrix} \varepsilon_{1n1} \\ \varepsilon_{1n2} \\ \varepsilon_{1n3} \\ \varepsilon_{2n1} \\ \varepsilon_{2n2} \\ \varepsilon_{2n3} \\ \varepsilon_{3n1} \\ \varepsilon_{3n2} \\ \varepsilon_{3n3} \end{pmatrix}.$$

The final relationship within the measurement model is the one that relates observed choice to the utilities. It is defined as:

$$d_{in} = \begin{cases} 1 & \text{if } u_{in} \geq u_{jn}, \quad j \neq i \\ 0 & \text{otherwise.} \end{cases} \tag{12}$$

ESTIMATION OF ICLV MODELS

We now discuss two estimation methods. The first method relies on maximizing the (simulated) log-likelihood of the sample. For efficiency reasons, we only focus on a full information solution. The second is a Bayesian Monte Carlo Markov Chain (MCMC) method that avoids calculating computationally complicated expressions such as multidimensional choice probabilities. Excellent background material on maximum simulated likelihood (MSL) and MCMC methods as they apply to discrete choice modelling may be found in Train (2002). Train shows that with large samples, both approaches should normally give similar results. Those estimation methods have been compared in traditional discrete choice model situations (see Bolduc *et al.*, 1997 or Train, 2001). One contribution of this paper is to provide feasibility to the estimation of ICLV model settings in the context of large choice sets.

Full information estimation of the ICLV model setting requires modelling jointly the parameters present in equations (1)-(4) and equations (10)-(12). Following Walker and Ben-Akiva (2002), full information estimation requires maximizing the following joint probability:

$$P(d_n, y_n \mid X_n, w_n, \delta),$$

where δ denotes the joint vector of parameters in $a, b, c, \lambda, \Sigma, \Psi, \Theta$. Assuming independence between the error terms in the choice model and in the measurement model, the decomposition $P(d_n, y_n \mid z_n^*) = P(d_n \mid z_n^*) \cdot f(y_n \mid z_n^*)$ and the relationship between w_n and z_n^* make it possible to write that joint probability as:

$$P(d_n, y_n \mid X_n, w_n, \delta) = \int_{z^*} P(d_n \mid z^*, X_n, w_n, \delta) f(y_n \mid \lambda, z^*) f(z^* \mid b, w_n) dz^*. \quad (13)$$

Assuming also that the v_n error term is composed of the two terms in equation (5), then because of equation (6), and writing z_n^* as $Bw_n + \Psi^{1/2}\eta$ and ξ_n as $\Sigma^{1/2}\omega_n$, this probability may be shown to be written as:

$$P(d_n, y_n \mid X_n, w_n, \delta) = \iint_{\omega\eta} \frac{\exp(X_{in}a + C_i Bw_n + C_i \Psi^{1/2}\eta + \Sigma_i^{1/2}\omega)}{\sum_{j=1}^{J} \exp(X_{jn}a + C_j Bw_n + C_j \Psi^{1/2}\eta + \Sigma_j^{1/2}\omega)} \quad (14)$$
$$\mathrm{n}(\varepsilon_n; y_n - \Lambda[Bw_n + \Psi^{1/2}\eta], \Theta)\mathrm{n}(\eta; 0, I_{JL})\mathrm{n}(\omega; 0, I_J) d\eta d\omega,$$

where $\Psi^{1/2}$ denotes the lower triangular Cholesky decomposition of Ψ and $\Sigma_h^{1/2}$ represents row h of the Cholesky matrix $\Sigma^{1/2}$ of Σ, where η and ω are standard normal variates.

Maximum Simulated Likelihood Estimation

In situations of reasonable size, the probability in equation (14) cannot be evaluated analytically. The logit kernel approach naturally replaces this probability with the following empirical mean:

$$\tilde{P}_n(i, y_n \mid X_n, w_n, \delta) = \quad (15)$$
$$\frac{1}{R}\sum_{r=1}^{R} \left\{ \frac{\exp(X_{in}a + C_i Bw_n + C_i \Psi^{1/2}\eta^r + \Sigma_i^{1/2}\omega^r)}{\sum_{j=1}^{J} \exp(X_{jn}a + C_j Bw_n + C_j \Psi^{1/2}\eta^r + \Sigma_j^{1/2}\omega^r)} \mathrm{n}(\varepsilon_n; y_n - \Lambda[Bw_n + \Psi^{1/2}\eta^r], \Theta) \right\},$$

where R draws η^r and ω^r are taken over the distributions of η and ω. When we consider the whole sample, and denoting as i_n the choice made by individual n, since the maximum likelihood method (ML) maximizes the log-likelihood of the whole sample:

$$L_n = \frac{1}{N}\sum_{n=1}^{N} \ln P_n(i_n, y_n),$$

the maximum simulated likelihood (MSL) approach maximizes the following expression:

$$\tilde{L}_n = \frac{1}{N}\sum_{n=1}^{N} \ln \tilde{P}_n(i_n, y_n). \tag{16}$$

In the past few years, a lot of progress has been made regarding MSL estimation. Train (2002) gives an in-depth analysis of the properties of MSL estimation. Recent results mostly attributable to Bhat (2001) suggest that simulations in equation (15) should exploit Halton draws or improvements such as shuffling and scrambling. Halton type sequences are known to produce simulators with a given level of accuracy using far fewer draws than using conventional uniform random draws. Simulated maximum likelihood is now well known and has been applied in numerous circumstances. The logit probability kernel in equation (15) makes the simulated log likelihood fairly well behaved. Asymptotically, the solution is identical to a solution arising from maximizing the usual log likelihood function. In the context of ICLV modelling, MSL is technically applicable. However in situations with a very large number of latent variables and large number of interdependent alternatives, the behaviour of MSL remains to be analyzed. From a practical standpoint, we believe that the next estimation method has a greater potential.

MCMC Estimation

In this section, we develop a methodology that allows one to estimate ICLV models using a Bayesian approach which exploits Gibbs sampling with data augmentation and Metropolis-Hastings (M-H) [see Koop (2003) for an introduction to those techniques]. We refer to it as a Monte Carlo Markov Chain (MCMC) setting. Bolduc *et al.* (1997) and Train (2001) demonstrate that it is possible to estimate discrete choice models using such a MCMC setting. An introduction to the subject may be found in Train (2002). In a Bayesian setting, parameters are assumed to have a statistical distribution. This distribution, called the *prior,* is used to carry the researcher beliefs about the values that the parameters can take in the model. Those anticipations are eventually revised as supplementary information comes in via the sample data (*likelihood function*). The combination of the *prior* distribution and the supplemental information brought in by the sample data determine the *posterior* distribution of the parameters. Technically, the density of the *posterior* distribution of δ is obtained applying the Bayes theorem:

$$p(\delta|X,d) = \frac{p(X,d|\delta)\times p(\delta)}{\int p(X,d|\delta)\times p(\delta)d\delta}, \tag{17}$$

where $p(.)$ is a given density or distribution function, δ represents the joint set of model parameters and X and d refer to the observed variables. As the denominator is a normalising constant, and not a function of δ, we can also write:

$$p(\delta|X,d) \propto l(\delta|X,d)\times p(\delta),$$

where $l(\delta|X,d)$ represents the likelihood of the sample and where $\propto$ denotes a sign of proportionality. The main difficulty in implementing a Bayesian approach is to characterize the *posterior* distribution of the parameters. MCMC methods permit to handle that problem in a straightforward way. The MCMC techniques that we use have been introduced by Geman and Geman (1984) and Tanner and Wong (1987). They have been used for the first time in the context of the estimation of traditional discrete choice models by Albert and Chib (1993).

In the following, we directly focus on the steps required to produce the estimation of the ICLV model discussed in the previous section where the choice model has a logit kernel (mixed logit) structure.

Our objective is to specifically deal with the following joint *posterior* distribution:

$$p(z_n^*,\xi_n,b,\gamma,\lambda,\Psi,\Sigma,\Theta \mid X_n,y_n,w_n,d_n), \tag{18}$$

where all terms were previously defined. In particular, recall that $\gamma = (a',c')'$. We include z_n^*,ξ_n as extra parameters to simplify the whole process through data augmentation. Tanner and Wong (1987) have shown that to consider the latent variables of the system as supplementary unknown parameters greatly simplifies the estimation process. For inference purposes, we then only focus on the $b,\gamma,\lambda,\Psi,\Sigma,\Theta$ terms. The basic algorithm depends on the properties of ergodic Markov chains.

Suppose that we can draw values for $z_n^*,\xi_n,b,\gamma,\lambda,\Psi,\Sigma,\Theta$ from an iterative sequence of "full conditional" *posterior* distributions also termed Gibbs layers:

$$\begin{aligned}
&\text{a) } z_n^* \sim p(z_n^* \mid \xi_n,b,\gamma,\lambda,\Psi,\Sigma,\Theta, y_n,X_n,w_n,d_n)\\
&\text{b) } \xi_n \sim p(\xi_n \mid z_n^*,b,\gamma,\lambda,\Psi,\Sigma,\Theta, y_n,X_n,w_n,d_n)\\
&\text{c) } b \sim p(b \mid z_n^*,\xi_n,\gamma,\lambda,\Psi,\Sigma,\Theta, y_n,X_n,w_n,d_n)\\
&\text{d) } \gamma \sim p(\gamma \mid z_n^*,\xi_n,b,\lambda,\Psi,\Sigma,\Theta, y_n,X_n,w_n,d_n)\\
&\text{e) } \lambda \sim p(\lambda \mid z_n^*,\xi_n,b,\gamma,\Psi,\Sigma,\Theta, y_n,X_n,w_n,d_n)
\end{aligned} \tag{19}$$

$$\text{f) } \Psi \sim p(\Psi \mid z_n^*, \xi_n, b, \gamma, \lambda, \Sigma, \Theta, y_n, X_n, w_n, d_n)$$
$$\text{g) } \Sigma \sim p(\Sigma \mid z_n^*, \xi_n, b, \gamma, \lambda, \Psi, \Theta, y_n, X_n, w_n, d_n)$$
$$\text{h) } \Theta \sim p(\Theta \mid z_n^*, \xi_n, b, \gamma, \lambda, \Psi, \Sigma, y_n, X_n, w_n, d_n).$$

Under fairly weak conditions, the sequence of random draws generated according to equations (19 a-h) forms an irreducible and ergodic Markov chain whose stable distribution corresponds to the joint posterior distribution of interest $p(z_n^*, \xi_n, a, b, c, \lambda, \Psi, \Sigma, \Theta \mid X_n, y_n, w_n, d_n)$. After the chain has converged, the mean of the sequence of draws generated by (19) is a consistent estimator for the posterior mean of the parameter of interest.

The Estimation Process

Consider the random variables $z_n^*, \xi_n, b, \gamma, \lambda, \Psi, \Sigma$ and Θ. They are assumed to come from the joint distribution denoted by $\pi(z_n^*, \xi_n, b, \gamma, \lambda, \Psi, \Sigma, \Theta)$. Consider an arbitrary set of starting values $z_n^{*(0)}, \xi_n^{(0)}, b^{(0)}, \gamma^{(0)}, \lambda^{(0)}, \Psi^{(0)}, \Sigma^{(0)}$ and $\Theta^{(0)}$. The MCMC sampler generates a sequence of random variables $\{z_n^{*(j)}, \xi_n^{(j)}, b^{(j)}, \gamma^{(j)}, \lambda^{(j)}, \Psi^{(j)}, \Sigma^{(j)}, \Theta^{(j)}\}_{j=1}^{\infty}$ by doing successive draws in the following conditional distributions:

1st draw:

$$z_n^{*(1)} \sim \pi(z_n^* \mid \xi_n^{(0)}, b^{(0)}, \gamma^{(0)}, \lambda^{(0)}, \Psi^{(0)}, \Sigma^{(0)}, \Theta^{(0)}, y_n, X_n, w_n, d_n)$$
$$\xi_n^{(1)} \sim \pi(\xi_n \mid z_n^{*(1)}, b^{(0)}, \gamma^{(0)}, \lambda^{(0)}, \Psi^{(0)}, \Sigma^{(0)}, \Theta^{(0)}, y_n, X_n, w_n, d_n)$$
$$b^{(1)} \sim \pi(b \mid z_n^{*(1)}, \xi_n^{(1)}, \gamma^{(0)}, \lambda^{(0)}, \Psi^{(0)}, \Sigma^{(0)}, \Theta^{(0)}, y_n, X_n, w_n, d_n)$$
$$\gamma^{(1)} \sim \pi(\gamma \mid z_n^{*(1)}, \xi_n^{(1)}, b^{(1)}, \lambda^{(0)}, \Psi^{(0)}, \Sigma^{(0)}, \Theta^{(0)}, y_n, X_n, w_n, d_n)$$
$$\lambda^{(1)} \sim \pi(\lambda \mid z_n^{*(1)}, \xi_n^{(1)}, b^{(1)}, \gamma^{(1)}, \Psi^{(0)}, \Sigma^{(0)}, \Theta^{(0)}, X_n, y_n, w_n, d_n)$$
$$\Psi^{(1)} \sim \pi(\Psi \mid z_n^{*(1)}, \xi_n^{(1)}, b^{(1)}, \gamma^{(1)}, \lambda^{(1)}, \Sigma^{(0)}, \Theta^{(0)}, y_n, X_n, w_n, d_n)$$
$$\Sigma^{(1)} \sim \pi(\Sigma \mid z_n^{*(1)}, \xi_n^{(1)}, b^{(1)}, \gamma^{(1)}, \lambda^{(1)}, \Psi^{(1)}, \Theta^{(0)}, y_n, X_n, w_n, d_n)$$
$$\Theta^{(1)} \sim \pi(\Theta \mid z_n^{*(1)}, \xi_n^{(1)}, b^{(1)}, \gamma^{(1)}, \lambda^{(1)}, \Psi^{(1)}, \Sigma^{(1)}, y_n, X_n, w_n, d_n)$$
$$\text{generated sequence}: \left\{z_n^{*(1)}, \xi_n^{(1)}, b^{(1)}, \gamma^{(1)}, \lambda^{(1)}, \Psi^{(1)}, \Sigma^{(1)}, \Theta^{(1)}\right\}$$

$$\vdots \qquad \vdots \qquad \vdots$$

j^{th} draw:

$$
\begin{aligned}
z_n^{*(j)} &\sim \pi(z_n^* \mid \xi_n^{(j-1)}, b^{(j-1)}, \gamma^{(j-1)}, \lambda^{(j-1)}, \Psi^{(j-1)}, \Sigma^{(j-1)}, \Theta^{(j-1)}, y_n, X_n, w_n, d_n) \\
\xi_n^{(j)} &\sim \pi(\xi_n \mid z_n^{*(j)}, b^{(j-1)}, \gamma^{(j-1)}, \lambda^{(j-1)}, \Psi^{(j-1)}, \Sigma^{(j-1)}, \Theta^{(j-1)}, y_n, X_n, w_n, d_n) \\
b^{(j)} &\sim \pi(b \mid z_n^{*(j)}, \xi_n^{(j)}, \gamma^{(j-1)}, \lambda^{(j-1)}, \Psi^{(j-1)}, \Sigma^{(j-1)}, \Theta^{(j-1)}, y_n, X_n, w_n, d_n) \\
\gamma^{(j)} &\sim \pi(\gamma \mid z_n^{*(j)}, \xi_n^{(j)}, b^{(j)}, \lambda^{(j-1)}, \Psi^{(j-1)}, \Sigma^{(j-1)}, \Theta^{(j-1)}, y_n, X_n, w_n, d_n) \\
\lambda^{(j)} &\sim \pi(\lambda \mid z_n^{*(j)}, \xi_n^{(j)}, b^{(j)}, \gamma^{(j)}, \Psi^{(j-1)}, \Sigma^{(j-1)}, \Theta^{(j-1)}, X_n, y_n, w_n, d_n) \\
\Psi^{(j)} &\sim \pi(\Psi \mid z_n^{*(j)}, \xi_n^{(j)}, b^{(j)}, \gamma^{(j)}, \lambda^{(j)}, \Sigma^{(j-1)}, \Theta^{(j-1)}, y_n, X_n, w_n, d_n) \\
\Sigma^{(j)} &\sim \pi(\Sigma \mid z_n^{*(j)}, \xi_n^{(j)}, b^{(j)}, \gamma^{(j)}, \lambda^{(j)}, \Psi^{(j)}, \Theta^{(j-1)}, y_n, X_n, w_n, d_n) \\
\Theta^{(j)} &\sim \pi(\Theta \mid z_n^{*(j)}, \xi_n^{(j)}, b^{(j)}, \gamma^{(j)}, \lambda^{(j)}, \Psi^{(j)}, \Sigma^{(j)}, y_n, X_n, w_n, d_n) \\
&\text{generated sequence} : \left\{ z_n^{*(j)}, \xi_n^{(j)}, b^{(j)}, \gamma^{(j)}, \lambda^{(j)}, \Psi^{(j)}, \Sigma^{(j)}, \Theta^{(j)} \right\}.
\end{aligned}
$$

After R iterations and if R is specified to be large enough, the sequence of draws $\{ z_n^{*(j)}, \xi_n^{(j)}, b^{(j)}, \gamma^{(j)}, \lambda^{(j)}, \Psi^{(j)}, \Sigma^{(j)}, \Theta^{(j)} \}$ obtained through those loops could be considered as directly coming from the joint distribution of interest in equation (18).

We now formally describe and explain each of the required steps. In the following, we develop on the conditional distributions involved in equations (a) to (h). In deriving the full conditional distributions in (19), we make use of well-known Bayesian results for the multivariate linear normal model [see, for example, Zellner (1971) or Koop (2003)]. We mention at this stage that the sequence in which the conditional distributions are visited within a given loop is irrelevant.

a) Conditional distribution of z_n^* given everything else.

This first case is fairly straightforward. Given that y_n and z_n^* are clearly inter-related through equations (1) and (10), Appendix 1 allows us to establish that:

$$(z_n^* \mid .) \sim N(E(z_n^* \mid .), \operatorname{Var}(z_n^* \mid .)) \tag{20}$$

where

$$
\begin{aligned}
E(z_n^* \mid .) &= Bw_n + \left[\Psi\Lambda'\right]\left[\Lambda\Psi\Lambda' + \Theta\right]^{-1}\left[y_n - \Lambda B w_n\right] \\
\operatorname{Var}(z_n^* \mid .) &= \Psi - \left[\Psi\Lambda'\right]\left[\Lambda\Psi\Lambda' + \Theta\right]^{-1}\left[\Lambda\Psi\right],
\end{aligned}
$$

where the period . replaces the following terms: $\xi_n, b, \gamma, \lambda, \Psi, \Sigma$, Θ, X_n, y_n, w_n and d_n. Implementation of this step corresponds to a data augmentation step which requires drawing normally distributed terms according to equation (20).

b) Conditional distribution of ξ_n given everything else.

Since by assumption, $\xi_n \sim MVN(0,\Sigma)$, draws for ξ_n are a function of Σ, the vector of utilities u_n and the choice depicted by d_n. Gilks (1996) shows how to derive the right full conditional distribution using a given prior and u_n. Generation of draws for u_n exploits the work of Scott (2003). This is discussed later in the text. This process constitutes an augmentation step for ξ_n which is then considered as known in the other steps of a given MCMC loop.

c) Conditional distribution of b given everything else.

Since b only relates to the latent variable model in equation (2), the conditional distribution of interest, given z_n^*, Ψ, W_n and d_n, becomes:

$$b \sim p(b|z_n^*, \Psi, W_n, d_n),$$

which could be obtained by combining the sample information and the *prior* information about b; i.e.

$$\begin{aligned} p(b|z_n^*, W_n, \Psi, d_n) &\propto l(b|z_n^*, W_n, \Psi, d_n).p(b) \\ &\propto p(z_n^*|b, W_n, \Psi, d_n).p(b) \end{aligned} \tag{21}$$

We use a multivariate normal distribution to characterize the *prior* distribution of b, because it is conjugate to the likelihood so that the *posterior* distribution belongs to the same distributional family as the *prior.* We assume that the *prior* distribution is as follows:

$$b \sim N(\bar{b}, \Sigma_b^{-1}),$$
$$p(b) = |\Sigma_b|^{0.5} \exp[-\frac{1}{2}(b-\bar{b})' \Sigma_b (b-\bar{b})],$$

where $\bar{b}$ and Σ_b^{-1} are pre-specified by the researchers.

The application of the Cholesky decomposition Q[3], where $QQ' = \Psi^{-1}$, to the relationship in equation (2) facilitates the calculations. In this case, we thus obtain:

$$Q'z_n^* = Q'W_n b + Q'\zeta_n, \qquad \text{where } Q'\zeta_n \sim N(0, I_{JL}),$$

which we rewrite as:

[3] If $QQ' = \Psi^{-1}$, then $(QQ')^{-1} = Q'^{-1}Q^{-1} = \Psi$ and $Q'\Psi Q = I$.

$$\tilde{z}_n^* = r_n b + \tilde{\zeta}_n, \qquad \text{where } \tilde{\zeta}_n \sim N(0, I_{JL})$$
$$\tilde{z}_n^* = Q' z_n^*$$
$$\tilde{\zeta}_n = Q' \tilde{\zeta}_n$$
$$r_n = Q' W_n.$$

After these transformations, the model becomes a simple regression model with unit variance, and it can then be shown that:

$$b | z_n^*, W_n, \Psi, d_n \sim N(\mu_b, \Xi_b) \tag{22}$$

where $\mu_b = \left[n \sum r_n' r_n + \Sigma_b \right]^{-1} \left[n \sum r_n' \tilde{z}_n^* + \Sigma_b \bar{b} \right]$ and $\Xi_b = \left[n \sum r_n' r_n + \Sigma_b \right]^{-1}$. Implementation of this step requires drawing normally distributed terms according to equation (22).

d) Conditional distribution of γ given everything else.

This step is the most computationally involved. Given that u_n is function of a mixture of normally distributed terms and of Gumbel terms, to draw from the correct distribution is not trivial. For situations like these, Metropolis-Hastings (M-H) should be used. Denoting the likelihood conditional on Z_n^{*s}, and ξ_n and denoting as i_n the choice made by agent *n,* we get:

$$L(Y | \gamma) = \prod_{n=1}^{N} \frac{\exp(X_{i_n n} a + Z_n^{*s} c + \xi_{i_n n})}{\sum_{j=1}^{J} \exp(X_{jn} a + Z_n^{*s} c + \xi_{jn})},$$
$$= \prod_{n=1}^{N} \frac{\exp([X_{i_n n} \vdots Z_n^{*s}]\gamma + \xi_{i_n n})}{\sum_{j=1}^{J} \exp([X_{jn} \vdots Z_n^{*s}]\gamma + \xi_{jn})}$$

where $\gamma = (a, c)'$, and where $Y = (y_1, \ldots, y_N)$, then the M-H step at iteration t should be implemented as follows:

compute

$$r = \frac{L(Y | \tilde{\gamma}^t)}{L(Y | \gamma^{t-1})},$$

where $\tilde{\gamma}^t = \gamma^{t-1} + \Sigma_\gamma^{-1/2} \varepsilon$, with $\varepsilon \sim N(0, I)$, where $\Sigma_\gamma^{-1/2}$ denotes the Cholesky decomposition of a pre-specified precision matrix for γ. Then, if $r > 1$, $\gamma^t - \tilde{\gamma}^t$, otherwise,

$$\gamma^t = \begin{cases} \tilde{\gamma}^t & \text{with probability } r \\ \gamma^{t-1} & \text{with probability } 1\text{-}r. \end{cases}$$

This ensures that only reasonable values for γ get retained. An alternative to this is to follow the approach of Scott (2003) which, given draws of the utilities u_n, produces a step that is very similar to the one developed in c). From a practical standpoint, the approach of Scott which leads to a standard Gibbs sampler with data augmentation should lead to a faster sampler. The test of this alternative sampler is left for future work.

e) Conditional distribution of λ given everything else.

Since λ is only present in equation (11) and since y_n is observed, the conditional distribution of interest, given z_n^*, y_n and Θ, becomes:

$$\lambda \sim p(\lambda | z_n^*, y_n, \Theta, d_n).$$

Using the Bayes theorem, we can characterize the conditional posterior distribution of λ as follows:

$$\begin{aligned} p(\lambda | y_n, z_n^*, \Theta, d_n) &\propto l(\lambda | y_n, z_n^*, \Theta, d_n).p(\lambda) \\ &\propto p(y_n | z_n^*, \lambda, \Theta, d_n).p(\lambda). \end{aligned} \tag{23}$$

Similarly to b and for identical reasons, we use the following multivariate normal distribution to characterize the *prior* distribution of λ:

$$\lambda \sim N(\bar{\lambda}, \Sigma_\lambda^{-1}),$$
$$p(\lambda) = |\Sigma_\lambda|^{0.5} \exp[-\frac{1}{2}(\lambda - \bar{\lambda})' \Sigma_\lambda (\lambda - \bar{\lambda})]$$

where $\bar{\lambda}$ and Σ_λ^{-1} are fixed by the researchers.

The application of the Cholesky decomposition Q, where $QQ' = \Theta^{-1}$, to the relationship (11) facilitates the calculations. In this case, we thus obtain:

$$Q' y_n = Q' Z_n^{*m} \lambda + Q' \varepsilon_n, \qquad \text{where } Q' \varepsilon_n \sim N(0, I_{JR}),$$

which we rewrite as:

$$\tilde{y}_n = r_n \lambda + \tilde{\varepsilon}_n, \text{ where } \tilde{\varepsilon}_n \sim N(0, I_{JR})$$
$$r_n = Q' Z_n^{*m}$$
$$\tilde{y}_n = Q' y_n$$
$$\tilde{\varepsilon}_n = Q' \varepsilon_n.$$

After these transformations, the model becomes a simple regression model with unit variance, and thus:

$$\lambda | y, z_n^*, \Theta, d_n \sim N(\mu_\lambda, \Xi_\lambda), \tag{24}$$

where $\mu_\lambda = \left[n \sum r_n' r_n + \Sigma_\lambda\right]^{-1} \left[n \sum r_n' \tilde{y}_n + \Sigma_\lambda \bar{\lambda}\right]$ and $\Xi_\lambda = \left[n \sum r_n' r_n + \Sigma_\lambda\right]^{-1}$. Implementation of this step requires drawing normally distributed terms according to equation (24).

f) Conditional distribution of Ψ given everything else.

To estimate Ψ, given the realisations of B and z_n^* and the observations of w_n and d_n, we can deduce from the relationship in equation (1), a realisation of ζ_n, where:

$$\zeta_n = z_n^* - B w_n, \qquad \text{and} \quad \zeta_n \sim N(0, \Psi).$$

So, the conditional distribution of interest is:

$$\Psi \sim p(\Psi | \zeta).$$

If we apply the Bayes theorem and we define $G = \Psi^{-1}$, the conditional posterior distribution of interest becomes:

$$p(G|\zeta) \propto l(G|\zeta).p(G)$$
$$\propto p(\zeta|G).p(G).$$

To allow an easy mixture of the *prior* distribution with the likelihood to get a tractable *posterior* distribution in the same family as the *prior*, we assume that the inverse of the covariance matrix Ψ follows a Wishart distribution:

$$\Psi^{-1} = G \sim W(\upsilon, V), \qquad \upsilon \geq JL,$$

where V is a positive definite matrix of dimension $(JL \times JL)$ and where $p(G) \propto |G|^{(\upsilon - JL)/2}$ $\exp[-\frac{1}{2} tr(GV)]$. By combining this *prior* distribution to the likelihood of the sample, we obtain the following *posterior* distribution:

$$G|\zeta \sim W(\eta, M)$$

where $M = V + \sum_{n=1}^{N} \zeta_n \zeta_n'$, $\eta = v + N$, V and v are fixed by the researcher and N represents the number of individuals in the sample. Once the matrix G is simulated[4], its inverse produces a simulated value for Ψ. For identification purpose, the terms on the diagonal of Ψ might be set to unity. The approach suggested in McCulloch *et al.* (2000) can be used in situations where constraints are imposed.

g) Conditional distribution of Σ given everything else.

To estimate Σ, given the realisations of ξ_n, we do as in the previous step. If we apply the Bayes theorem and if we assume $G = \Theta^{-1}$, the posterior distribution of interest becomes:

$$p(G|\varepsilon) \propto l(G|\varepsilon).p(G)$$
$$\propto p(\varepsilon|G).p(G).$$

As for Ψ, we postulate that the inverse of the covariance matrix Σ follow a Wishart distribution:

$$\Sigma^{-1} = G \sim W(K, k), \qquad k \geq J,$$

where K is a positive definite matrix of dimension $(J \times J)$ and $p(G) \propto |G|^{(k-J)/2}$ $\exp[-\frac{1}{2}tr(GK)]$. By combining this *prior* distribution to the likelihood of the sample, we obtain the following *posterior* distribution:

$$G|\varepsilon \sim W(\eta, M)$$

where $M = K + \sum_{n=1}^{N} \xi_n \xi_n'$, $\eta = k + N$, K and k are determined by the researchers and N is the number of individuals in the sample. Once the matrix G is simulated, its inverse produces a simulated value for Σ. A simpler version with a diagonal Σ would instead involve making draws from a chi-square distribution.

h) Conditional distribution of Θ given everything else.

To estimate Θ, given the realisations of Λ and z_n^* and the observations of y_n and d_n, we can

---

[4]See Appendix 2 for more details on how to simulate matrix G from a Wishart distribution.

obtain a realisation of ε_n, from the relationship in equation (10), where:

$$\varepsilon_n = y_n - \Lambda z_n^*, \qquad \text{and } \varepsilon_n \sim N(0,\Theta).$$

So, the conditional distribution of interest is:

$$\Theta \sim p(\Theta|\varepsilon).$$

If we apply the Bayes theorem to the previous relationship and we define $G = \Theta^{-1}$, the posterior distribution of interest becomes :

$$\begin{aligned} p(G|\varepsilon) &\propto l(G|\varepsilon).p(G) \\ &\propto p(\varepsilon|G).p(G). \end{aligned}$$

As for Ψ, we postulate that the inverse of the covariance matrix Θ follow a Wishart distribution:

$$\Theta^{-1} = G \sim W(K,k), \qquad k \geq JR,$$

where K is a positive definite matrix of dimension $(JR \times JR)$ and $p(G) \propto |G|^{(k-JR)/2}$ $\exp[-\frac{1}{2}tr(GK)]$. By combining this *prior* distribution to the likelihood of the sample, we obtain the following *posterior* distribution:

$$G|\varepsilon \sim W(\eta, M)$$

where $M = K + \sum_{n=1}^{N} \varepsilon_n \varepsilon_n'$, $\eta = k + N$, K and k are determined by the researchers and N is the number of individuals in the sample. Once the matrix G is simulated, its inverse produces a simulated value for Θ. A simpler version with a diagonal Θ would instead involve making draws from a chi-square distribution.

SUMMARY

This paper presents the technical details regarding the estimation of the integrated discrete choice model formulations where choices are described by a multinomial specification and where a large number of latent variables are permitted. Simulation based techniques such as MSL and a MCMC approach are considered. We show how to implement each of the approaches. Based on our experience, the Bayesian MCMC technique should be more satisfactory in terms of computation time savings in situations where the choice set and/or the number of latent variables are large. In situations with limited number of latent variables and with choice set of reasonable size, MSL is likely to be a very good alternative.

Empirical evidence regarding the implementations recommended here are considered in Bolduc and Godbout (2004).

APPENDICES

Appendix 1

Dropping the n index for convenience, we now show that z^* and y have the following joint distribution:

$$\begin{bmatrix} z^* \\ y \end{bmatrix} \sim N\left(\begin{bmatrix} Bw \\ \Lambda Bw \end{bmatrix}, \begin{bmatrix} \Psi & \Psi\Lambda' \\ \Lambda\Psi & \Lambda\Psi\Lambda' + \Theta \end{bmatrix} \right).$$

Proof:

Using the notation in equations (1) and (10), we thus have:

$$\begin{aligned} z^* &= Bw + \zeta, \quad \zeta \sim MVN(0, \Psi) \\ y &= \Lambda z^* + \varepsilon, \quad \varepsilon \sim MVN(0, \Theta), \end{aligned}$$

which might also be written as:

$$\underbrace{\begin{bmatrix} I_{JL} & 0 \\ -\Lambda & I_{JR} \end{bmatrix}}_{\Gamma} \underbrace{\begin{bmatrix} z^* \\ y \end{bmatrix}}_{\omega} = \underbrace{\begin{bmatrix} Bw \\ 0 \end{bmatrix}}_{\kappa} + \underbrace{\begin{bmatrix} \zeta \\ \varepsilon \end{bmatrix}}_{\xi}. \tag{25}$$

This thus leads to:

$$\begin{aligned} \omega &= \Gamma^{-1}\kappa + \Gamma^{-1}\xi \\ \begin{bmatrix} z^* \\ y \end{bmatrix} &= \begin{bmatrix} I_{JL} & 0 \\ \Lambda & I_{JR} \end{bmatrix} \begin{bmatrix} Bw \\ 0 \end{bmatrix} + \begin{bmatrix} I_{JL} & 0 \\ \Lambda & I_{JR} \end{bmatrix} \begin{bmatrix} \zeta \\ \varepsilon \end{bmatrix}, \end{aligned} \tag{26}$$

and therefore:

$$\begin{bmatrix} z^* \\ y \end{bmatrix} \sim MVN\left(\begin{bmatrix} Bw \\ \Lambda Bw \end{bmatrix}, \begin{bmatrix} \Psi & \Psi\Lambda' \\ \Lambda\Psi & \Lambda\Psi\Lambda' + \Theta \end{bmatrix} \right), \tag{27}$$

where $|\Gamma| = \prod_{i=1}^{N} \gamma_{ii} = 1$ (*triangular.*)

Appendix 2

We describe how to generate G from a Wishart distribution. Let

$G|\varepsilon_n \sim W(\Phi, \phi)$, where ε_n represents an error term with cov matrix Φ.

First step:

Factor G so that

$$G = CAC',$$

where C, is a lower triangular matrix which Cholesky decomposes Φ, and matrix A, of same size as Φ, is from a standard Wishart distribution (i.e. $A \sim W(I, \phi)$).

Second step:

Factor A so that

$$A = TT'$$

where T is a lower triangular matrix generated as follows:

i) terms not on the diagonal are standard normal;

$$T_{ij} \sim N(0,1),$$

ii) elements on the diagonal are chi squared distributed;

$$T_{ii} \sim \sqrt{\chi^2_{\phi-i+1}}.$$

Third step:

Generate G as follows:

$$G = (CT)(CT)' = CTT'C'.$$

REFERENCES

Albert, J. H. and S. Chib (1993). Bayesian Analysis of Binary and polychotomous Response Data. *JASA*, **88**, 669-679.

Ben-Akiva, M. and B. Boccara (1987). Integrated Framework for Travel Behavior Analysis. IATBR Conference, Aix-en-provence, France.

Ben-Akiva, M. and B. Boccara (1995). Discrete Choice Models with Latent Choice Sets. *International Journal of Research in Marketing*, **12**, 9-24.

Ben-Akiva, M., D. Bolduc and J. Walker (2001). Specification, Estimation & Identification of the Logit Kernel (or Continuous Mixed Logit) Model. Working paper, Massachusetts Institute of Technology.

Ben-Akiva, M. and S. R. Lerman (1985). *Discrete Choice Analysis: Theory and Application to Travel Demand*. MIT Press, MA.

Ben-Akiva, M., D. McFadden, T. Gärling, D. Gopinath, J. Walker, D. Bolduc, A. Börsch-Supan, P. Delquié, O. Larichev, T. Morikawa and V. Rao (1999). Extended Framework for Modeling Choice Behavior. *Marketing Letters*, **10**(3), 187-203.

Ben-Akiva, M., D. McFadden, K. Train, J. Walker, C. Bhat, M. Bierlaire, D. Bolduc, A. Börsch-Supan, D. Brownstone, D. S. Bunch, A. Daly, A. de Palma, D. Gopinath, A. Karlstrom and M. A. Munizaga (2002). Hybrid Choice Models: Progress and Challenges. *Marketing Letters*, **13**(3), 163-175.

Ben-Akiva, M., J. Walker, A. T. Bernardino, D. A. Gopinath, T. Morikawa, A. Polydoropoulou (2002). Integration of Choice and Latent Variable Models. International Association of Traveler Behaviour Research (IATBR) book from the 1997 Conference in Austin, Texas.

Bentler, P. M. (1980). Multivariate Analysis with Latent Variables. *Annual Review of Psychology*, **31**, 419-456.

Bhat, C. R. (2001). Simulation Estimation of Mixed Discrete Choice Models Using Randomized and Scrambled Halton Sequences. Working Paper, University of Texas at Austin.

Bolduc, D., B. Fortin and M. A. Fournier (1996). The Impact of Incentive Policies to Influence Practice Location of General Practioners: A Multinomial Probit Analysis. *Journal of Labor Economics*, **14**, 703-732.

Bolduc, D., B. Fortin, and S. Gordon (1997). Multinomial Probit Estimation of Spatially Interdependent Choices: An Empirical Comparison of Two New Techniques. *International Regional Science Review*, **20**(1-2), 77-101.

Bolduc, D. and M. H. Godbout (2004). Empirical Tests of the Classical and Bayesian Simulation Based Estimations of Hybrid Choice Models. Working paper in progress.

Geman, S. and D. Geman (1984). Stochastic Relaxation, Gibbs Distribution and the Bayesian Restoration of Images. *IEEE Transaction of Pattern Analysis and Machine Intelligence*, **6**, 721-741.

Gilks, W. R. (1996). Full conditional distributions. In: *Markov Chain Monte Carlo in Practice* (Gilks, W. R., S. Richardson, D. J. Spiegelhalter, eds.), pp. 75-88. Chapman and Hall, London.

Jöreskog, K. G. (1973). A General Method for Estimating a Linear Structural Equation System. *Structural Models in the Social Sciences* (A. S. Goldberger and O. D. Duncan, eds.). Academic Press, New York.

Jöreskog, K. G. and D. Sörbom (1984). *LISREL VI: Analysis of linear structural relations by maximum likelihood, instrumental variables and least squares methods.* User's guide, Department of Statistics, University of Uppsala, Uppsala, Sweden.

Keesling, J. W. (1972). *Maximum Likelihood Approaches to Causal Analysis.* Ph.D. Thesis, University of Chicago.

Koop, G. (2003). Bayesian Econometrics. John Wiley and Sons Ltd.

Manski, C. (1977). The Structure of Random Utility Models. *Theory and Decision,* **8**, 229-254.

McCulloch, R. E., N. G. Polson and P. E. Rossi (2000). Bayesian Analysis of the Multinomial Probit Model with Fully Identified Parameters. *Journal of Econometrics*, **99**, 173-193.

McFadden, D. (1973). Conditional Logit Analysis of Qualitative Choice Behavior. In: *Frontiers in Econometrics* (P. Zarembka, ed.). Academic Press, New York.

McFadden, D. (1986). The Choice Theory Approach to Marketing Research. *Marketing Science,* **5**(4), 275-297.

McFadden, D. (1999). Rationality for Economists. *Journal of Risk and Uncertainty*, **19**(1/3), Kluwer, pp. 73-105.

McFadden, D. and K. Train (2000). Mixed MNL Models for Discrete Response. *Journal of Applied Econometrics,* **15**(5), 447-470.

Morikawa, T., M. Ben-Akiva and D. McFadden (2002). Discrete Choice Models Incorporating Revealed Preferences and Psychometric Data. *Econometric Models in Marketing,* **16**, 27-53.

Scott, S. L. (2003). Data Augmentation for the Bayesian Analysis of Multinomial Logit Models. *Proceedings of the American Statistical Association.* Section on Bayesian Statistical Science, Alexandria, VA: American Statistical Association.

Tanner, T., W. H. Wong (1987). The Calculation of the Posterior Distribution by Data Augmentation. *JASA,* **82**, 528-549.

Train, K. (2001). A comparison of hierarchical bayes and maximum simulated likelihood for mixed logit. Working Paper, Department of Economics, University of California, Berkeley.

Train, K. E. (2002). *Discrete Choice Methods with Simulation.* Cambridge University Press.

Walker, J. L. (2001). *Extended Discrete Choice Models: Integrated Framework, Flexible Error Structures, and Latent Variables.* Ph.D. Dissertation, Massachusetts Institute of Technology, Cambridge, USA.

Walker, J. L. and M. Ben-Akiva (2002). Generalized Random Utility Model. *Mathematical Social Science,* **43**(3), 303-343.

Wiley, D. E. (1973). The Identification Problem for Structural Equation Models with Unmeasured Variables. *Structural Models in the Social Science* (A. S. Goldberger and O. D. Duncan, eds.). Academic Press, New York.

Zellner, A. (1971). An Introduction to Bayesian Inference in Econometrics. Wiley, New York.

ACKNOWLEDGEMENTS

This paper was supported by the Social Sciences and Humanities Research Council of Canada, through a Major Collaborative Research Initiative, 2000-2005: *Access to activities and services in urban Canada: behavioural processes that condition equity and sustainability.*

INDEX

A

B

C

Printed and bound by CPI Group (UK) Ltd, Croydon, CR0 4YY
16/09/2026
14963403-0001